# Antidepressants

# Contemporary Neuroscience

# Antidepressants

## *New Pharmacological Strategies*

Edited by

## Phil Skolnick

*National Institutes of Health, Bethesda, MD*

Humana Press ✳ Totowa, New Jersey

Library of Congress Cataloging in Publication Data

Antidepressants: new pharmacological strategies/edited by Phil Skolnick.
    p. cm.—(Contemporary neuroscience)
  Includes bibliographical references and index.
  ISBN 0-89603-469-0 (alk. paper)
  1. Antidepressants.   I. Skolnick, Phil.   II. Series.
  [DNLM: 1. Antidepressive agents.   QV 77.5 A62897 1997]
RM332.A579   1997
616.85'27061—dc21
DNLM/DLC
for Library of Congress                                           97-16900
                                                                  CIP

Forty years ago, the first reports appeared describing the antidepressant properties of iproniazid and imipramine. Equal measures of serendipity and astute clinical observation led to the discovery of these "first-generation" antidepressants. Thus, the antidepressant properties of iproniazid (an analog of isoniazid) were first recognized in patients being treated for tuberculosis, while imipramine was synthesized in an attempt to improve on the therapeutic properties of another tricyclic molecule, chlorpromazine.

The demonstration that these first-generation antidepressants affect the disposition and metabolism of biogenic amines led to a revolution in biological psychiatry, resulting in new theories and testable hypotheses about the origins of affective disorders. These early studies led in the synthesis of new compounds in an attempt to improve the therapeutic qualities of first-generation agents. The fruits of these efforts include a group of "second-generation" antidepressants, such as the serotonin-specific reuptake inhibitors (SSRIs) and reversible inhibitors of monoamine oxidase (RIMAs). A "third generation" of antidepressants is now available, possessing both serotonin and norepinephrine reuptake blocking properties, but apparently devoid of many other pharmacological properties common to tricyclic antidepressants. The principal advantage of these second- and third-generation agents is an improved side-effect profile compared to first-generation antidepressants. These newer agents are clearly superior to placebo, but SSRIs (for example) appear to be no more efficacious antidepressants than imipramine and, though still controversial, may be less effective than traditional therapies in severely depressed (melancholic) patients *(1–4)*.

In order to achieve a clear improvement over current antidepressants, two major therapeutic objectives must be met. First, a reduction in the rate of nonresponders from the ~30% reported in most carefully controlled studies. Second, a reduction (and in the ideal, elimination) of the "therapeutic lag" encountered with traditional antidepressants. This therapeutic lag remains one of the great enigmas of antidepressant therapy. Aside from the obvious benefit to the patient, therapies that reduce the latency between onset of treatment and significant remission of symptoms will undoubtedly result in a better understanding of the molecular mechanisms of antidepressant action.

Despite the indisputable evidence that most clinically effective antidepressants perturb monoaminergic pathways, there has been a resurgence of interest in the mechanisms responsible for antidepressant action. This interest stems from the inability of monoaminergic mechanisms to provide a heuristic framework capable of adequately explaining the significant proportion of patients who do not respond to current antidepressants and the time required for a therapeutic response. The very strategy of targeting compounds to block biogenic uptake and metabolism may result in the two most significant limitations common to currently used antidepressants. Thus, we may be compelled to focus on alternative mechanisms and strategies in order to overcome these limitations. This objective is both worthy and timely, since it has been estimated that the current cost of depression (from absenteeism, lost productivity, lost earnings, treatment, and rehabilitation) exceeds $40 billion in the United States. Moreover, the market for second-generation antidepressants (e.g., SSRIs such as sertraline, paroxetine, and fluoxetine) has been estimated at $4 billion. The objective of this book is to bring together novel concepts of antidepressant therapy that are either not grounded on biogenic amine-based hypotheses or attempt to exploit new insights from the laboratory to improve current, biogenic amine-based therapies.

## REFERENCES

1. Danish University Antidepressant Group (1986) Citalopram: clinical effect profile in comparison with clomipramine. A controlled multicenter study. *Psychopharmacol.* **90,** 131–138.
2. Danish University Antidepressant Group (1990) Paroxetine: a selective serotonin reuptake inhibitor showing better tolerance, but weaker efficacy than clomipramine in a controlled multicenter study. *J. Affect. Dis.* **18,** 289–299.
3. Roose, S. P., Glassman, A. H., Attia, E., and Woodring, S. (1994) Comparative efficacy of selective serotonin reuptake inhibitors and tricyclics in the treatment of melancholia. *Amer. J. Psychiat.* **151,** 1735–1739.
4. Kasper, S. and Heiden, A. (1995) Do SSRIs differ in their antidepressant efficacy. *Human Psychopharmacol.* **10,** S163–S172.

*Phil Skolnick*

# Contents

# Contributors

FRANCESC ARTIGAS • *Department of Neurochemistry, Instituto de Investigaciones Biomédicas de Barcelona, Spain*

MICHAEL BRILEY • *Centre de Recherche Pierre Fabre, France*

JOSEP Mª CASANOVAS • *Department of Neurochemistry, Instituto de Investigaciones Biomédicas de Barcelona, Spain*

ISABELLE CLOËZ-TAYARANI • *Unité de Pharmacologie Neuro-Immuno-Endocrinienne, Institut Pasteur, Paris Cedex, France*

ROSER CORTÉS • *Department of Neurochemistry, Instituto de Investigaciones Biomédicas de Barcelona, Spain*

RONALD S. DUMAN • *Laboratory of Molecular Psychiatry and Pharmacology, Yale University School of Medicine, Connecticut Mental Health Center, New Haven, CT*

GILLES FILLION • *Unité de Pharmacologie Neuro-Immuno-Endocrinienne, Institut Pasteur, Paris Cedex, France*

MARIE-PAULE FILLION • *Unité de Pharmacologie Neuro-Immuno-Endocrinienne, Institut Pasteur, Paris Cedex, France*

EDWARD I. GINNS • *Clinical Neuroscience Branch, NIMH, National Institute of Health, Bethesda, MD*

VIVETTE GLOVER • *Department of Paediatrics, Queen Charlotte's and Chelsea Hospital, London, UK*

BRIGITTE GRIMALDI • *Unité de Pharmacologie Neuro-Immuno-Endocrinienne, Institut Pasteur, Paris Cedex, France*

ILDEFONS HERVÁS • *Department of Neurochemistry, Instituto de Investigaciones Biomédicas de Barcelona, Spain*

NUO-YU HUANG • *Laboratory of Neuroscience, National Institutes of Health, Bethesda, MD*

RICHARD T. LAYER • *Laboratory of Neuroscience, National Institutes of Health, Bethesda, MD*

BRIAN E. LEONARD • *Pharmacology Department, University College, Galway, Ireland*

YANBIN LIANG • *Lilly Research Laboratories, Eli Lilly and Company, Indianapolis, IN*

OLIVIER MASSOT • *Unité de Pharmacologie Neuro-Immuno-Endocrinienne, Institut Pasteur, Paris Cedex, France*

CHANTAL MORET • *Centre de Recherche Pierre Fabre, France*

MASASHI NIBUYA • *Laboratory of Molecular Psychiatry and Pharmacology, Yale University School of Medicine, Connecticut Mental Health Center, New Haven, CT*

MARIUSZ PAPP • *Institute of Pharmacology, Polish Academy of Sciences, Kwakow, Poland*

IAN A. PAUL • *Laboratory of Neurobehavioral Pharmacology and Immunology, Division of Neurobiology and Behavior Research, Department of Psychiatry, University of Mississippi Medical Center, Jackson, MS*

STEVEN M. PAUL • *Lilly Research Laboratories, Eli Lilly and Company, Indianapolis, IN; Clinical Neuroscience Branch, NIMH, National Institute of Health, Bethesda, MD*

NICOLE PRUDHOMME • *Unité de Pharmacologie Neuro-Immuno-Endocrinienne, Institut Pasteur, Paris Cedex, France*

OLGIERD PUCILOWSKI • *Department of Psychiatry, Temple University Hospital, Philadelphia, PA*

LUZ ROMERO • *Department of Neurochemistry, Instituto de Investigaciones Biomédicas de Barcelona, Spain*

S. PAUL ROSSBY • *Department of Psychiatry, Vanderbilt University School of Medicine, Nashville, TN*

JEAN-CLAUDE ROUSSELLE • *Unité de Pharmacologie Neuro-Immuno-Endocrinienne, Institut Pasteur, Paris Cedex, France*

LAURE SEGUIN • *Unité de Pharmacologie Neuro-Immuno-Endocrinienne, Institut Pasteur, Paris Cedex, France*

JEAN-CHRISTOPHE SEZNEC • *Unité de Pharmacologie Neuro-Immuno-Endocrinienne, Institut Pasteur, Paris Cedex, France*

PHIL SKOLNICK • *Laboratory of Neuroscience, National Institutes of Health, Bethesda, MD*

FRIDOLIN SULSER • *Departments of Psychiatry and Pharmacology, Vanderbilt University School of Medicine, Nashville, TN*

RAMON TRULLAS • *Neurobiology Unit, Institut d'Investigaciones Biomèdiques de Barcelona, Consejo Superior de Investigaciones Científicas, Barcelona, Spain*

VIDITA A. VAIDYA • *Laboratory of Molecular Psychiatry and Pharmacology, Yale University School of Medicine, Connecticut Mental Health Center, New Haven, CT*

PAUL WILLNER • *Department of Psychology, University of Wales, Swansea, UK*

XIN WU • *Lilly Research Laboratories, Eli Lilly and Company, Indianapolis, IN*

# 1

# Strategies to Optimize the Antidepressant Action of Selective Serotonin Reuptake Inhibitors



**Luz Romero, Josep Mª Casanovas, Ildefons Hervás, Roser Cortés, and Francesc Artigas**



## 1. INTRODUCTION

### 1.1. The Brain Serotonergic System as Target for Antidepressant Drugs

It is beyond doubt that major depression can be treated by selectively manipulating the function of different brain neurotransmitters *(1,2)*. Yet, of the many brain neuronal systems, the serotonin (5-hydroxytryptamine, 5-HT) system is the most common neurobiological target for such treatments. Tricyclic antidepressants (TCAs) act on 5-HT and noradrenergic (NE) neurons by inhibiting, with different potencies, transmitter reuptake *(3,4)*, and monoamine (MAO) oxidase inhibitors (MAOIs) increase 5-HT and NE transmission by preventing their metabolism. Yet, it was not until the advent of the selective serotonin reuptake inhibitors (SSRIs) that the antidepressant potential of 5-HT transporter blockade was fully appreciated *(5,6)*. The clinically useful SSRIs are chemically dissimilar, but share the property of selectively inhibiting the 5-HT reuptake process (Fig. 1). Unlike TCAs, the SSRIs display little affinity for aminergic receptors *(7)* and therefore lack the severe side effects associated with the use of the former agents. This results in both an improved quality of life for the patients and greater treatment compliance, which is compromised in some instances by the use of TCAs. It is generally recognized that the antidepressant efficacy of SSRIs is comparable to that of TCAs, although several studies have shown that the latter are more effective in severely depressed inpatients *(5,6,8,9)*.

Given the ability of SSRIs to selectively inhibit the 5-HT transporter thereby increasing the synaptic concentration of 5-HT, their clinical actions can thus be attributed to an enhanced activation of one or several postsynaptic 5-HT receptors. This view is supported by clinical data showing that the administration of tryptophan-free amino acid mixtures to recovered major depressive patients receiving either SSRIs or MAOIs transiently abolishes their antidepressant effects *(10)*. This procedure causes a very marked reduction of neuronal 5-HT release in rats subjected to repeated treat-



*From:* Antidepressants: *New Pharmacological Strategies*
*Edited by: P. Skolnick, Humana Press Inc., Totowa, NJ*



*1*

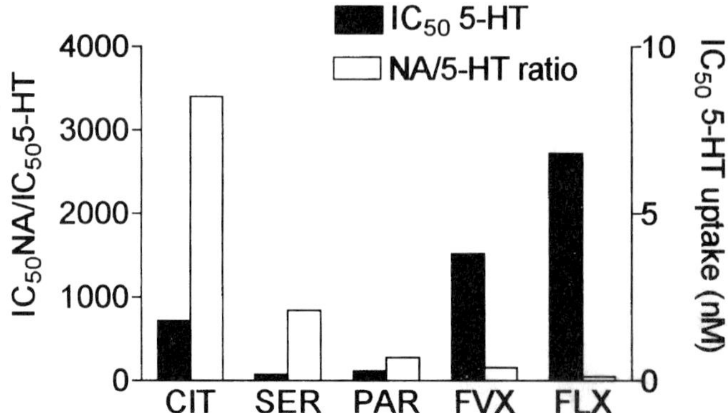



**Fig. 1.** Bar diagram showing the affinity of the SSRIs for the 5-HT transporter (filled bars, right *y*-axis) and their selectivity, as measured by the ratio between the affinities for the 5-HT and NE transporters (open bars, left *y*-axis). Most selective compound is citalopram; that with a higher affinity for the 5-HT transporter is paroxetine. CIT, citalopram; FLX, fluoxetine; FVX, fluvoxamine; PAR, paroxetine; SER, sertraline. Redrawn from ref. *4*.

ment with the SSRI fluvoxamine *(11)*, thus supporting the association between recovery from depression and enhancement of 5-HT activity. It is unclear how many of the 14 or so 5-HT receptor subtypes identified so far mediate the antidepressant effects of the SSRIs *(12)*. There is evidence that hippocampal postsynaptic 5-HT$_{1A}$ receptors participate in the action of several types antidepressant drugs *(13–15)*. However, given the plethora of symptoms exhibited by depressed patients, it is likely that the effects of antidepressants involve the activation of receptors in more than one brain structure.

## 1.2. Serotonergic Autoreceptors and Onset of Antidepressant Action

Despite the ability of the SSRIs to block the 5-HT transporter soon after their administration, significant clinical improvement of depressed patients requires prolonged administration. This suggests the existence of neurobiological adaptative mechanisms responsible for their clinical action. This delay cannot be attributed to a downregulation of the cortical β-adrenoceptor-coupled cAMP generating system, because most SSRIs do not induce such effect after chronic treatment *(16–19)*. More likely, the slow onset of clinical action and the limited efficacy of antidepressant drugs (less than two-thirds of patients usually respond to the first drug administered) may be partly ascribed to the inhibition of 5-HT release by forebrain serotonergic nerve terminals after the administration of drugs that inhibit 5-HT uptake or MAO activity *(20,21)*. This negative feedback, which involves the activation of somatodendritic 5-HT$_{1A}$ autoreceptors, is also responsible for the attenuation of cell firing observed after a single administration of antidepressant drugs *(22,23)*. The present chapter will critically review this evidence and provide new data in support of a differential inhibition of the serotonergic pathways of the dorsal and median raphe nuclei (DRN and MRN, respectively) during the acute phase of the treatment with SSRIs.

## 2. METHODOLOGICAL ASPECTS

### 2.1. Autoradiographic and In Situ Hybridization Studies

Receptor autoradiography and *in situ* hybridization (ISH) histochemistry were used to examine the localization and density of 5-HT receptor and transporter proteins and the mRNAs encoding them in fresh-frozen rat and human brain. Human brain samples were provided by Dr. F. Cruz-Sánchez from the Human Brain Bank of Barcelona. Specimens were obtained at autopsy from control subjects aged 40–70 years, after postmortem delays ranging from 4 to 22 h. The brains were dissected into blocks, quickly frozen on dry ice and later sectioned into 14-µm-thick slices using a microtome-cryostat and mounted onto microscope glass slides. Labeling of 5-HT$_{1A}$ receptors was performed, as previously described *(24)*, using 2 n*M* [$^3$H]8-OH-DPAT ([$^3$H]8-hydroxy-2-[di-*n*-propylamino]tetralin) as ligand. To detect the mRNA encoding the rat 5-HT$_{1A}$ receptor and the rat and human 5-HT transporter by ISH, tissue sections were incubated with synthetic oligonucleotide probes complementary to the rat 5-HT$_{1A}$ receptor cDNA *(25)* and the rat transporter cDNA *(26)*, respectively, both end-labeled with [$^{35}$S]α-dATP, using published protocols *(27)*. Incubated sections were exposed to Hyperfilm-$^3$H or Hyperfilm βmax films (Amersham, UK). After film exposure, sections processed for ISH were dipped into NTB2 nuclear track emulsion (Kodak, Rochester, NY) to allow the cellular detection of hybridization signal. Optical densities were measured on film autoradiograms using an image analysis system.

### 2.2. Microdialysis Studies and the Serotonergic System

A large percentage of the data presented in this chapter has been obtained using in vivo microdialysis. This technique enables monitoring of neurotransmitter changes in the interstitial brain space of the living, freely-moving rat. Given the anatomical and functional similarities between the serotonergic system in the rat and human brains, it is likely that many of the presynaptic adaptative changes triggered by antidepressants observed in the rat brain can also take place in the human brain. In both species, the ascending serotonergic axons originate in several neuronal groups located in the midbrain, the dorsal and median raphe nuclei (DRN, and MRN, respectively; B5–B8 groups in the classification of Dahlström and Fuxe for the rat brain *[28]*). Figure 2 shows the localization of 5-HT cells in the DRN of the rat and human brainstem, as labeled by ISH of the mRNA encoding the 5-HT transporter. These neurons have long, extremely arborized axons, which, together with the noradrenergic system, constitute the most expansive neuronal transmitter system in the mammalian brain. The number of serotonergic neurons is low (a total of approx 11,000 and 250,000 in the rat and human DRN, respectively), but innervate the diencephalon and telencephalon with a very high density of nerve terminals (approx 5 × 10$^6$ varicosities/mm$^3$ cortex or hippocampus) *(29–31)*. Thus, it has been estimated that each serotonergic neuron gives rise to approx 5 × 10$^6$ varicosities *(31)*.

The cell bodies and dendrites of serotonergic neurons in both species contain somatodendritic 5-HT$_{1A}$ autoreceptors *(24,32,33)*, which inhibit electric and metabolic activity when activated, and a very large density of the 5-HT transporter that internalizes 5-HT from the extracellular space *(34,35)*. Figure 3 shows the density of

　　　　　　　　　　　　　　　　　　　　　　　　*Romero et al.*

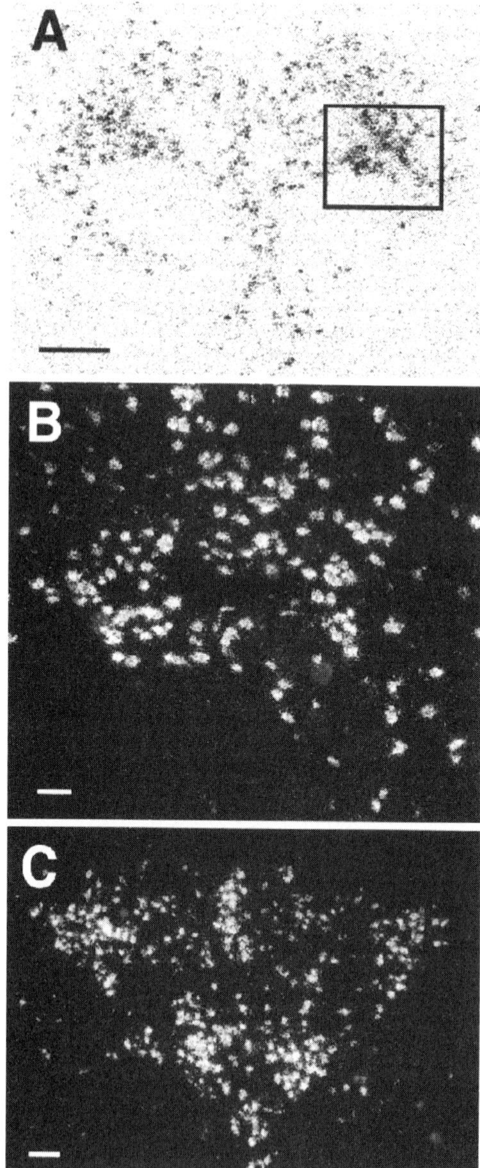

**Fig. 2.** Detection of 5-HT transporter mRNA by *in situ* hybridization. (**A**) Film autoradiogram obtained from a section through the human midbrain showing the presence of 5-HT transporter mRNA hybridization signal over the region of the DRN. (**B**) Dark-field photomicrograph obtained from the section shown in A (boxed area), after dipping into nuclear track emulsion at a higher magnification. Specific hybridization signal is observed as accumulations of silver grains over cell bodies. (**C**) Dark-field photomicrograph showing 5-HT transporter mRNA hybridization signal over cells of the rat DRN. Bars: A = 1 mm; B and C = 100 μm.

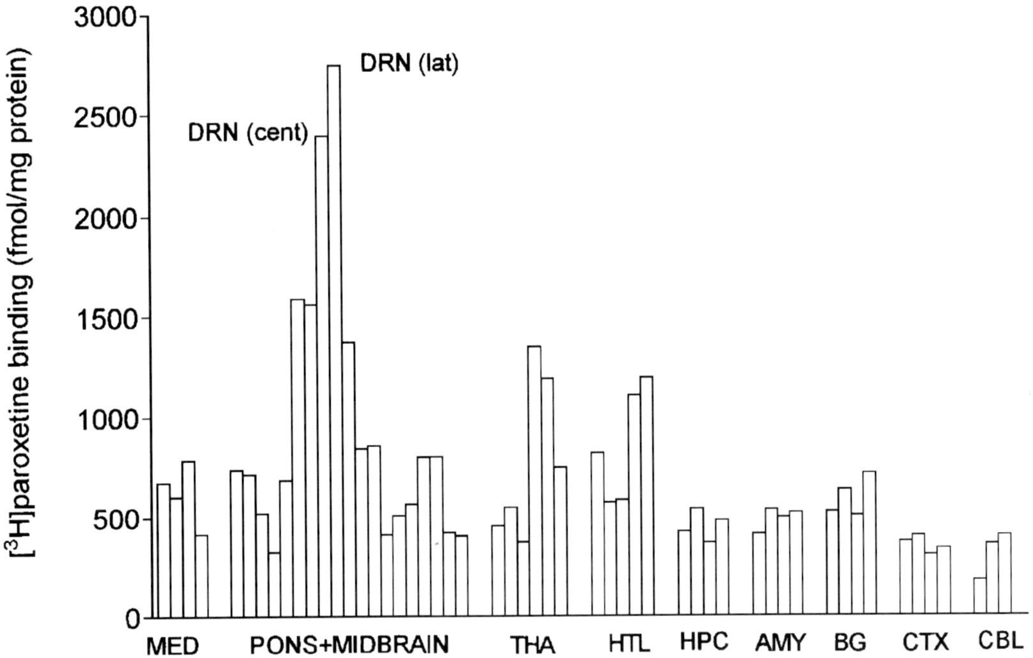



**Fig. 3.** Mean density of the 5-HT transporter, as indicated by [3H]paroxetine binding, in several brain regions of the human brain. Note the presence of a very large density in central and lateral parts of the DRN, as well as in the nucleus centralis superior and nucleus raphe linearis. Redrawn from ref. *34.*

the 5-HT transporter, as labeled by [3H]paroxetine, in the human brain. The large density found in the various raphe nuclei, and also observed in the rat brain, is rather remarkable. Although some of this protein may correspond to newly synthesized 5-HT transporter present in cell bodies before its transport to axonal varicosities in forebrain, experimental data support the observation that the somatodendritic transporter is fully operational in rat brain, actively removing 5-HT from the extracellular space (*see* Section 3.1.). This suggests that the cell bodies of serotonergic neurons are key targets in the action of SSRIs, because of the presence of such a large density of uptake sites.

Finally, but of no less importance, a large percentage of serotonergic terminals (approx 60–80%) do not make synaptic contacts in neocortex *(36,37).* This implies that most 5-HT acts in a paracrine manner, activating receptors relatively distant from its release site. These characteristics have two important corollaries. First, the 5-HT concentration found in the interstitial brain space is representative of that seen by high-affinity 5-HT receptors. Second, the 5-HT transporter located in the neuronal—and perhaps glial *(38)*—membrane is the main mechanism that controls the extracellular concentration of 5-HT (5-HT$_{ext}$), and, therefore, the extent of the activation of such receptors. Thus, 5-HT$_{ext}$, as measured by the microdialysis technique, is representative of the active concentration of 5-HT in brain. Intrasynaptic 5-HT concentrations cannot be assessed with microdialysis; therefore, drug-induced changes in axons making synaptic contacts are unlikely to be properly evaluated with this technique.

Drugs were locally (dissolved in the artificial cerebrospinal fluid [CSF] used to perfuse the probes) or systemically administered. Local administration of antidepressants permits an examination of the effects in restricted areas of brain tissue; the changes of 5-HT$_{ext}$ found after systemic administration correspond to an integrated response of the CNS, because effects of drugs in distal parts of the brain may affect the response in the area sampled by the probe. The modifications of 5-HT$_{ext}$ after local and systemic administration of antidepressants have been assessed principally in the midbrain raphe nuclei (DRN and MRN) and in forebrain projection areas of both (frontal cortex and striatum for the DRN; dorsal and ventral hippocampus for the MRN). Also, dual probe implants have been extensively used. Drugs were then locally applied in the DRN or DRN + MRN and 5-HT release was monitored in selective projection areas. This enabled us to assess the effects of 5-HT$_{ext}$ changes at the somatodendritic level on 5-HT release by forebrain nerve terminals. The interested reader is referred to publications by this and other groups listed at the end of the chapter for a detailed description of microdialysis procedures. Table 1 lists the stereotaxic coordinates used in the experiments described herein. The size of the probes was adjusted to that of the brain region examined. Thus, small (1.5-mm-long) probes were used for the DRN and MRN; larger (4-mm-long) probes were used for frontal cortex or striatum. Data from larger probes have been corrected for size, using an empirical plot of recovery vs length, and expressed as for 1.5-mm probes. This allows a direct comparison of dialysate values obtained with probes of different sizes.

All experiments described in the present chapter and in publications by this group have been obtained in unanesthetized, freely moving rats. This is an important methodological aspect, because of the marked differences in the discharge rate of serotonergic neurons between waking (active) and sleep (inactive) periods. Furthermore, we noticed that pentobarbital anesthesia caused a profound (−60%) reduction of somatodendritic and terminal 5-HT release (Abellán and Artigas, unpublished results), another compelling reason for the use of unanesthetized animals in the study of a fully functional serotonergic system.

## 3. EFFECTS OF SSRIs ON 5-HT$_{EXT}$ IN THE RAT BRAIN

### 3.1. Preferential Increases in 5-HT$_{ext}$ in the Raphe Nuclei After a Single SSRI Treatment

An early study reported large (i.e., four- to sixfold) increases in 5-HT$_{ext}$ in the midbrain raphe after either local or systemic administration of the nonselective 5-HT uptake inhibitor clomipramine *(20)*. By contrast, local administration increased 5-HT$_{ext}$ in frontal cortex to a similar extent, but systemic administration failed to do so. Such regional selectivity after systemic treatment with antidepressant drugs was subsequently demonstrated for SSRIs, thus showing that it was not an unspecific effect of clomipramine *(39–42)*. In all cases examined, the increases in 5-HT$_{ext}$ elicited in the DRN or DRN + MRN by these drugs were greater than in the frontal cortex, hippocampus, or striatum. More importantly, single treatment with low doses of antidepressants significantly elevated 5-HT$_{ext}$ in the raphe nuclei, but not in frontal cortex, thus raising doubts about the enhancement of serotonergic transmission produced by clinically relevant antidepressant doses. By contrast, repeated treatment for 2 wk with

**Table 1**
**Length of Microdialysis Probes and Stereotaxical Coordinates of Implants**

| Region | Length, mm | AP | DV | L | Vertical Angle (degrees) |
|---|---|---|---|---|---|
| Dorsal raphe nucleus | 1.5 | −7.8 | −7.5 | −3.1 | 30 |
| Median raphe nucleus | 1.5 | −7.8 | −8.9 | −2.0 | 13 |
| Dorsal + median raphe nuclei | 4.0 | −7.8 | −9.0 | −0.5 | 0 |
| Dorsal hippocampus | 1.5 | −3.8 | −4.0 | −2.0 | 0 |
| Ventral hippocampus | 4.0 | −5.8 | −8.0 | −5.0 | 0 |
| Dorsal striatum | 4.0 | 0.2 | −8.0 | −3.0 | 0 |
| Frontal cortex | 4.0 | 3.4 | −6.0 | −2.5 | 0 |
| Amygdala | 1.5 | −4.3 | −9.0 | −4.1 | 0 |

Coordinates (AP, DV, L) are given in mm, with respect to bregma and duramater, according to the rat brain atlas of Paxinos and Watson (1986). Probes in the DRN and MRN were implanted with lateral angles of 30 and 13 degrees, respectively, to avoid obstruction of the cerebral aqueduct.

the same daily doses (e.g., 1 mg/kg/d fluvoxamine, 4 mg/kg/d imipramine, or 0.5 mg/kg/d tranylcypromine) significantly increased 5-HT$_{ext}$ in frontal cortex *(43–45)*. Repeated treatment with higher SSRI doses also elevated baseline 5-HT$_{ext}$ in frontal cortex or potentiated the effects of a further dose of the SSRI *(46–48)*. The reasons for the cortical 5-HT$_{ext}$ elevations after prolonged treatment with SSRIs are not fully determined, but may involve a reduction of the activity of somatodendritic or terminal 5-HT autoreceptors *(13,14,46–49)* *(see below)*, as well as of the efficacy of the 5-HT transporter *(50)*.

Recent results indicate that the increase in 5-HT$_{ext}$ elicited in the MRN by single fluoxetine administration is even greater than that in the DRN *(51)*. Similarly, basal dialysate 5-HT values in the MRN obtained in presence of 1 $\mu M$ of the SSRI citalopram doubled those in the DRN, despite comparable values in both regions when the dialysis probes were perfused with artificial CSF without the uptake inhibitor. Both observations indicate that the increment of 5-HT$_{ext}$ produced by SSRIs is greater in the MRN than in the DRN, and both are greater than those produced in projection areas (Fig. 4). This difference cannot be accounted for by a greater density of 5-HT transporters or number of neurons in the MRN vs the DRN (in fact, the opposite is true in both instances). Possibly, the greater 5-HT$_{ext}$ concentration in the MRN after blockade of the 5-HT transporter may indicate a higher 5-HT release by dendrites or efferent 5-HT fibers within the boundaries of the MRN. Under normal conditions, dialysate 5-HT is representative of the equilibrium between release and reuptake. When the latter is blocked, dialysate 5-HT can be taken as a measure of release by the brain tissue surrounding the probe.

The factors controlling the activity of 5-HT neurons and 5-HT release in the raphe nuclei are still poorly understood. Electron microscopy and retrograde tracing studies have identified a large number of afferents to the DRN *(31,52–54)*, a finding somewhat paradoxical in view of the relative insensitivity of the firing of 5-HT cells to many physiological and environmental stimuli in awake cats *(31,55)* (although the activity of subgroups of DRN 5-HT neurons is extremely sensitive to some movements and sen-

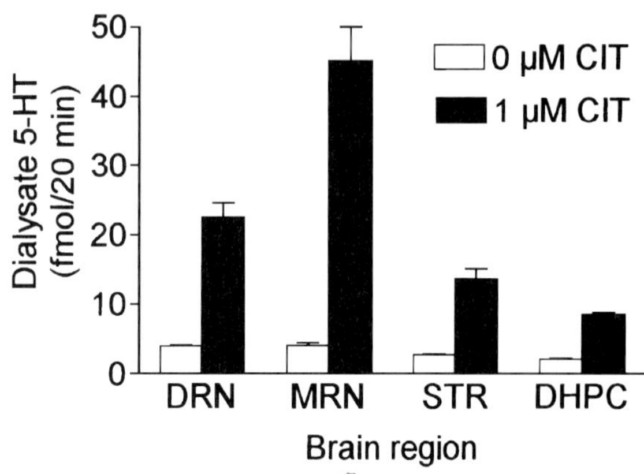



**Fig. 4.** Dialysate 5-HT concentrations in the DRN and MRN, and in selective projection areas of both (dorsal striatum, STR, and dorsal hippocampus, DHPC) in absence or presence of 1 µ*M* citalopram in the perfusion fluid. The absolute and percent increments of dialysate 5-HT produced by citalopram addition are greater in the MRN than in any other brain area examined so far, including the DRN. Data from 15 to 26 rats/region, without citalopram in the dialysis fluid, and from 23 to 38 rats/region, with citalopram.

sory stimuli *[56]*). Catecholaminergic afferents are relevant for the control of serotonergic function, because the electric activity of 5-HT neurons is dependent on the activation of $\alpha_1$-adrenoceptors *(57)* and dopamine $D_2$-like receptors in the DRN that increase somatodendritic 5-HT release *(58)*. Moreover, glutamatergic and GABA-ergic inputs are likely to figure extensively in the control of serotonergic activity *(59,60)*. These aspects are particularly important, not only to understand the regulation of 5-HT neurons, but also from a pharmacological point of view. Thus, given the presence of a high density of inhibitory 5-HT$_{1A}$ autoreceptors in cell bodies and dendrites, physiological or drug-induced increments in somatodendritic 5-HT release by certain neuronal groups within the raphe nuclei may precipitate a profound reduction of the electric and metabolic activity in neighboring 5-HT neurons.

Thus, the marked raphe–forebrain differences in the effects of SSRIs and MAOIs, as well as the unchanged forebrain 5-HT$_{ext}$ after the single administration of clinically relevant doses of these agents, might be somehow related to their delayed onset of action. Consequently, we reasoned that a study of the mechanisms responsible for the limitation of antidepressant effects in forebrain might provide some clues to shorten their latency and to improve their efficacy.

### 3.2. SSRIs Reduce 5-HT Release in Forebrain Through a Raphe-Based Mechanism

The limited increase in 5-HT$_{ext}$ elicited by SSRIs in forebrain areas is the consequence of the activation of somatodendritic (5-HT$_{1A}$) autoreceptors by the excess 5-HT$_{ext}$ produced in the vicinity of cell bodies and dendrites after their systemic administration. This reduces cell firing and terminal release, thus attenuating the elevations produced by blockade of the 5-HT reuptake in forebrain. This inhibitory process is exemplified in Fig. 5, which shows the reduction of cortical 5-HT release induced by

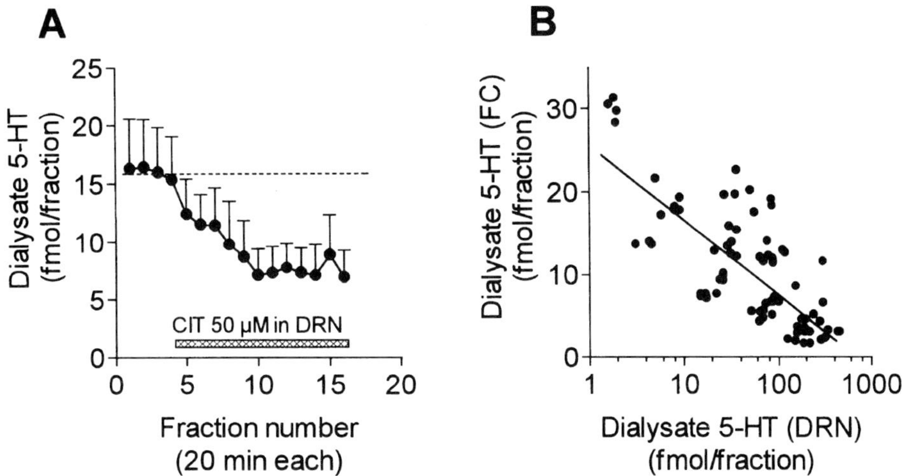

**Fig. 5.** (**A**) Reduction of the 5-HT release in frontal cortex by the application of citalopram (50 $\mu M$) in the DRN (shown by a crosshatched horizontal bar). The probe placed in frontal cortex was perfused with artificial CSF containing 1 $\mu M$ citalopram. Under this condition, dialysate 5-HT values are representative of the 5-HT release. (**B**) Correlation between individual dialysate 5-HT values in frontal cortex and DRN of the same rats during the experiment depicted in A (16 fractions/rat; $n = 5$ rats). Best fit (semilog) yielded a correlation coefficient of 0.582.

the SSRI citalopram. The local application of citalopram in the DRN elicited a profound reduction of 5-HT release in frontal cortex. There was a significant inverse correlation ($r^2 = 0.582$; $p < 0.0001$) between the 5-HT$_{ext}$ concentration in the DRN and that in frontal cortex, thus indicating that the release in the latter area is tightly controlled by the activation of 5-HT$_{1A}$ receptors by raphe 5-HT$_{ext}$. Similarly, the administration of paroxetine (3 mg/kg ip) reduced 5-HT$_{ext}$ in frontal cortex when the microdialysis probe was perfused with a physiological solution containing 1 $\mu M$ citalopram *(61)*. Using this procedure, the systemic administration of SSRIs causes little additional blockade of the 5-HT transporter in the tissue surrounding the dialysis probe, because this is already locally blocked *(62)*, but the elevation of 5-HT$_{ext}$ produced in the somatodendritic region by the SSRIs activates 5-HT$_{1A}$ autoreceptors, thus reducing 5-HT release by nerve terminals in forebrain.

### 3.3. Regional Selectivity of the Effects of SSRIs in Forebrain: Dorsal vs Median Raphe Neuronal Pathways

The serotonergic innervation of forebrain originates from the DRN and MRN, which give rise to six different ascending tracts *(63)*. Dorsal raphe neurons innervate preferentially frontal cortex and basal ganglia; those from the MRN project mainly to limbic areas *(63,64)*. With few exceptions (e.g., the dorsal striatum), most forebrain structures are innervated by both nuclei.

Neurons from the DRN and MRN are morphologically dissimilar *(65)* and have been reported to display a differential sensitivity to the activation of 5-HT$_{1A}$ autoreceptors. Thus, the local or systemic administration of the prototypical 5-HT$_{1A}$ agonist

8-OH-DPAT preferentially inhibited serotonergic cell firing and forebrain 5-HT synthesis in the DRN pathway *(66–68)*. However, a microdialysis study in anesthetized rats reported similar reductions of 5-HT release in DRN- and MRN-innervated areas after systemic 8-OH-DPAT treatment *(69)*. Moreover, a recent study using extracellular recordings reported a comparable reduction of 5-HT neuronal firing rate in the DRN and MRN to the inhibitory actions of paroxetine and 8-OH-DPAT *(70)*. Indeed, this is a controversial but important issue, because a lesser sensitivity of MRN neurons to the self-inhibitory actions of SSRIs would imply that serotonergic transmission in limbic areas would be less affected by this negative feedback than that in frontal cortex or striatum.

The administration of single maximal doses of two different SSRIs, fluoxetine (10 mg/kg) and paroxetine (3 mg/kg), elevated 5-HT$_{ext}$ to a similar extent in cortex or hippocampus *(51,61)*. Yet, the net effects of an SSRI in a given brain area depend on the affinity of the drug for the 5-HT transporter, the density of serotonergic innervation, and the extent of self-inhibition of the fibers innervating that particular area.

We have assessed the latter point using two different approaches. First, by examining the effects of selective 5-HT$_{1A}$ agonists on the 5-HT release in forebrain structures preferentially innervated by the DRN and the MRN. Three different agents, 8-OH-DPAT, ipsapirone, and alnespirone (S-20499), induced a clear regional pattern of action at all doses examined, with greater reductions of 5-HT release in frontal cortex and striatum (the latter innervated exclusively by 5-HT neurons of the DRN) vs dorsal or vental hippocampus, receiving afferents mainly from the MRN *(71,72)*. Figure 6 shows the reduction of 5-HT$_{ext}$ in striatum elicited by a low dose of 8-OH-DPAT (25 µg/kg)—greater than that produced in dorsal hippocampus. Higher doses (0.1 and 0.3 mg/kg) reduced 5-HT$_{ext}$ in dorsal hippocampus, but always less than in striatum.

We have also examined the reduction of 5-HT release in DRN- and MRN-innervated areas induced by the systemic administration of paroxetine during local blockade of the 5-HT transporter. As previously outlined, this procedure reduced terminal 5-HT release in frontal cortex because of the activation of 5-HT$_{1A}$ autoreceptors by the excess 5-HT$_{ext}$ caused by paroxetine in the somatodendritic region. Figure 7 shows that the administration of 3 mg/kg paroxetine markedly reduced 5-HT$_{ext}$ in striatum and amygdala (both areas receiving a prominent 5-HT innervation from the DRN), but not in dorsal or ventral hippocampus, which are innervated mainly by the MRN. The participation of 5-HT$_{1A}$ receptors in this effect is illustrated by the reversal of the paroxetine-induced reduction by the selective 5-HT$_{1A}$ receptor antagonist WAY-100635 (1 mg/kg sc). This dose of WAY-100635 did not alter baseline 5-HT$_{ext}$ when given alone, but elevated 5-HT$_{ext}$ in amygdala, striatum, and ventral hippocampus to above pre-paroxetine level. This increment over baseline may be related to the additional blockade of the 5-HT transporter produced by the administration of paroxetine.

These results clearly indicate that direct (8-OH-DPAT) or indirect agonists (SSRIs) of 5-HT$_{1A}$ autoreceptors attenuate 5-HT more markedly in DRN-innervated areas. It is unclear whether these differences are attributable to a distinct sensitivity of the discharge rate of DRN and MRN neurons to the action of 5-HT$_{1A}$ receptor agonists or to local factors affecting 5-HT release in a regionally selective manner.

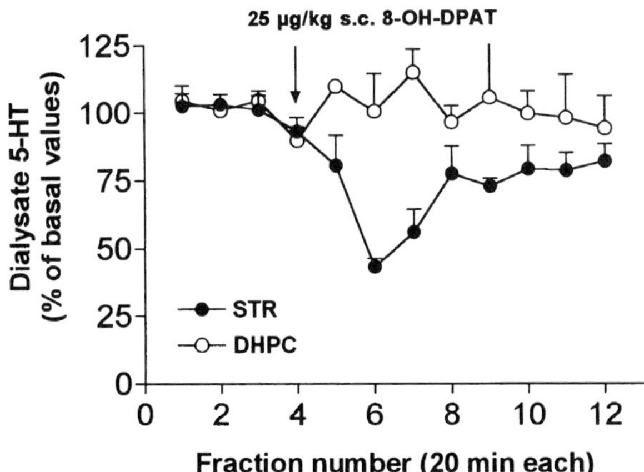


**Fig. 6.** Effects of the single administration of 8-OH-DPAT (25 µg/kg sc, shown by an arrow) on 5-HT release in dorsal striatum (STR, filled circles; $n = 6$), innervated by serotonergic neurons of the DRN, and in dorsal hippocampus (DHPC, open circles; $n = 5$), mainly innervated by those from the MRN. 8-OH-DPAT induced a very significant reduction ($p < 0.001$, ANOVA for repeated measures) of the striatal 5-HT release, and did not significantly affect that in dorsal hippocampus.

Thus, it can be speculated that local depolarizations produced by excitatory heteroreceptors located on hippocampal 5-HT terminals might perhaps compensate the inhibition of firing-dependent 5-HT release caused by activation of 5-HT$_{1A}$ receptors in the MRN.

## 4. ENHANCEMENT OF THE EFFECTS OF SSRIs BY 5-HT$_{1A}$ AUTORECEPTOR ANTAGONISTS

### 4.1. Changes of 5-HT$_{1A}$ Autoreceptors During Antidepressant Treatments

The above neurochemical data, i.e., large increases in 5-HT$_{ext}$ in the raphe nuclei after single treatment with 5-HT uptake or MAO inhibitors, are fully consistent with earlier reports on the ability of these agents to inhibit the discharge of identified serotonergic neurons of the DRN *(14,22,23)*. Prolonged treatments with serotonergic antidepressants induced a progressive recovery of the firing rate of serotonergic neurons *(13,14)*, a finding attributed to a reduced ability of the somatodendritic autoreceptor (5-HT$_{1A}$) controlling cell firing to inhibit the discharge of 5-HT neurons following prolonged antidepressant treaments. Further attempts to reveal the desensitization of somatodendritic 5-HT$_{1A}$ receptors using various experimental paradigms have yielded somewhat contradictory results *(73–75)*, although recent data gives additional support to this view *(47–49,76–78)*. Such a loss in efficacy of 5-HT$_{1A}$ autoreceptors does not appear to be accounted for by concurrent reduction of their number. Figures 8 and 9 show the unchanged density of somatodendritic and postsynaptic 5-HT$_{1A}$ receptors and of its mRNA in rats treated for 2 wk with the 5-HT uptake inhibitors clomipramine ($2 \times 10$ mg/kg/d, 24 h washout) and fluoxetine ($2 \times 5$ mg/kg/d, 48 h washout). It is thus evident that the changes in the sensitivity of 5-HT$_{1A}$ receptors

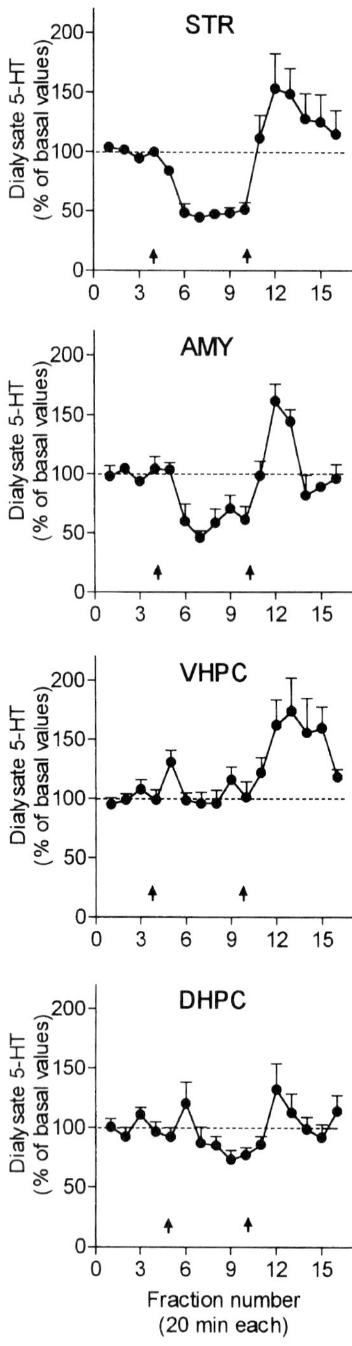


**Fig. 7.** Similar to the regional effects on 5-HT release elicited by 8-OH-DPAT, the systemic administration of paroxetine (3 mg/kg ip, first arrow) significantly reduced ($p < 0.001$, ANOVA for repeated measures) the striatal (STR) 5-HT release, but not that in dorsal or ventral hippocampus (DHPC and VHPC, respectively). The 5-HT release in the amygdala, which receives a dense input from the DRN, was also reduced by paroxetine administration. The sc administration of 1 mg/kg of the selective 5-HT$_{1A}$ receptor antagonist WAY-100635 (second arrow) reversed the attenuation produced by paroxetine in dorsal striatum and further elevated 5-HT. Data from 5–6 rats/group (except for the amygdala, $n = 3$).

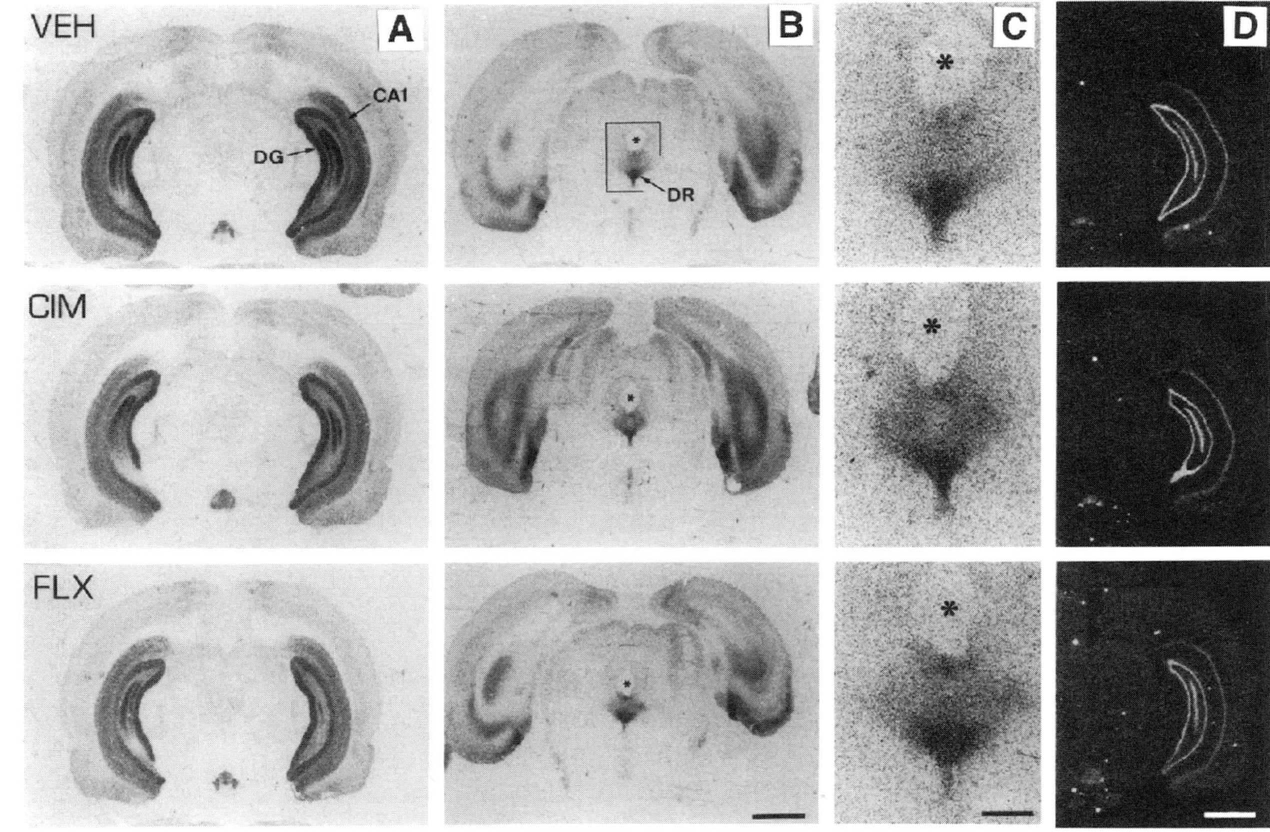



**Fig. 8.** (**A–C**) Autoradiograms showing the density of the 5-HT$_{1A}$ receptor in coronal brain sections (A, hippocampal level; B and C, midbrain level; C corresponds to an enlargement of the area marked in B, showing the DRN with greater detail), as labeled by the selective agonist [$^3$H]8-OH-DPAT. Asterisk marks the cerebral aqueduct. (**D**) *In situ* hybridization analysis of the density of the mRNA encoding the 5-HT$_{1A}$ receptor in hippocampus. Note the exclusive localization in the dentate gyrus (DG) and the CA1 hippocampal field at this level. Upper, middle, and lower rows correspond to rats treated with vehicle (saline), clomipramine (CIM, $2 \times 10$ mg/kg/d; 24 h washout), and fluoxetine (FLX, $2 \times 5$ mg/kg/d; 48 h washout) for 2 wk. Bars: A, B, and D = 2 mm; C = 500 µm.

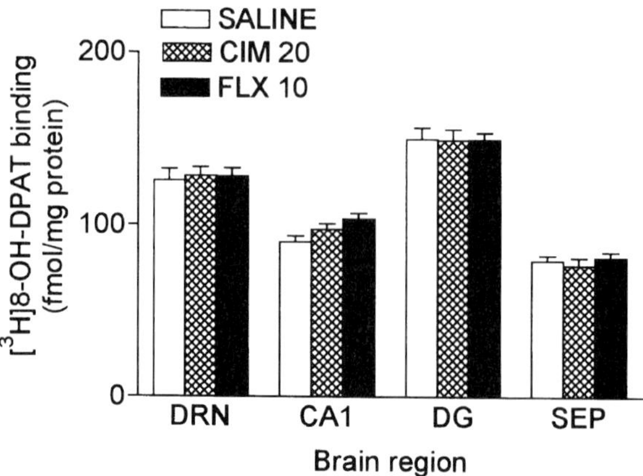


**Fig. 9.** Density of pre- and postsynapic 5-HT$_{1A}$ receptors in the DRN, dentate gyrus (DG), and CA1 hippocampal areas, and in septum (SEP). One-way ANOVA revealed the absence of changes of the density of 5-HT$_{1A}$ receptors in these brain areas after the treatment of rats with 20 mg/kg/d clomipramine (CIM 20) and 10 mg/kg/d fluoxetine (FLX 10). Data from 6 rats/group.

revealed by behavioral, neurochemical, and electrophysiological means must be accounted for by other factors, such as, for example, variations in the efficacy of the receptor–effector coupling.

### 4.2. Rationale for the Use of 5-HT$_{1A}$ Autoreceptor Antagonists

In view of the above observations on the attenuation of 5-HT release by SSRIs, it becomes evident that the 5-HT-enhancing action of antidepressant drugs is offset by the activation of somatodendritic 5-HT$_{1A}$ receptors. It was then hypothesized that the treatment with SSRIs (or MAOIs) and 5-HT$_{1A}$ receptor antagonists would accelerate and enhance the antidepressant effects of the former *(21)*. The normalization of cell firing and release produced by 5-HT$_{1A}$ receptor antagonists would enable SSRIs to increase 5-HT$_{ext}$ to a greater extent than when administered alone. The accumulated experimental evidence fully supports this working hypothesis. Earlier work indicated that the application in the DRN of the nonselective 5-HT$_1$ antagonist methiothepin enabled a low dose of citalopram to significantly increase 5-HT$_{ext}$ in frontal cortex *(40)*. Yet, methiothepin reduced 5-HT neuronal firing by itself *(79)*, which somewhat complicates the interpretation of these results. Further reports using local or systemic administration of either mixed β-adrenoceptor/5-HT$_{1A}$ antagonists (such as (−)penbutolol or (−)pindolol) or selective 5-HT$_{1A}$ antagonists (such as UH-301 or WAY 100635) have documented greater increments of 5-HT$_{ext}$ when the SSRIs are concurrently administered with these agents *(41,61,80–84)*.

The greater enhancement of 5-HT$_{ext}$ by the combination of SSRIs and 5-HT$_{1A}$ antagonists is convincingly accounted for by the prevention of the inhibitory effects of SSRIs on 5-HT neuronal firing and release by the latter *(41,84–87)*. Our own data indicate that the elevations in 5-HT$_{ext}$ produced by four SSRIs (citalopram, fluoxetine, fluvoxamine, and paroxetine), the TCA clomipramine, and the nonselective MAOI phenelzine can be further enhanced by the combined administration of 1 mg/kg WAY-

100635 *(61,88,89)*. Moreover, the application of WAY-100635 in the DRN markedly potentiated the elevation of 5-HT$_{ext}$ produced by an injection of 3 mg/kg paroxetine, thus supporting a crucial role for somatodendritic 5-HT$_{1A}$ autoreceptors in the action of WAY-100635 *(61)*. Figure 10 shows the maximal increments of dialysate 5-HT after the administration of 10 mg/kg of the 5-HT uptake inhibitors clomipramine, fluoxetine, and fluvoxamine in combination with 1 mg/kg WAY-100635. By contrast, this agent did not modify the modest change of 5-HT$_{ext}$ induced by the injection of the TCA desipramine (which neither inhibits 5-HT cell firing nor blocks the 5-HT transporter), thus supporting the exclusive participation of serotonergic neurons in the action of WAY-100635.

Also, (–)pindolol (15 mg/kg sc) enhanced the elevation of 5-HT$_{ext}$ produced by the SSRIs citalopram and paroxetine in dorsal striatum *(84)*. Yet, the extent of the potentiation was much lower than that produced by 1 mg/kg WAY-100635 in all brain areas examined (Fig. 11). The enhancement induced by WAY-100635 was maximal in striatum and minimal in dorsal hippocampus, in agreement with previous observations of a greater inhibition of 5-HT release by SSRI treatment in DRN-innervated areas. By contrast, (–)pindolol potentiated the effects of paroxetine in striatum *(84)* and hypothalamus *(82)*, but not in frontal cortex or ventral hippocampus. These differences between the effects of (–)pindolol and WAY-100635 cannot be accounted for by a differential action at the somatodendritic level. At the doses used, both agents fully prevented the paroxetine-induced inhibition of firing of DRN neurons *(41,84,85)*, which suggests that the lower potentiation induced by (–)pindolol is not caused by an insufficient antagonism of somatodendritic 5-HT$_{1A}$ receptors. Instead, this may be because of the unspecific nature of its action, possibly derived from the marked antagonism of β-adrenoceptors by the large doses required to block the somatodendritic 5-HT$_{1A}$ receptor. Terminal 5-HT$_{1B}$ autoreceptors are unlikely to play a role, because local application of (–)pindolol through the dialysis probe did not alter paroxetine effects (in any case, an opposite effect should be observed, given the functional antagonism of [–]pindolol at 5-HT$_{1B}$ receptors). At a behavioral level, rats administered 15 mg/kg (–)pindolol (but not saline) and paroxetine (3 mg/kg, ip) or citalopram (1 mg/kg, ip) were extremely sedated, with minimal spontaneous motor activity. By contrast, those treated with the combination of SSRIs and WAY-100635 had motor activities comparable to those administered SSRIs alone, but, unlike these animals, displayed a prominent pattern of sexual self-stimulation lasting for about 1 h after administration of WAY-100635 (an effect not produced by WAY-100635 alone). This behavior appears difficult to reconcile with the increase of male sexual activity produced by the 5-HT$_{1A}$ agonist 8-OH-DPAT *(90)*, since the administration of WAY-100635 should prevent any increment of sexual activity mediated by the activation of postsynaptic 5-HT$_{1A}$ receptors. This may indicate the presence of postsynaptic 5-HT receptors other than 5-HT$_{1A}$, whose activation by the enhanced 5-HT$_{ext}$ concentration promotes sexual behavior.

### 4.2. Paradoxical Reductions in 5-HT$_{ext}$ in Frontal Cortex by Mixed β-Adrenoceptor/5-HT$_{1A}$ Antagonists

The presence of an action of (–)pindolol unrelated to its antagonism at 5-HT$_{1A}$ receptors is also supported by the striking reduction in the effects of 3 mg/kg paroxetine elicited by this agent in frontal cortex (Figs. 11 and 12). The dose used (15 mg/kg)

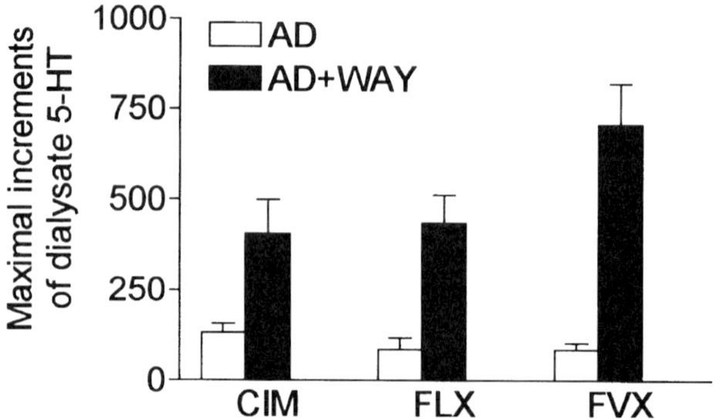



**Fig. 10.** Maximal increments over basal (in percentage) of dialysate 5-HT in frontal cortex of rats treated with 10 mg/kg of the antidepressants clomipramine (CIM), fluoxetine (FLX), and fluvoxamine (FVX) before (AD, open bars) or after (AD+WAY, filled bars) the treatment with 1 mg/kg sc of WAY-100635. The latter was injected when the 5-HT uptake inhibitors had already elicited a maximal or nearly maximal increment of dialysate 5-HT. The potentiation elicited by WAY-100635 was statistically significant in all cases ($p < 0.01$; paired Student's $t$-test). Data from 4 to 6 rats/group.

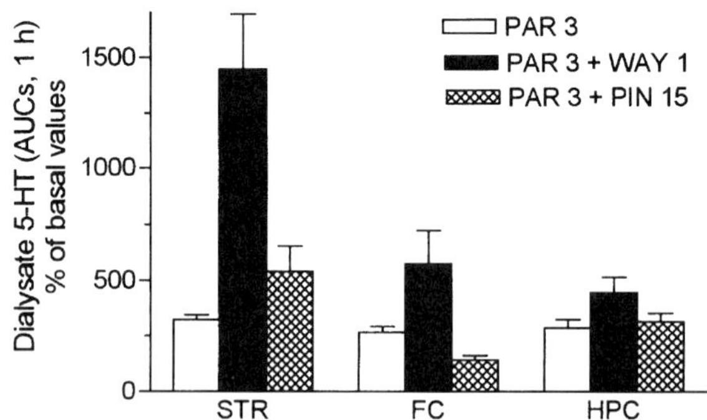



**Fig. 11.** AUC values of 1-h periods corresponding to the time of the maximal potentiation of the effects of 3 mg/kg ip, paroxetine on 5-HT$_{ext}$ by the nonselective 5-HT$_{1A}$ antagonist (−) pindolol (15 mg/kg sc) and by the selective antagonist WAY-100635 (1 mg/kg sc). WAY-100635 induced a significant potentiation of paroxetine effects in the three regions examined (though more markedly in STR and FC, predominantly innervated by the DRN). By contrast, (−)pindolol potentiated the effect of paroxetine in striatum but not in hippocampus or frontal cortex. Data from 5–18 rats/group.

did not alter 5-HT release in frontal cortex (Fig. 12) or striatum *(84)* nor reduced 5-HT discharge in the DRN when administered alone *(84)*. This suggests that the attenuation in the elevation of 5-HT$_{ext}$ produced by paroxetine was not caused by a suppression of the firing activity of DRN serotonergic neurons, despite the potential for pindolol to reduce 5-HT synthesis in limbic areas *(91)*.

The attenuation of the paroxetine-induced elevation of 5-HT$_{ext}$ in frontal cortex by (−)pindolol appears to be entirely independent of its 5-HT$_{1A}$ antagonistic properties. Thus, WAY-100635, lacking affinity for other aminergic receptors *(92)*, markedly poten-

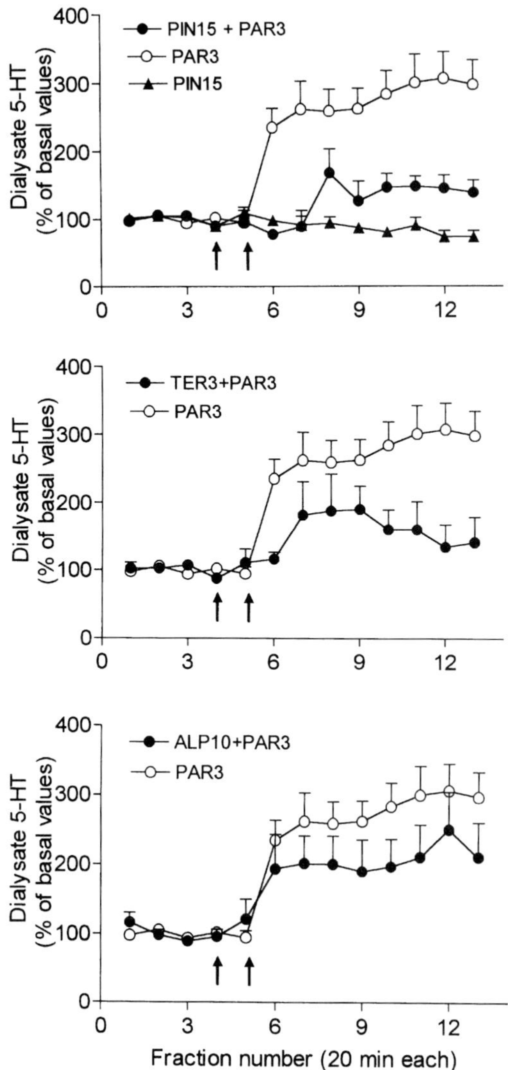

**Fig. 12.** Attenuation of the 5-HT$_{ext}$-enhancing effects of paroxetine (3 mg/kg ip, open circles; injection indicated by the second arrow) in frontal cortex by the previous treatment with the mixed β-adrenoceptor/5-HT$_{1A}$ receptor antagonists (filled circles) (–)pindolol (15 mg/kg sc; PIN 15), (–)tertatolol (3 mg/kg sc TER 3), and alprenolol (10 mg/kg sc, ALP 10). Significant reductions were observed for (–)pindolol and (–)tertatolol ($p < 0.01$, significant effects of the treatment and drug factors, and of the interaction between both). The effect of alprenolol did not reach statistical significance. Note that this dose of (–)pindolol did not alter 5-HT$_{ext}$ by itself (filled triangles in upper panel). Also, at the dose used, (–)tertatolol elevated 5-HT$_{ext}$ dorsal striatum. Data from 4 to 12 rats/group.

tiated the effects of various SSRIs, including paroxetine, in frontal cortex. By contrast, a comparable attenuation of cortical 5-HT$_{ext}$ was observed after the combined treatment with 3 mg/kg paroxetine and 3 mg/kg (–)tertatolol or 10 mg/kg alprenolol, two other β-adrenoceptors antagonists with moderate affinities for the 5-HT$_{1A}$ receptor. It is remarkable that, at the same dose, (–)tertatolol increased 5-HT release in striatum by itself *(89)*, in agreement with its enhancing action on 5-HT neuronal discharge in the DRN *(93)*. Since the serotonergic innervation of frontal cortex also originates in the DRN for the most part, it is difficult to conceive that both actions of (–)tertatolol have a common substrate. More likely, the attenuation of paroxetine effects in frontal cortex by mixed β-adrenoceptor/5-HT$_{1A}$ antagonists may derive from a blockade of β-adrenergic transmission. The antagonism of 5-HT$_{1A}$ receptors may predominate in neurons projecting to the striatum; 5-HT release would be inhibited by an unknown mechanism in fibers projecting to frontal cortex after paroxetine treatment.

A large body of data in the literature indicates the existence of functional interrelationships between 5-HT and NE. In addition to their parallel widespread innervation of forebrain, serotonergic neurons of the anterior raphe nuclei (dorsal and median) and NE neurons of the locus ceruleus display a similar pattern of activity, with higher firing rates during behavioral and motor activation and lower firing rates during sleep *(31,94)*. There is a reciprocal innervation of 5-HT and NE cell groups in the brain stem *(31,52–54)*. Thus, the firing rate of serotonergic neurons of the DRN is under the tonic excitatory influence of $\alpha_1$-adrenoceptors *(57)* and terminal 5-HT release is inhibited by $\alpha_2$-adrenergic heteroreceptors *(95–97)*. Little is known about the effects of the blockade of β-adrenoceptors on 5-HT release in brain. In enterochromaffin cells of the gut, stimulation of β-adrenoceptors promotes 5-HT release *(98)*. On the other hand, glutamate-induced excitation of locus ceruleus NE neurons is attenuated by 5-HT, and NE release is inhibited by presynaptic 5-HT$_{1D}$ heteroreceptors in peripheral tissues and cerebral vasculature *(99,100)*. This complex network of interactions facilitates the existence of a crosstalk between both systems at multiple levels, which may be involved in the attenuation of the effects of paroxetine in frontal cortex by mixed β-adrenoceptor/5-HT$_{1A}$ antagonists. These results further emphasize the need for using selective compounds to assess drug-induced changes on serotonergic transmission.

## 5. CLINICAL STUDIES

At present, several open-labeled and controlled clinical studies have been or are being carried out to assess the possible use of the mixed β-adrenoceptor/5-HT$_{1A}$ receptor antagonist pindolol, in combination with SSRIs, to accelerate and enhance the antidepressant effects of the latter. An early study reported rapid improvements of several previously untreated major depressive patients when administered a combination of 20 mg/d paroxetine and 7.5 mg/d of racemic pindolol (2.5 mg, three times daily) *(101)*. Also, the addition of this pindolol regime to the medication (paroxetine, fluvoxamine, imipramine, and phenelzine), taken by several treatment-resistant patients, induced a rapid (<1 wk) improvement in about two-thirds of them. The conclusions of this study were fully confirmed in a second open-labeled trial using the combination of paroxetine (20 mg/d) and pindolol (7.5 mg/d). Likewise, several depressed patients treated with (but unresponsive to) SSRIs and a MAO inhibitor (moclobemide) benefited from the addition of pindolol *(102)*.

The results of a double-blind, placebo-controlled study with the combination of fluoxetine and pindolol, conducted by this group in Barcelona in 111 major depressive outpatients, are also consistent with the view that pindolol enhances the antidepressant action of SSRIs *(103)*. After 1 wk on single-blind placebo (to exclude those individuals with a high placebo response) patients were randomly assigned to each of two treatment arms. Fifty-six patients received 20 mg/d fluoxetine + placebo three times daily. The rest of the patients received 20 mg/d fluoxetine + 2.5 mg pindolol three times daily. At the end point (6 wk of active treatment), a significantly greater percentage of the patients treated with fluoxetine + pindolol had experienced a therapeutic response (75 vs 59% in the fluoxetine + placebo group). The same difference (60 vs 44%) was noted in the percentage of patients who experienced a remission (a reduction in score of the Hamilton Depression Scale of 17 items to 8 or less), although in that case the difference was only marginally significant ($p < 0.055$). Moreover, the median time needed to achieve a sustained, 50% reduction of the symptoms was 19 d in the fluoxetine + pindolol group and 29 d in the fluoxetine + placebo group ($p < 0.01$). Such differences were markedly reduced when considering only responder patients in both groups, which suggests that the main effect of pindolol was to increase the response to fluoxetine, thus reducing the median response time of the whole group. The enhancement of the antidepressant effects of fluoxetine was achieved without a concurrent increment of the number of side effects associated with its use (mainly gastrointestinal).

Another controlled study carried out in London, with the combination of paroxetine (20 mg/d) and pindolol (7.5 mg/d), reported a significant reduction of the time to onset of antidepressant action. By d 4, 23% of the patients taking paroxetine + pindolol and 5% of those taking paroxetine + placebo showed a decrease in severity of at least 50% from baseline. A significant difference was noted between both treatment groups until d 21, but no further enhancement of the efficacy of paroxetine was noted (M. Isaac, personal communication).

## 6. OPEN QUESTIONS

### 6.1. Does Pindolol Block 5-HT$_{1A}$ Receptors in Depressed Patients?

The above clinical data are consistent with the hypothesis derived from experimental studies *(21)*, i.e., that the prevention of the SSRI-induced self-attenuation of serotonergic neurons by 5-HT$_{1A}$ antagonists might result in enhanced effects of the SSRIs. Indeed, there are still many open questions that will be addressed at both the experimental and clinical level in forthcoming years. The first concerns the mechanism of action of pindolol, as well as the dose used, given its lack of selectivity for 5-HT$_{1A}$ receptors. Until the recent availability of selective 5-HT$_{1A}$ antagonists, (−)pindolol was the prototypical agent used to block the actions of selective 5-HT$_{1A}$ agonists in preclinical studies. In the rat, 4–8 mg/kg of the active (−) enantiomer were typically employed. At much lower doses (single, 20–30 mg oral doses of the racemic compound), pindolol blocked various 5-HT$_{1A}$-mediated responses in humans, like the hypothermia or hormonal secretions induced by selective 5-HT$_{1A}$ agonists *(104–107)*. Recent data on sleep architecture in depressed patients treated with 5-HT uptake inhibitors plus pindolol (2.5 mg, three times daily) is fully consistent with an enhancement of serotonergic transmission by pindolol *(108)*. However, the dose used in clinical studies raises reasonable doubts about the neurobiological substrate of pindolol action in humans.

This difference might be accounted for by a greater affinity for human vs rat 5-HT$_{1A}$ receptors. To our knowledge, there are no comparative data on the affinity of pindolol for 5-HT$_{1A}$ receptors in both species. Indirect evidence suggests that the affinity of pindolol for human 5-HT$_{1A}$ receptors might be greater than for rat 5-HT$_{1A}$ receptors. It was found that 5-HT displaced by 40–50% the high-affinity (subnanomolar) binding of [$^{125}$I]cyanopindolol—used to label β-adrenoceptors in human brain—in DRN and hippocampus, two brain areas with a very high density of 5-HT$_{1A}$ receptors. The displacement by 5-HT was absent in areas lacking 5-HT$_{1A}$ sites, which suggests that low concentrations of cyanopindolol labeled these sites in human brain *(109)*. It remains to be established whether pindolol mimics the ability of its cyano derivative to interact with these sites with such high affinity. Also, the distribution of [$^{125}$I]pindolol binding sites in the rat hippocampal formation is markedly different from that in the human brain, which suggests that it labels different subpopulations of receptors in both species *(110)*. Moreover, the affinity of (−)pindolol for 5-HT$_{1A}$ receptors (but not those of 5-HT or other serotonergic agents) is reduced two orders of magnitude by single-point mutation in the seventh transmembrane domain of the 5-HT$_{1A}$ receptor *(111)*. This supports the existence of distinct binding sites for 5-HT and (−)pindolol within the receptor protein. Then, small differences in the structure of the human and rat receptors might set the basis for affinity differences. Hence, although direct evidence is lacking, a differential interaction of pindolol with rat and human 5-HT$_{1A}$ receptors would not be surprising, in view of the above findings.

A second aspect concerns the dose used. So far, all clinical studies have employed the dose first reported (2.5 mg, three times daily) *(101)*. However, this may be lower than optimal. Circumstantial evidence in open trials suggests that some resistant patients may improve at a higher (15 mg/d) dose, which emphasizes the need for conducting dose-efficacy studies.

### 6.2. Is Antagonism at β-adrenoceptors Involved in the Pindolol-Induced Potentiation of SSRIs?

The dose of pindolol used in clinical studies (7.5 mg/d) is several times lower than that used to treat hypertension—up to 45 mg/d *(112)*—and, therefore, a low peripheral β-blocking activity may be predicted. This is confirmed by the very moderate drop of pulse rate found in the 56 patients treated with fluoxetine + pindolol (−5 beats/min, compared to fluoxetine + placebo). No difference in blood pressure was noted between treatment groups *(103)*.

Despite the ability of TCAs and MAOIs to downregulate cortical β-adrenoceptors, antagonists at these sites have never demonstrated an antidepressant activity by themselves *(113)*. On the contrary, their use has been associated with a higher consumption of antidepressants *(114)*. Recently, it has been documented that pindolol did not potentiate the effects of desipramine and trimipramine, two TCAs without affinity for either the 5-HT transporter or 5-HT receptors *(115)*, thus suggesting that pindolol cannot augment the antidepressant efficacy of compounds without a primary serotonergic mechanism (despite the fact that desipramine downregulates β-adrenoceptors). Moreover, the antidepressant efficacy of SSRIs at clinically relevant levels is not associated with downregulation of β-adrenoceptors *(16–19)*. Therefore, it is unclear how an additional blockade of cortical β-adrenoceptors might improve their efficacy.

Furthermore, the increased efficacy and shorter latency of the fluoxetine + pindolol treatment cannot be ascribed to a hypothetical action of pindolol on anxiety items of the Hamilton scale used to evaluate the severity, because the change in these items was not different in both treatment groups. On the contrary, significant differences were noted in the subscales of core symptoms and retardation, in support of a true antidepressant effect of pindolol addition.

Hence, the available evidence indicates that the beneficial action of pindolol is not a result of its antagonistic properties at β-adrenoceptors. Yet, the relatively lower potentiation induced by (–)pindolol of paroxetine's effects at an experimental level (compared to that elicited by WAY-100635) indicates that blockade of β-adrenoceptors may alter 5-HT transmission in forebrain (particularly in frontal cortex, where several mixed β-adrenoceptor/5-HT$_{1A}$ antagonists reduced, rather than potentiated, the 5-HT$_{ext}$-enhancing effects of paroxetine).

The full confirmation of the current working hypothesis on the mechanisms underlying the enhancement of antidepressant effects by pindolol will require the clinical demonstration that β-adrenoceptor blockers devoid of affinity for 5-HT$_{1A}$ receptors do not potentiate the antidepressant effects of SSRIs, and also that selective 5-HT$_{1A}$ autoreceptor antagonists enhance the antidepressant effects of SSRIs. The first option may be feasible in the near future, given the clinical availability of selective β-adrenoceptor antagonists devoid of 5-HT$_{1A}$ activity (atenolol, betaxolol, and so on). However, the second option must be delayed, because 5-HT$_{1A}$ receptor antagonists currently used at experimental levels are still in early phases of development.

### 6.3. Effects of Pindolol on Pre- and Postsynaptic 5-HT$_{1A}$ Receptors

A third important issue involving the mechanism of action of pindolol concerns the possible blockade of postsynaptic 5-HT$_{1A}$ receptors in hippocampus, which mediate, at least in part, antidepressant effects *(13–15)*. Should the effects of SSRIs be mediated exclusively through hippocampal 5-HT$_{1A}$ receptors, their blockade by pindolol would abolish the potential benefits of increasing the activity of the presynaptic component of serotonergic transmission *(116)*. However, although (–)pindolol blocked several 5-HT$_{1A}$-mediated responses in the rat *(104,117,118)*, it did not antagonize the 5-HT$_{1A}$ receptor-mediated hyperpolarization of pyramidal neurons in the CA$_3$ field of dorsal hippocampus induced by the microiontophoretic application of 5-HT and 8-OH-DPAT *(84)*. By contrast, the same dose of (–)pindolol prevented the reduction in firing of dorsal raphe 5-HT neurons induced by paroxetine and LSD *(84,88)*. This differential activity at pre- and postsynaptic 5-HT$_{1A}$ receptors is analogous to that found for spiperone *(119)*, and cannot be explained by the higher receptor reserve of 5-HT$_{1A}$ receptors in the raphe *(120,121)* (which would produce an opposite situation). Other 5-HT$_{1A}$ antagonists, like BMY-7378, NAN-190, or SDZ-216525, behave as antagonists at hippocampal 5-HT$_{1A}$ receptors and as partial agonists at somatodendritic 5-HT$_{1A}$ autoreceptors *(122–124)*. Moreover, chemically dissimilar high-affinity 5-HT$_{1A}$ agonists elicit different behavioral patterns that are supposedly mediated by postsynaptic 5-HT$_{1A}$ receptors *(122)*. To add a further degree of complexity, microiontophoretically applied (–)pindolol attenuated the excitation (but not the inhibition) produced by 8-OH-DPAT in neurons of the dorsal vagal nucleus. Perhaps more surprising, WAY-100635 and 8-OH-DPAT produced the same inhibitory effects in some neurons *(125)*.

The reasons for these pharmacologial differences among 5-HT$_{1A}$ receptors are unclear, and do not seem to be explained by molecular differences as subtypes of the 5-HT$_{1A}$ receptor have not been reported. Raphe and hippocampal 5-HT$_{1A}$ receptors display similar biochemical characteristics *(126)*. It is thus possible that some of these differences may be accounted for by the different neuronal phenotypes on which pre- and postsynaptic 5-HT$_{1A}$ receptors are located. Differences in the type of G protein, or its coupling to either potassium channels or local factors in the recognition of compounds, may also be involved. Also, the 5-HT$_{1A}$ receptor appears to be exclusively coupled to a potassium channel in raphe *(127,128)*, but three different effector mechanisms have been described in hippocampus: a potassium channel *(129)*, inhibition of forskolin-stimulated adenylate cyclase *(130)*, and activation of adenylate cyclase *(131,132)* (the latter response may perhaps correspond to 5-HT$_7$ receptors, for which 8-OH-DPAT displays moderate affinity) *(133)*. Differences in G-receptor coupling and sensitivity to spiperone have been reported between the so-called intra- and extrasynaptic hippocampal 5-HT$_{1A}$ receptors *(119,134)*. This astonishing functional complexity may encompass a pharmacological distinction of pre- and postsynaptic 5-HT$_{1A}$ receptors, even if both are identical molecular entities.

On the other hand, despite the well-known actions of antidepressants via 5-HT$_{1A}$ receptors, the involvement of other 5-HT receptor subtypes cannot be discarded *a priori*, considering the variety of psychological and physical signs and symptoms exhibited by depressed patients and the large number of 5-HT receptor subtypes. Earlier binding data indicated the presence of adaptative changes of 5-HT$_1$ and 5-HT$_2$ sites *(1,16)*. More recently, a role for 5-HT$_{1B/1D}$ receptors in the mechanism of several antidepressants has been suggested *(135)*. Also, the blockade of 5-HT actions at cortical 5-HT$_2$ receptors has been proposed to contribute to the actions of antidepressant drugs *(136)*. Indeed, the stimulation of 5-HT$_{2C}$ receptors by the nonselective agonist m-chorophenyl piperazine (mCPP) increases anxiety in panic patients and healthy controls *(137)*, and various chronic antidepressant treatments reduce the function of 5-HT$_{2B/2C}$ receptors in brain *(138)*. Also, very little is known about the role of other 5-HT receptors localized in cortical and limbic structures in the effects of antidepressants. Recent data suggested a downregulation of hypothalamic 5-HT$_7$ receptors by chronic treatment with the SSRI fluoxetine *(139)*. Therefore, the view expressed that antidepressant effects are solely mediated via postsynaptic 5-HT$_{1A}$ receptors may be incomplete, as judged from the large number of other 5-HT receptors and the little current knowledge on its relevance for the treatment of major depression and other psychiatric conditions.

## 7. FUTURE PERSPECTIVES

The greater effects of antidepressants in combination with 5-HT$_{1A}$ antagonists clearly demonstrate the dramatic importance that 5-HT$_{1A}$-mediated self-inhibition of 5-HT neurons have in their mechanism of action. The scheme in Figure 13 summarizes the current views on the acute effects of SSRIs on the presynaptic component of serotonergic transmission. The blockade of the transporter at the somatodendritic level prevents the reuptake of 5-HT released by dendrites (and possibly by axonal varicosities of neighboring 5-HT fibers). The excess 5-HT$_{ext}$ is sufficient to activate somatodendritic 5-HT$_{1A}$ receptors, leading to a reduction of cell firing and release by (mainly through DRN) axon terminals in forebrain. Terminal autoreceptors also exert a local control of 5-HT synthesis and release, but its quantitative importance during acute SSRI treatment is less than that played by 5-HT$_{1A}$ autoreceptors, because SSRIs increase 5-HT$_{ext}$ more in the raphe (rich in 5-HT$_{1A}$ receptors) than in regions rich in nerve terminals.

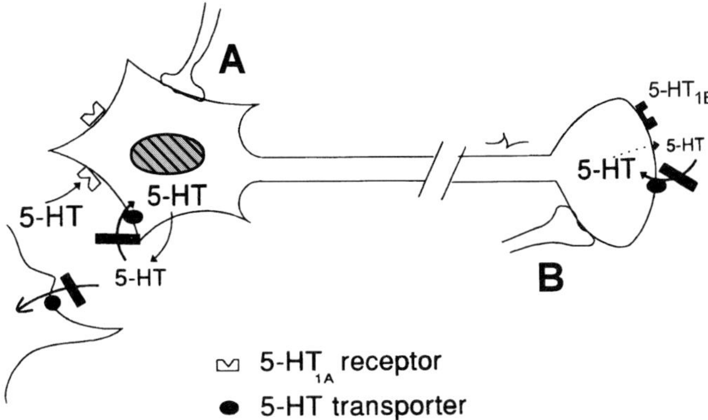


**Fig. 13.** Schematic representation of the effects of 5-HT uptake inhibitors on serotonergic neurons. The blockade of the 5-HT transporter at the level of the raphe nuclei elevates 5-HT$_{ext}$ to an extent that activates somatodendritic autoreceptors (5-HT$_{1A}$). This causes the opening of potassium channels, a reduction of the discharge rate and the subsequent reduction of 5-HT release by forebrain terminals. Heteroregulation of the activity of serotonergic neurons is exerted by afferents at somatodendritic (**A**) and terminal (**B**) levels.

The physiological role played by somatodendritic 5-HT$_{1A}$ autoreceptors remains unclear. Recently, it has been reported that the application of 8-OH-DPAT into the DRN reduced somatodendritic release and markedly increased the time spent in REM sleep in the cat *(140)*. Given its precise anatomical localization and the presence of a somatodendritic release of 5-HT *(141–144)*, it is evident that 5-HT$_{1A}$ autoreceptors are important elements in the regulation of serotonergic activity. Hence, physiologically induced increments in the activity of excitatory afferents to 5-HT neurons may increase somatodendritic 5-HT release, which, by acting on 5-HT$_{1A}$ autoreceptors of neighboring neurons, may attenuate their firing activity and reduce 5-HT release in target forebrain areas. The attenuation of terminal 5-HT release observed after SSRI treatment is possibly a pharmacological manifestation of a more general physiological mechanism used by 5-HT neurons to maintain homeostasis.

Yet, on the basis that enhancement of serotonergic activity leads to clinical improvement in major depression, pharmacological interventions increasing terminal 5-HT release are advisable. It is therefore important to obtain a deeper understanding of the mechanisms that determine 5-HT$_{ext}$ concentrations at the somatodendritic and terminal levels, encompassing both self-control mechanisms (i.e., transporter and autoreceptors) and heterocontrol mechanisms exerted by excitatory and inhibitory afferents at somatodendritic and terminal level (A and B, respectively, in Fig. 13).

Another aspect that deserves current and future attention concerns the postsynaptic changes elicited by enhanced 5-HT$_{ext}$ after chronic SSRI treatments, and the identification of the receptors involved. Indeed, the administration of drugs acting selectively at certain postsynaptic receptors would induce more rapid and effective improvements by avoiding the self-inhibition of 5-HT neurons during early treatment stages with serotonergic antidepressants (SSRIs, MAOIs, and some TCAs). The existence of adaptative changes in second and third messengers by chronic antidepressant treatments in target neurons has been documented *(145–147; see also* Chapters 10 and 11). However, it is uncertain if the clinical improvement is the immediate consequence of an enhanced activation of certain 5-HT receptors, or whether postsynaptic adaptive changes are also required (in the first case, only the presynaptic component would be responsible for the delayed onset of action). The inability of pindolol to fully suppress the delay in antidepressant action  suggests that postsynaptic adaptive mechanisms are required for obtaining a full remission of symptoms. Yet, in view of the data obtained in rat brain, more potent $5\text{-HT}_{1A}$ antagonists could greatly accelerate and potentiate the effects of SSRIs. Clinical observations indicate the presence of rapid mood changes in depressed patients that are inconsistent with the longer periods required for gene regulation or changes in protein synthesis. For instance, some melancholic patients display dramatic diurnal mood changes, with a much greater severity of symptoms in the morning. Also, the rapid relapse and subsequent recovery of depressed patients treated with SSRIs, when administered a tryptophan-free amino acid mixture, indicates that, in certain conditions, there may be a rapid switch between euthymic and depressive states (and vice versa). Finally, severely depressed patients experience a rapid remission when subjected to sleep deprivation, a procedure with an unknown neurobiological substrate, but which may be related to an overall increase in the activity of aminergic neurons (otherwise silent during sleep) during this period. Unlike those antidepressant drugs, the effects of sleep deprivation are rapid and transient. This may lead to speculation that there are two different neurobiological components to the action of antidepressant treatments, related to the initial remission of symptoms and the persistence of the clinical improvement, respectively. The first one might involve essentially presynaptic components and its efficacy could be enhanced by preventing self-inhibition of 5-HT (and perhaps NE) neurons. The second one would involve changes at the level of gene expression after prolonged additional activation of certain postsynaptic receptors, once an increased extracellular concentration of 5-HT has been attained. Failure of the treatments to induce a recovery might then depend on the first, i.e., presynaptic component (the inability to produce a sufficient increase of $5\text{-HT}_{ext}$ in specific forebrain areas). Thus, patients resistant to SSRI treatment might be viewed as those individuals with, e.g., a greater somatodendritic 5-HT release, hypersensitive $5\text{-HT}_{1A}$ autoreceptors, or hyperactive 5-HT transporter, all leading to a greater-than-average self-inhibition of 5-HT neurons during SSRI treatment. Those patients would have additional difficulty increasing $5\text{-HT}_{ext}$ in forebrain, despite an effective blockade of the 5-HT transporter in nerve terminals. Addition of $5\text{-HT}_{1A}$ antagonists to the therapeutic regime would bring about a therapeutic response by disrupting this negative feedback and allowing 5-HT neurons to normalize their firing and terminal release. Although this hypothesis lacks experimental support, it is consistent with the greater response observed with the combination fluoxetine + pindolol in a controlled study, and with the effects of pindolol in resistant patients.

In summary, the combined administration of 5-HT uptake inhibitors (either tricyclic or selective) and $5\text{-HT}_{1A}$ antagonists results in an enhanced increase in forebrain $5\text{-HT}_{ext}$. This is a result of the blockade of self-inhibition of 5-HT neurons elicited by the activation of somatodendritic $5\text{-HT}_{1A}$ receptors during these treatments. In the rat, this effect is more marked in forebrain areas innervated by serotonergic neurons of the DRN. The clinical data obtained in pilot and controlled studies using the nonselective $5\text{-HT}_{1A}$ antagonist pindolol is in agreement with these experimental findings. Additional sites of action of pindolol cannot be fully excluded, but are unlikely. Future work should be aimed at gaining a better knowledge of the control of serotonergic activity by afferents to the somatodendritic and terminal areas of 5-HT neurons as a way to selectively modulate 5-HT release in areas of interest. Further studies should also address the relative role of pre- and postsynaptic adaptative processes in the mechanism of action of SSRIs.

## ACKNOWLEDGMENTS

This work has been supported by a grant from the Fondo de Investigación Sanitaria (FIS 95/266). L. R. and I. H. are recipients of predoctoral fellowships from the CIRIT (Generalitat de Catalunya). The financial support from pharmaceutical companies (Bayer, CIBA-Geigy, Eli Lilly, and Servier) in some of the cited studies is gratefully acknowledged. Thanks are given to Lundbeck, Smith Kline Beecham, and Wyeth-Ayerst for the supply of citalopram, paroxetine, and WAY-100635, respectively, and to F. Cruz-Sánchez for supply of brain samples. We thank Letizia Campa for excellent technical assistance.

## REFERENCES

1. Baker, G. B. and Greenshaw, A. J. (1989) Effects of long-term administration of antidepressants and neuroleptics on receptors in the central nervous system. *Cel. Mol. Neurobiol.* **9**, 1–44.
2. Caldecott-Hazard, S., Morgan, D. G., De Leon-Jones, F., Overstreet, D. H., and Janowsky, D. (1991) Clinical and biochemical aspects of depressive disorders: II. Transmitter/receptor theories. *Synapse,* **9**, 251–301.
3. Richelson, E. and Pfenning, M. (1984) Blockade by antidepressants and related compounds of biogenic amine uptake into rat brain synaptosomes: most antidepressants selectively block norepinephrine uptake. *Eur. J. Pharmacol.* **104**, 277–286.
4. Hyttel, J. (1994) Pharmacological characterization of selective serotonin reuptake inhibitors (SSRIs). *Int. Clin. Psychopharmacol.* **7**, 866–867.
5. Tollefson, G. D. and Holman, S. L. (1994) How long to onset of antidepressant action: a meta-analysis of patients treated with fluoxetine or placebo. *Int. Clin. Psychopharmacol.* **9**, 245–250.
6. Anderson, I. M. and Tomenson, B. M. (1994) The efficacy of selective serotonin reuptake inhibitors in depression: a meta-analysis of studies against tricyclic antidepressants. *J. Psychopharmacol.* **8**, 238–249.
7. Cusack, B., Nelson, A., and Richelson, E. (1994) Binding of antidepressants to human brain receptors: focus on newer generation compounds. *Psychopharmacology* **114**, 559–565.
8. Danish University Antidepressant Group (1986) Citalopram: clinical effect profile in comparison with clomipramine. A controlled multicenter study. *Psychopharmacology* **90**, 131–138.
9. Danish University Antidepressant Group (1990) Paroxetine: a selective serotonin reuptake inhibitor showing better tolerance, but weaker antidepressant effect than clomipramine in a controlled multicenter study. *J. Affect. Disord.* **18**, 289–299.
10. Delgado, P. L., Charney, D. S., Price, L. H., Aghajanian, G. K., Landis, H., and Heninger, G. R. (1990) Serotonin function and the mechanism of antidepressant action. Reversal of antidepressant-induced remission by rapid depletion of plasma tryptophan. *Arch. Gen. Psychiatry* **47**, 411–418.

11. Bel, N. and Artigas, F. (1996) Reduction of serotonergic function in rat brain by tryptophan depletion. Effects in control and fluvoxamine-treated rats. *J. Neurochem.* **67,** 669–676.

12. Hoyer, D., Clarke, D. E., Fozard, J. R., Hartig, P. R., Martin, G. R., Mylecharane, E. J., Saxena, P. R., and Humphrey, P. P. A. (1994) International union of pharmacology classification of receptors for 5-hydroxytryptamine (serotonin). *Pharmacol. Rev.* **46,** 157–203.

13. Blier, P., De Montigny, C., and Chaput, Y. (1987) Modifications of the serotonin system by antidepressant treatment: implications for the therapeutic response in major depression. *J. Clin. Psychopharmacol.* **7,** 24S–35S.

14. Blier, P. and De Montigny, C. (1994) Current advances and trends in the treatment of depression. *Trends Pharmacol. Sci.* **15,** 220–226.

15. De Vry, J. (1995) 5-HT$_{1A}$ receptor agonists: recent developments and controversial issues. *Psychopharmacology* **121,** 1–26.

16. Peroutka, S. J. and Snyder, S. H. (1980) Long-term antidepressant treatment decreases spiroperidol-labeled serotonin receptor binding. *Science* **210,** 88–90.

17. Dechant, K. L. and Clissold, S. P. (1991) Paroxetine. A review of its pharmacodynamic and pharamcokinetic properties, and therapeutic potential in depressive illness. *Drugs* **41,** 225–253.

18. Hyttel, J., Overo, K. F., and Arnt, J. (1984) Biochemical effects and drug levels in rats after long-term treatment with the specific 5-HT-uptake inhibitor, citalopram. *Psychopharmacology* **83,** 20–27.

19. Hrdina, P. D. and Vu, T. B. (1993) Chronic fluoxetine treatment upregulates 5-ht uptake sites and 5-ht2 receptors in rat brain—an autoradiographic study. *Synapse* **14,** 324–331.

20. Adell, A. and Artigas, F. (1991) Differential effects of clomipramine given locally or systemically on extracellular 5-hydroxytryptamine in raphe nuclei and frontal cortex. An in vivo microdialysis study. *Naunyn-Schmiedeberg's Arch. Pharmacol.* **343,** 237–244.

21. Artigas, F. (1993) 5-HT and antidepressants: new views from microdialysis studies. *Trends Pharmacol. Sci.* **14,** 262.

22. Aghajanian, G. K., Graham, A. W., and Sheard, M. H. (1970) Serotonin-containing neurons in brain: depression of firing by monoamine oxidase inhibitors. *Science* **169,** 1100–1102.

23. Scuvée-Moreau, J. and Dresse, A. (1979) Effect of various antidepressant drugs on the spontaneous firing rate of locus coeruleus and dorsal raphe neurons of the rat. *Eur. J. Pharmacol.* **57,** 219–225.

24. Pazos, A. and Palacios, J. M. (1985) Quantitative autoradiographic mapping of serotonin receptors in the rat brain. I. Serotonin-1 receptors. *Brain Res.* **346,** 205–230.

25. Kobilka, B. K., Frielle, T., Collins, S., Yang-Feng, T., Kobilka, T. S., Francke, U., Lefkowitz, R. J., and Caron, M. G. (1987) An intronless gene encoding a potential member of the family of receptors coupled to guanine nucleotide regulatory proteins. *Nature* **329,** 75–79.

26. Blakely, R. D., Berson, H. E., Fremeau R. T., Caron, M. G., Peek, M. M., Prince, H. K., and Bradley, C. C. (1991) Cloning and expression of a functional serotonin transporter from rat brain. *Nature* **354,** 66–70.

27. Cortés, R., Mengod, G., Celada, P., and Artigas, F. (1993) p-Chlorophenylalanine increases tryptophan-5-hydroxylase mRNA levels in the rat dorsal raphe: a time course study using in situ hybridization. *J. Neurochem.* **60,** 761–764.

28. Dahlström, A. and Fuxe, K. (1964) Evidence for the existence of monoamine-containing neurons in the central nervous system. I. Demonstration of monoamines in cell bodies of brain neurons. *Acta Physiol. Scand.* **62(Suppl 232),** 1–55.

29. Descarries, L., Watkins, K. C., Garcia, S., and Beaudet, A. (1982) The serotonin neurons in nucleus raphe dorsalis of adult rat: a light and electron microscope radioautographic study. *J. Comp. Neurol.* **207,** 239–254.

30. Oleskevich, S. and Descarries, L. (1990) Quantified distribution of the serotonin innervation in adult rat hippocampus. *Neuroscience* **34,** 19–33.
31. Jacobs, B. L. and Azmitia, E. C. (1992) Structure and function of the brain serotonin system. *Physiol. Rev.* **72,** 165–229.
32. Marcinkiewicz, M., Vergé, D., Gozlan, H., Pichat, L., and Hamon, M. (1984) Autoradiographic evidence for the heterogeneity of 5-ht1 sites in the rat brain. *Brain Res.* **291,** 159–163.
33. Hoyer, D., Pazos, A., Probst, A., and Palacios, J. M. (1986) Serotonin receptors in the human brain. I. Characterization and autoradiographic localization of 5-HT$_{1A}$ recognition sites. Apparent absence of 5-HT$_{1B}$ recognition sites. *Brain Res.* **376,** 85–96.
34. Cortés, R., Soriano, E., Pazos, A., Probst, A., and Palacios, J. M. (1988) Autoradiography of antidepressant binding sites in the human brain: localization using [$^3$H]imipramine and [$^3$H]paroxetine. *Neuroscience* **27,** 473–496.
35. Hrdina, P. D., Foy, B., Hepner, A., and Summers, R. J. (1990) Antidepressant binding sites in brain: autoradiographic comparison of [$^3$H]paroxetine and [$^3$H]imipramine localization and relationship to serotonin transporter. *J. Pharmacol. Exp. Ther.* **252,** 410–418.
36. Seguela, P., Watkins, K. C., and Descarries, L. (1989) Ultrastructural relationships of axon terminals in the cerebral cortex of the adult rat. *J. Comp. Neurol.* **289,** 129–142.
37. Smiley, J. F. and Goldman-Rakic, P. S. (1996) Serotonergic axons in monkey prefrontal cortex synapse predominantly on interneurons as demonstrated by serial section electron microscopy. *J. Comp. Neurol.* **367,** 431–433.
38. Artigas, F., Bel, N., Figueras, G., Suñol, C., Vilaró, M. T., and Mengod, G. (1995) Antidepressants inhibit the neuronal 5-HT transporter in glial cells. In vivo and in vitro studies. *Soc. Neurosci. Abstracts* **21,** 1369.
39. Bel, N. and Artigas, F. (1992) Fluvoxamine preferentially increases extracellular 5-hydroxytryptamine in the raphe nuclei: an in vivo microdialysis study. *Eur. J. Pharmacol.* **229,** 101–103.
40. Invernizzi, R., Belli, S., and Samanin, R. (1992) Citalopram's ability to increase the extracellular concentration of serotonin in the dorsal raphe prevents the drug's effect in frontal cortex. *Brain Res.* **584,** 322–324.
41. Gartside, S. E., Umbers, V., Hajos, M., and Sharp, T. (1995) Interaction between a selective 5-HT$_{1A}$ receptor antagonist and an SSRI in vivo: effects on 5-HT cell firing and extracellular 5-HT. *Br. J. Pharmacol.* **115,** 1064–1070.
42. Malagié, I., Trillat, A. C., Jacquot, C., and Gardier, A. (1996) Effects of acute fluoxetine on extracellular serotonin levels in the raphe: an in vivo microdialysis study. *Eur. J. Pharmacol.* **286,** 213–217.
43. Bel, N. and Artigas, F. (1993) Chronic treatment with fluvoxamine increases extracellular serotonin in frontal cortex but not in raphe nuclei. *Synapse* **15,** 243–245.
44. Bel, N. and Artigas, F. (1996) In vivo effects of the simultaneous blockade of serotonin and norepinephrine transporters on serotonergic function. Microdialysis studies. *J. Pharmacol. Exp. Ther.,* **15,** 349–360.
45. Ferrer, A. and Artigas, F. (1994) Effects of single and chronic treatment with tranylcypromine on extracellular serotonin in rat brain. *Eur. J. Pharmacol.* **263,** 227–234.
46. Rutter, J. J., Gundlah, C., and Auerbach, S. B. (1994) Increase in extracellular serotonin produced by uptake inhibitors is enhanced after chronic treatment with fluoxetine. *Neurosci. Lett.* **171,** 183–186.
47. Invernizzi, R., Bramante, M., and Samanin, R. (1994) Chronic treatment with citalopram facilitates the effect of a challenge dose on cortical serotonin output: role of presynaptic 5-HT$_{1A}$ receptors. *Eur. J. Pharmacol.* **260,** 243–246.

48. Invernizzi, R., Bramante, M., and Samanin, R. (1995) Extracellular concentrations of serotonin in the dorsal hippocampus after acute and chronic treatment with citalopram. *Brain Res.* **696,** 62–66.

49. Kreiss, D. S. and Lucki, I. (1995) Effects of acute and repeated administration of antidepressant drugs on extracellular levels of 5-hydroxytryptamine measured in vivo. *J. Pharmacol. Exp. Ther.* **274,** 866–876.

50. Piñeyro, G., Blier, P., Dennis, T., and de Montigny, C. (1994) Desensitization of the neuronal 5-HT carrier following its long-term blockade. *J. Neurosci.* **14,** 3036–3047.

51. Hervás, I. and Artigas, F. (1996) Regional effects of fluoxetine on extracellular 5-HT in DRN- and MRN-innervated areas of the rat brain. *J. Neurochem.* **86(Suppl 2),** 36S.

52. Sakai, K., Salvert, D., Touret, M., and Jouvet, M. (1977) Afferent connections of the nucleus raphe dorsalis in the cat as visualized by the horseradish peroxidase technique. *Brain Res.* **137,** 11–35.

53. Baraban, J. M. and Aghajanian, G. K. (1981) Noradrenergic innervation of serotonergic neurons in the dorsal raphe: demonstration by electronic microscopic autoradiography. *Brain Res.* **24,** 1–11.

54. Peyron, C., Luppi, P. H., Fort, P., Rampon, C., and Jouvet, M. (1996) Lower brainstem catecholamine afferents to the rat dorsal raphe nucleus. *J. Comp. Neurol.* **364,** 402–413.

55. Jacobs, B. L. and Fornal, C. A. (1991) Activity of brain serotonergic neurons in the behaving animal. *Pharmacol. Rev.* **43,** 563–578.

56. Fornal, C. A., Metzler, C. W., Marrosu, F., Ribiero do valle, L. E., and Jacobs, B. L. (1996) A subgroup of dorsal raphe serotonergic neurons in the cat is strongly activated during oral-buccal movements. *Brain Res.* **716,** 123–133.

57. Baraban, J. M. and Aghajanian, G. K. (1980) Suppression of serotonergic neuronal firing by alpha-adrenoceptor antagonists: evidence against GABA mediation. *Eur. J. Pharmacol.* **66,** 287–294.

58. Ferré, S., Cortés, R., and Artigas, F. (1994) Dopaminergic regulation of the sertonergic raphe-striatal pathway: microdialysis studies in freely moving rats. *J. Neurochem.* **14,** 4839–4846.

59. Levine, E. S. and Jacobs, B. L. (1992) Neurochemical afferents controlling the activity of serotonergic neurons in the dorsal raphe nucleus: microiontophoretic studies in the awake cat. *J. Neurosci.* **12,** 4037–4044.

60. Tao, R. and Auerbach, S. B. (1996) Differential effect of NMDA on extracellular serotonin in rat midbrain raphe and forebrain sites. *J. Neurochem.* **66,** 1067–1075.

61. Artigas, F. and Romero, L. (1996) Enhancement of SSRI effects by 5-HT$_{1A}$ antagonists. Regional selectivity and role of somatodendritic autoreceptors. *Soc. Neurosci. Abstracts* **22,** 611.

62. Rutter, J. J. and Auerbach, S. B. (1993) Acute uptake inhibition increases extracellular serotonin in the rat forebrain. *J. Pharmacol. Exp. Ther.* **265,** 1319–1324.

63. Azmitia, E. C. and Segal, M. (1978) An autoradiographic analysis of the differential ascending projections of the dorsal and median raphe nuclei in the rat. *J. Comp. Neurol.* **179,** 641–668.

64. Imai, H., Steindler, D. A., and Kitai, S. T. (1986) The organization of divergent axonal projections from the midbrain raphe nuclei in the rat. *J. Comp. Neurol.* **243,** 363–380.

65. Kosofsky, B. E. and Molliver, M. E. (1987) The serotoninergic innervation of cerebral cortex: different classes of axon terminals arise from dorsal and median raphe nuclei. *Synapse* **1,** 153–168.

66. Sinton, C. M. and Fallon, S. L. (1988) Electrophysiological evidence for a functional differentiation between subtypes of the 5-HT1 receptor. *Eur. J. Pharmacol.* **157,** 173–181.

67. Blier, P., Serrano, A., and Scatton, B. (1990) Differential responsiveness of the rat dorsal and median raphe 5-HT systems to 5-HT1 receptor agonists and p-chloroamphetamine. *Synapse* **5**, 120–133.

68. Invernizzi, R., Carli, M., Di Clemente, A., and Samanin, R. (1991) Administration of 8-hydroxy-2-(Di-n-propylamino)tetralin in raphe nuclei dorsalis and medianus reduces serotonin synthesis in rat brain: differences in potency and regional sensitivity. *J. Neurochem.* **56**, 243–247.

69. Hjorth, S. and Sharp, T. (1991) Effect of the 5-HT$_{1A}$ receptor agonist 8-OH-DPAT on the release of 5-HT in dorsal and median raphe-innervated rat brain regions as measured by in vivo microdialysis. *Life Sci.* **48**, 1779–1786.

70. Hajos, M., Gartside, S. E., and Sharp, T. (1995) Inhibition of median and dorsal raphe neurons following administration of the selective serotonin reuptake inhibitor paroxetine. *Naunyn-Schmiedeberg's Arch. Pharmacol.* **351**, 624–629.

71. Lesourd, M., Mocaër, E., Casanovas, J. M., and Artigas, F. (1995) Effects of the full 5-HT$_{1A}$ agonist alnespirone (S-20499) on 5-HT release in DRN- and MRN-innervated areas in rat brain. *Soc. Neurosci. Abstracts* **21**, 1368.

72. Casanovas, J. M. and Artigas, F. (1996) Differential effects of ipsapirone on 5-HT release in the dorsal and median raphe neuronal pathways. *J. Neurochem.* **67**, 1945–1952.

73. Welner, S. A., de Montigny, C., Desroches, J., Desjardins, P., and Suranyi-Cadotte, B. E. (1989) Autoradiographic quantification of serotonin$_{1A}$ receptors in rat brain following antidepressant drug treatment. *Synapse* **4**, 347–352.

74. Hensler, J. G., Covachich, A., and Frazer, A. (1991) A quantitative autoradiographic study of serotonin 1A receptor regulation. Effect of 5,7-dihydroxytryptamine and antidepressant treatments. *Neuropsychopharmacology* **4**, 131–144.

75. Hjorth, S. and Auerbach, S. B. (1994) Lack of 5-HT(1A) autoreceptor desensitization following chronic citalopram treatment, as determined by in vivo microdialysis. *Neuropharmacology* **33**, 331–334.

76. Jolas, T., Haj-Dahmane, S., Kidd, E. J., Langlois, X., Lanfumey, L., Fattaccini, C. M., Vantalon, V., Laporte, A. M., Adrien, J., Gozlan, H., and Hamon, M. (1994) Central pre- and postsynaptic 5-HT$_{1A}$ receptors in rats treated chronically with a novel antidepressant, cericlamine. *J. Pharmacol. Exp. Ther.* **268**, 1432–1443.

77. Lepoul, E., Laaris, N., Doucet, E., Laporte, A. M., Hamon, M., and Lanfumey, L. (1995) Early desensitization of somato-dendritic 5-HT$_{1A}$ autoreceptors in rats treated with fluoxetine or paroxetine. *Naunyn-Schmiedeberg's Arch. Pharmacol.* **352**, 141–148.

78. Martin, K. F., Phillips, I., Hearson, M., Prow, M. R., and Heal, D. J. (1992) Characterization of 8-OH-DPAT-induced hypothermia in mice as a 5-HT$_{1A}$ autoreceptor response and its evaluation as a model to selectively identify antidepressants. *Br. J. Pharmacol.* **107**, 15–21.

79. Gallager, D. W. and Aghajanian, G. K. (1976) Effect of antipsychotic drugs on the firing of dorsal raphe cells. I. Role of adrenergic system. *Eur. J. Pharmacol.* **39**, 341–355.

80. Hjorth, S. (1993) Serotonin 5-HT$_{1A}$ autoreceptor blockade potentiates the ability of the 5-HT reuptake inhibitor citalopram to increase nerve terminal output of 5-HT in vivo: a microdialysis study. *J. Neurochem.* **60**, 776–779.

81. Artigas, F., Romero, L., Celada, P., and Bel, N. (1994) 5-HT$_{1A}$ antagonists and terminal 5-HT release. Effects of the combined treatment with 5-HT uptake inhibitors. *Soc. Neurosci. Abstracts* **20**, 1541.

82. Dreshfield, L. J., Wong, D. T., Perry, K. W., and Engleman, E. A. (1996) Enhancement of fluoxetine-dependent increase of extracellular serotonin (5-HT) levels by (–)-pindolol, an antagonist at 5-HT$_{1A}$ receptors. *Neurochem. Res.* **21**, 557–562.

83. Arborelius, L., Nomikos, G. G., Hertel, P., Salmi, P., Grillner, P., Hook, B. B., Hacksell, U., and Svensson, T. H. (1996) The 5-HT$_{1A}$ receptor antagonist (S)-UH-301 augments the increase in extracellular concentrations of 5-HT in the frontal cortex produced by both acute and chronic treatment with citalopram. *Naunyn-Schmiedeberg's Arch. Pharmacol.* **353**, 630–640.

84. Romero, L., Bel, N., Artigas, F., De Montigny, C., and Blier, P. (1996) Effect of pindolol on the function of pre- and postsynaptic 5-HT$_{1A}$ receptors: in vivo microdialysis and electrophysiological studies in the rat brain. *Neuropsychopharmacology* **15**, 349–360.

85. Blier, P., Seletti, B., Bouchard, C., Artigas, F., and De Montigny, C. (1994) Functional evidence for the differential responsiveness of pre- and postsynaptic 5-HT$_{1A}$ receptors in the rat brain. *Soc. Neurosci. Abstracts* **20**, 1540.

86. Arborelius, L., Nomikos, G. G., Grillner, P., Hertel, P., Hook, B. B., Hacksell, U., and Svensson, T. H. (1995) 5-HT$_{1A}$ receptor antagonists increase the activity of serotonergic cells in the dorsal raphe nucleus in rats treated acutely or chronically with citalopram. *Naunyn-Schmiedeberg's Arch. Pharmacol.* **352**, 157–165.

87. Hjorth, S. and Auerbach, S. B. (1994) Further evidence for the importance of 5-HT$_{1A}$ autoreceptors in the action of selective serotonin reuptake inhibitors. *Eur. J. Pharmacol.* **260**, 251–255.

88. Artigas, F., Romero, L., De Montigny, C., and Blier, P. (1996) Acceleration of the effect of selected antidepressant drugs in major depression by 5-HT$_{1A}$ antagonists. *Trends Neurosci.*, in press.

89. Romero, L., Bel, N., Casanovas, J. M., and Artigas, F. (1996) Two actions are better than one: avoiding self-inhibition of serotonergic neurones enhances the effects of serotonin uptake inhibitors. *Int. Clin. Psychopharmacol.* **11(Suppl. 4)**, 1–8.

90. Ahlenius, S. and Larsson, K. (1987) Evidence for a unique pharmacological profile of 8-OH-DPAT by evaluation of its effects on male rat sexual behavior, in *Brain 5-HT$_{1A}$ Receptors* (Dourish, C. T., Ahlenius, S., and Hutson, P. H., eds.). Ellis Horwood, Chichester, UK, pp. 185–198.

91. Hjorth, S. and Carlsson, A. (1985) (−)-Pindolol stereospecifically inhibits rat brain serotonin (5-HT) synthesis. *Neuropharmacology* **24**, 1143–1146.

92. Forster, E. A., Cliffe, I. A., Bill, D. J., Dover, G. M., Jones, D., Reilly, Y., and Fletcher, A. (1995) A pharmacological profile of the selective silent 5-HT$_{1A}$ receptor antagonist, WAY-100635. *Eur. J. Pharmacol.* **281**, 81–88.

93. Jolas, T., Hajdahmane, S., Lanfumey, L., Fattaccini, C. M., Kidd, E. J., Adrien, J., Gozlan, H., Guardiola-Lemaitre, B., and Hamon, M. (1993) (−)Tertatolol is a potent antagonist at presynaptic and postsynaptic serotonin 5-HT(1A) receptors in the rat brain. *Naunyn-Schmiedeberg's Arch. Pharmacol.* **347**, 453–463.

94. Sakai, K. (1991) Physiological properties and afferent connections of the locus coeruleus and adjacent tegmental neurons involved in the generation of paradoxical sleep in the rat. *Prog. Brain Res.* **88**, 31–46.

95. Starke, K., Göthert, M., and Kilbinger, H. (1989) Modulation of neurotransmitter release by presynaptic autoreceptors. *Physiol. Rev.* **69**, 864–989.

96. Tao, R. and Hjorth, S. (1992) $\alpha_2$-adrenoceptor modulation of rat ventral hippocampal 5-hydroxytryptamine release in vivo. *Naunyn-Schmiedeberg's Arch. Pharmacol.* **345**, 137–143.

97. Bel, N. and Artigas, F. (1996) In vivo effects of the simultaneous blockade of serotonin and norepinephrine transporters on serotonergic function. Microdialysis studies. *J. Pharmacol. Exp. Ther.* **268**, 1064–1072.

98. Racké, K., Reimann, A., Schwörer H., and Kilbinger, H. (1995) Regulation of 5-HT release from enterochromaffin cells. *Behav. Brain Res.* **73**, 83–87.

99. Aston-Jones, G., Shipley, M. T., Chovet, T. G., Ennis M., van Bockstaele, E., Pieribone, V., Shiekhattar, R., Akaoka, H., Drolet, G., Astier, B., Charléty, P., Valentino, R. J., and Williams, J. T. (1991) Afferent regulation of locus coeruleus neurons: anatomy, physiology and pharmacology. *Prog. Brain. Res.* **88**, 47–75.

100. Molderings, G. J., Frolich, D., Likungu, J., and Göthert, M. (1996) Inhibition of noradrenaline release via presynaptic 5-HT$_{1D}$ alpha receptors in human atrium. *Naunyn-Schmiedeberg's Arch. Pharmacol.* **353**, 272–280.

101. Artigas, F., Pérez, V., and Alvarez, E. (1994) Pindolol induces a rapid improvement of depressed patients treated with serotonin reuptake inhibitors. *Arch. Gen. Psychiatry* **51**, 248–251.

102. Blier, P. and Bergeron, R. (1995) Effectiveness of pindolol with selected antidepressant drugs in the treatment of major depression. *J. Clin. Psychopharmacol.* **15**, 217–222.

103. Pérez, V., Gilaberte, I., Faries, D., Alvarez, E., and Artigas, F. (1996) Pindolol augments the antidepressant efficacy of fluoxetine. Results of a double-blind, randomized. *Lancet*, in press.

104. Middlemiss, D. N., Neill, J., and Tricklebank, M. D. (1985) Subtypes of the 5-HT receptor involved in hypothermia and forepaw treading induced by 8-OH-DPAT. *Br. J. Pharmacol.* **85**, 25.

105. Lesch, K. P., Poten, B., Sohnle, K., and Schulte, H. M. (1990) Pharmacology of the hypothermic response to 5-HT$_{1A}$ receptor activation in humans. *Eur. J. Clin. Pharmacol.* **39**, 17–19.

106. Meltzer, H. Y. and Maes, M. (1994) Effect of pindolol on the L-5-HTP-induced increase in plasma prolactin and cortisol concentrations in man. *Psychopharmacology* **114**, 635–643.

107. Park, S. B. G. and Cowen, P. J. (1995) Effect of pindolol on the prolactin response to d-fenfluramine. *Psychopharmacology* **118**, 471–474.

108. Wilson, S. J., Bell, C. J., Coupland, N. J., and Nutt, D. J. (1996) SSRIs, TCAs and augmentation therapies. Exploring mechanisms by acute and chronic sleep effects. *Eur. Neuropsychopharmacol.* **6(Suppl 3)**, 32.

109. Pazos, A., Gonzalez-Gil, J., and Castillo, M. J. (1994) Increased affinity of beta-blockers for 5-HT$_{1A}$ receptors in the human brain: an autoradiographic study. *IUPHAR Satellite Meeting on Serotonin* **3**, 99.

110. Duncan, G. E., Little, K. Y., Koplas, P. A., Kirkman, J. A., Breese, G. R., and Stumpf, W. E. (1991) Beta adrenergic receptor distribution in human and rat hippocampal formation: marked species differences. *Brain Res.* **561**, 84–92.

111. Guan, X. M., Peroutka, S. J., and Kobilka, B. K. (1992) Identification of a single amino acid residue responsible for the binding of a class of beta-adrenergic receptor antagonists to 5-hydroxytryptamine1A receptors. *Mol. Pharmacol.* **41**, 695–698.

112. Martindale (1993) *The Extra Pharmacopoeia*. Pharmaceutical Press, London.

113. Bailly, D. (1996) The role of beta adrenoceptor blockers in the treatment of psychiatric disorders. *CNS Drugs* **5**, 115–136.

114. Avron, J., Everitt, D. E., and Weiss, D. (1986) Increased antidepressant use in patients prescribed beta-blockers. *J. Am. Med. Assoc.* **255**, 357–360.

115. Blier, P., Bergeron, R., and de Montigny, C. (1996) Present and future tools to improve the antidepressant response based on the differential properties of 5-HT receptor subtypes. *Eur. Neuropsychopharmacol.* **6(Suppl 3)**, 17.

116. Deakin, J. F. W., Graeff, F. G., and Guimaraes, F. S. (1993) Acute reuptake inhibitors do not increase 5-HT availability. *Trends Pharmacol. Sci.* **14**, 398.

117. Aulakh, C. S., Wozniak, K. M., Haas, K., Hill, J. L., Zohar, J., and Murphy, D. L. (1988) Food intake, neuroendocrine and temperature effects of 8-OH-DPAT in the rat. *Eur. J. Pharmacol.* **146,** 253–259.

118. Scott, P. A., Chou, J. M., Tang, H., and Frazer, A. (1994) Differential induction of 5-HT$_{1A}$-mediated responses in vivo by three chemically dissimilar 5-HT$_{1A}$ agonists. *J. Pharmacol. Exp. Ther.* **270,** 198–208.

119. Blier, P., Lista, A., and de Montigny, C. (1993) Differential properties of presynaptic and postsynaptic 5-hydroxytryptamine(1A) receptors in the dorsal raphe and hippocampus. 1. Effect of spiperone. *J. Pharmacol. Exp. Ther.* **265,** 7–15.

120. Cox, R. F., Meller, E., and Waszczak, B. L. (1993) Electrophysiological evidence for a large receptor reserve for inhibition of dorsal raphe neuronal firing by 5-HT(1A) agonists. *Synapse* **14,** 297–304.

121. Meller, E., Goldstein, M., and Bohmaker, K. (1990) Receptor reserve for 5-hydroxytryptamine-mediated inhibition of serotonin synthesis: possible relationship to anxiolyticproperties of 5-hydroxytriptamine agonists. *Mol. Pharmacol.* **37,** 231–237.

122. Hjorth, S. and Sharp, T. (1990) Mixed agonist/antagonist properties of NAN-190 at 5-HT$_{1A}$ receptors: behavioural and in vivo brain microdialysis studies. *Life Sci.* **46,** 955–963.

123. Sharp, T., Backus, L. I., Hjorth, S., Bramwell, S. R., and Grahame-Smith, D. G. (1990) Further investigation of the in vivo pharmacological properties of the putative 5-HT$_{1A}$ antagonist, BMY-7378. *Eur. J. Pharmacol.* **176,** 331–340.

124. Sharp, T., Mcquade, R., Fozard, J. R., and Hoyer, D. (1993) The novel 5-HT(1A)-receptor antagonist, SDZ 216-525 decreases 5-HT release in rat hippocampus in vivo. *Br. J. Pharmacol.* **109,** 699–702.

125. Wang, Y., Jones, J. F. X., Ramage, A. G., and Jordan, D. (1995) Effects of 5-HT and 5-HT$_{1A}$ receptor agonists and antagonists on dorsal vagal preganglionic neurons in anesthetized rats an ionophoretic study. *Br. J. Pharmacol.* **116,** 2291–2297.

126. Radja, F., Daval, G., Hamon, M., and Vergé, D. (1992) Pharmacological and physico-chemical properties of pre- versus postsynaptic 5-hydroxytryptamine$_{1A}$ receptor binding sites in the rat brain: a quantitative autoradiographic study. *J. Neurochem.* **58,** 1338–1346.

127. Aghajanian, G. K. and Lakoski, J. M. (1984) Hyperpolarization of serotonergic neurons by serotonin and LSD: studies in brain slices showing increased K+ conductance. *Brain Res.* **305,** 181–185.

128. Clarke, W. P., Yocca, F. D., and Maayani, S. (1996) Lack of 5-hydroxytryptamine(1A)-mediated inhibition of adenylyl cyclase in dorsal raphe of male and female rats. *J. Pharmacol. Exp. Ther.* **277,** 1259–1266.

129. Andrade, R., Malenka, R. C., and Nicoll, R. A. (1986) A G protein couples serotonin and GABA-B receptors to the same channel in hippocampus. *Science* **234,** 1261–1265.

130. De Vivo, M. and Maayani, S. (1986) Characterization of the 5-hydroxytryptamine 1A receptor-mediated inhibition of forskolin-stimulated adenylate cyclase activity in guinea-pig and rat hippocampal membranes. *Pharmacol. Exp. Ther.* **238,** 248–253.

131. Markstein, R., Hoyer, D., and Engel, G. (1986) 5-HT$_{1A}$ receptors mediate stimulation of adenylate cyclase in rat hippocampus. *Naunyn-Schmiedeberg's Arch. Pharmacol.* **333,** 335–341.

132. Sijbesma, H., Schipper, J., Molewijk, H. E., Bosch, A. I., and de Kloet, E. R. (1991) 8-hydroxy-2-(di-N-propylamino)tetralin increases the activity of adenylate cyclase in the hippocampus of freely-moving rats. *Neuropharmacology* **30,** 967–975.

133. Ruat, M., Traiffort, E., Leurs, R., Tardivellacombe, J., Diaz, J., Arrang, J. M., and Schwartz, J. C. (1993) Molecular cloning, characterization, and localization of a high-

affinity serotonin receptor (5-HT[7]) activating cAMP formation. *Proc. Natl. Acad. Sci. USA* **90,** 8547–8551.

134. Blier, P., Lista, A., and de Montigny, C. (1993) Differential properties of pre- and postsynaptic 5-hydroxytryptamine receptors in the dorsal raphe and hippocampus: II. Effect of pertussis and cholera toxins. *J. Pharmacol. Exp. Ther.* **265,** 16–23.
135. O'Neill, M. F., Fernandez, A. G., and Palacios, J. M. (1996) GR 127935 blocks the locomotor and antidepressant-like effects of RU 24969 and the action of antidepressants in the mouse tail suspension test. *Pharmacol. Biochem. Behav.* **53,** 535–539.
136. Borsini, F. (1994) Balance between cortical 5-HT$_{1A}$ and 5-HT2 receptor fuction: hypothesis for a faster antidepressant action. *Pharmacol. Res.* **30,** 1–11.
137. Charney, D. S., Woods, S. W., Goodman, W. K., and Heninger, G. R. (1987) Serotonin function in anxiety. II. Effects of the serotonin agonist mCPP in panic disorder patients and healthy subjects. *Psychopharmacology* **92,** 14–24.
138. Kennett, G. A., Lightowler, S., Debiasi, V., Stevens, N. C., Wood, M. D., Tulloch, I. F., and Blackburn, T. P. (1994) Effect of chronic administration of selective 5-hydroxytryptamine and noradrenaline uptake inhibitors on a putative index of 5-HT$_{2C/2B}$ receptor function. *Neuropharmacology* **33,** 1581–1588.
139. Sleight, A. J., Carolo, C., Petit, N., Zwingelstein, C., and Bourson, A. (1995) Identification of 5-hydroxytryptamine(7) receptor binding sites in rat hypothalamus sensitivity to chronic antidepressant treatment. *Mol. Pharmacol.* **47,** 99–103.
140. Portas, C. M., Thakkar, M., Rainnie, D., and Mccarley, R. W. (1996) Microdialysis perfusion of 8-hydroxy-2-(di-n-propylamino)tetralin (8-OH-DPAT) in the dorsal raphe nucleus decreases serotonin release and increases rapid eye movement sleep in the freely moving cat. *J. Neurosci.* **16,** 2820–2828.
141. Héry, F., Faudon, M., and Ternaux, J. P. (1982) In vivo release of serotonin in two raphe nuclei (raphe dorsalis and magnus) of the cat. *Brain Res. Bull.* **8,** 123–129.
142. Adell, A., Carceller, A., and Artigas, F. (1993) In vivo brain dialysis study of the somatodendritic release of serotonin in the raphe nuclei of the rat. Effects of 8-hydroxy-2-(di-N-propylamino)tetralin. *J. Neurochem.* **60,** 1673–1681.
143. Bosker, F., Klompmakers, A., and Westenberg, H. (1994) Extracellular 5-hydroxytryptamine in median raphe nucleus of the conscious rat is decreased by nanomolar concentrations of 8-hydroxy-2-(di-n-propylamino)tetralin and is sensitive to tetrodotoxin. *J. Neurochem.* **63,** 2165–2171.
144. Davidson, C. and Stamford, J. A. (1995) Evidence that 5-hydroxytryptamine release in rat dorsal raphe nucleus is controlled by 5-HT$_{1A}$, 5-HT$_{1B}$ and 5-HT$_{1D}$ autoreceptors. *Br. J. Pharmacol.* **114,** 1107–1109.
145. Nestler, E. J., Terwilliger, R. Z., and Duman, R. S. (1989) Chronic antidepressant administration alters the subcellular distribution of cyclic AMP-dependent protein kinase in rat frontal cortex. *J. Neurochem.* **53,** 1644–1647.
146. Popoli, M., Vocaturo, C., Pérez, J., Smeraldi, E., and Racagni, G. (1995) Presynaptic Ca$^{2+}$/calmodulin-dependent protein kinase. II: Autophosphorylation and activity increase in the hippocampus after long-term blockade of serotonin reuptake. *Mol. Pharmacol.* **48,** 623–629.
147. Nibuya, M., Nestler, E. J., and Duman, R. S. (1996) Chronic antidepressant administration increases the expression of cAMP response element binding protein (CREB) in rat hippocampus. *J. Neurosci.* **16,** 2365–2372.

# 2

# Antidepressant Properties of Specific Serotonin–Noradrenaline Reuptake Inhibitors

**Michael Briley and Chantal Moret**

## 1. INTRODUCTION

Despite 30 yr of intensive research and development of antidepressant drugs, about 30% of patients do not respond satisfactorily to treatment, and no antidepressant starts to produce its therapeutic effect in less than 2–3 wk. Although considerable progress has been made in improving the tolerability of antidepressant drugs, the tricyclic antidepressants (TCAs), developed in the early 1960s, are still the "gold standard" for antidepressant efficacy. The monoamine hypothesis of depression, which dates from the same period *(1)*, is still the most widely accepted, even though it has undergone a series of minor modifications over the years. The symptoms of depression are generally considered to be related to a decreased content and/or activity of cerebral monoamines, particularly noradrenaline (NA) and serotonin (5-hydroxytryptamine, 5-HT) *(2,3)*. TCAs inhibit the reuptake of both NA and 5-HT to varying extents. In addition, they interact directly, with affinities from 10 to 100 n$M$, with various synaptic receptors, notably the cholinergic muscarinic receptor, $\alpha_1$-adrenoceptors, and histamine $H_1$ receptors *(4)*. These interactions are responsible for the major side effects of these compounds: dry mouth, constipation, blurred vision and other cholinergic side effects, orthostatic hypotension, and sedation. In addition to the discomfort they cause, these side effects diminish antidepressant efficacy either by causing treatment to be abandoned or by reducing the compliance and encouraging the use of low, therapeutically suboptimal doses.

The considerable research efforts that have been expended over the last few decades have thus centered on the search for compounds devoid of these receptor interactions and their associated side effects. The first major advance since the TCAs, the selective 5-HT reuptake inhibitors (SSRI), in fact went one step further, because they had not only lost the direct receptor interactions, but also the ability to inhibit the reuptake of NA.

Compared to the tricyclic antidepressants (TCAs), the SSRIs are better tolerated. Their efficacy in major depression is, however, no greater, and their onset of action no more rapid, than that of the TCAs. Indeed, some studies have shown that the efficacy of TCAs, such as clomipramine, may be significantly greater than that of SSRIs, such

*From: Antidepressants: New Pharmacological Strategies*
*Edited by: P. Skolnick, Humana Press Inc., Totowa, NJ*

as citalopram or paroxetine *(5,6)*. A recent meta-analysis *(7)* shows that, although globally the efficacy of SSRIs and TCAs are comparable, there is a significant superiority of the TCAs over the SSRIs in hospitalized patients and in patients with severe depression, as judged by a high Hamilton Depression Rating Score (HDRS).

There is considerable evidence to suggest that increasing NA function has an antidepressant effect. Relatively selective NA reuptake inhibitors, such as desipramine and lofepramine, have clear antidepressant efficacy. Increasing noradrenergic neurotransmission via blockade of $\alpha_2$-adrenoceptors, with compounds such as mianserine, mirtazapine *(8)*, or idazoxan *(9)*, improves depressive symptoms; electroconvulsive therapy (ECT) has been shown to increase the release of NA *(10)*. Evidence that simultaneous effects on the NA system and the 5-HT system are probably supplementary is provided by combination studies of SSRIs and selective NA reuptake inhibitors *(11,12)*. Consequently, there has been a recent tendency toward the development of new antidepressants that selectively and simultaneously inhibit the reuptake of both 5-HT and NA, with no affinity for the synaptic receptors responsible for the adverse effects of the TCAs. These compounds are referred to as the specific 5-HT/NA reuptake inhibitors (SNRIs).

This chapter will review the pharmacological and clinical profiles of the two principal compounds of this class, milnacipran (Ixel®, F2207, 1-phenyl-1-diethyl-aminocarbonyl-2-aminomethyl-cyclopropane [Z] hydrochloride [Pierre Fabre Medicament]) and venlafaxine (Effexor® WY-45,030, [Δ]1-[2-{dimethylamino}-1-{4-methoxyphenyl} ethyl] cyclohexanol hydrochloride [Wyeth]) (Fig. 1). A third compound, duloxetine (LY248686, [+]-*N*-methyl-3-[1-naphthalenyloxy]-2-thiophenepropanamine hydrochloride [Eli Lilly]) *(13)*, has entered early clinical trials, but its development as an antidepressant has apparently been suspended (except in Japan) for reasons of intolerance.

## 2. PHARMACOLOGICAL PROPERTIES

### 2.1. Effects on Neurotransmitter Systems

#### 2.1.1. Milnacipran

Milnacipran inhibits the uptake of radiolabeled 5-HT and NA into rat hypothalamic slices with similar potency *(14*; Table 1), but not that of dopamine into striatal slices. The compounds, H75/12 and H 77/77, are taken up into cells by the selective 5-HT and NA transporter systems, respectively, where they displace endogenous monoamines. This massive release of monoamines leads to a hyperthermic reaction in rats. The reduction of this hyperthermic effect can thus be used as an in vivo indicator of the inhibition of the respective monoamine transporter system. Milnacipran inhibits, virtually equipotently, the hyperthermia produced by the two monoamine displacers, thus confirming the inhibition of both the uptake of 5-HT and NA by milnacipran in vitro *(14)*.

The direct in vivo examination of extracellular neurotransmitter levels has become possible since the advent of intracerebral microdialysis, a technique that allows direct sampling and measurement of brain neurotransmitters and their metabolites in the extracellular fluid of either anesthetized or freely moving animals. The extracellular levels of both 5-HT and NA, measured by microdialysis in hypothalamus of freely moving

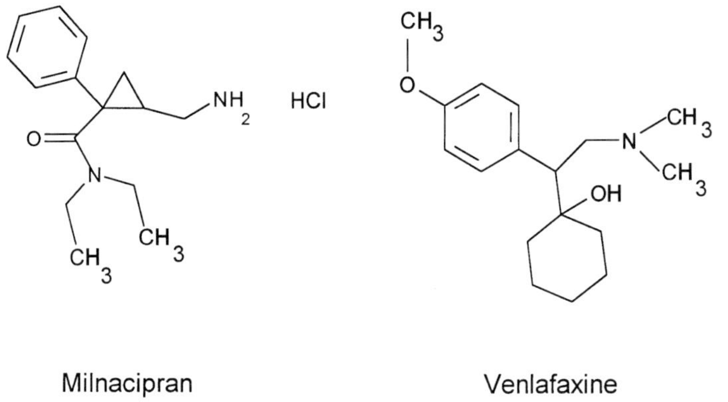

Milnacipran

Venlafaxine

**Fig. 1.** Chemical structures of milnacipran (F2207, 1-phenyl-1-diethyl-aminocarbonyl-2-aminomethyl-cyclopropane [Z] hydrochloride) and venlafaxine (WY-45,030, [Δ]1-[2-{dimethylamino}-1-{4-methoxyphenyl}ethyl] cyclohexanol hydrochloride).

guinea pigs, were increased by up to 300% after the administration of milnacipran at 10 and 40 mg/kg ip *(15)*. The 5-HT metabolite, 5-hydroxyindoleacetic acid (5-HIAA), and the NA metabolite, 4-hydroxy-3-methoxyphenyl-glycol (MHPG), were decreased by about 50%, as expected, since the inhibition of reuptake prevents intracellular production of metabolites. These results confirm that, in vivo, milnacipran massively increases the extracellular levels of both 5-HT and NA, as expected from its in vitro affinity for monoamine uptake sites *(14*; Table 1).

Milnacipran does not interact with any neurotransmitter receptor tested. Binding affinities, expressed as $IC_{50}$ values, were greater than $10^{-5}M$ for the 40 receptors studied *(14*; and unpublished data). In particular, and in contrast to the TCAs, there was no affinity for $\alpha_1$-adrenoceptors, or muscarinic or histaminergic $H_1$ receptors, which are considered to be responsible for the orthostatic hypotension, anticholinergic effects (dry mouth, constipation, blurred vision) and sedation seen with TCAs *(14)*.

A number of studies from various laboratories *(16)* have confirmed the original observations *(14,17)* that, unlike many antidepressant drugs, repeated administration of milnacipran does not result in downregulation of β-adrenoceptors. This absence of effect on radioligand binding has also been extended to an absence of effect on the β-adrenoceptor-linked adenylate cyclase and on β-adrenoceptor-mediated behavior, such as salbutamol-induced hypoactivity in mice *(18)*. Since milnacipran inhibits the reuptake of NA and increases the extracellular levels of NA in vivo, as measured by microdialysis *(15)*, it is rather surprising that there is no influence on β-adrenoceptor sensitivity. For the moment, there is no clear explanation. In addition, when administered repeatedly, milnacipran produced no alterations in $\alpha_1$- or $\alpha_2$-adrenoceptors, $5\text{-HT}_1$ or $5\text{-HT}_2$ receptors, or benzodiazepine binding sites, which are modified by certain antidepressants. Moreover, after chronic administration, uptake and accumulation of 5-HT and NA were unmodified, and the potency of milnacipran to inhibit monoamine uptake in vitro in the cortex remained unaltered *(17)*.

The rate-limiting enzymes for the synthesis of 5-HT and NA are tryptophan and tyrosine hydroxylase, respectively. These enzymes are highly sensitive to feedback con-

**Table 1**
**Inhibition of Monoamine Reuptake into Rat Brain Tissue by TCAs,
SSRIs, and SNRIs**

| | NA, n$M$ | 5-HT, n$M$ |
|---|---|---|
| Tricyclic antidepressants (TCAs) | | |
| Imipramine | 41 | 14 |
| Amitriptyline | 43 | 40 |
| Clomipramine | 60 | 18 |
| Selective serotonin reuptake inhibitors (SSRIs) | | |
| Fluoxetine | 250 | 12 |
| Fluvoxamine | 550 | 3.1 |
| Paroxetine | 175 | 0.6 |
| Citalopram | 2000 | 1.3 |
| Selective serotonin/noradrenaline reuptake inhibitors (SNRIs) | | |
| Duloxetine[a] | 7 | 2.6 |
| Milnacipran[b] | 100 | 203 |
| Venlafaxine[c] | 640 | 210 |
| *O*-Demethylvenlafaxine[c] | 1160 | 180 |
| *N*-Demethylvenlafaxine[c] | 4700 | 1600 |
| *N,O*-Didemethylvenlafaxine[c] | >10000 | 2800 |

Data taken from refs. [a]13, [b]14, and [c]41. Other values are those found typically in the literature.

trol, so that any increase in the extracellular concentrations of the corresponding monoamine produces a significant inhibition of their enzymatic activity. The effects of milnacipran on tryptophan and tyrosine hydroxylase were determined ex vivo by measuring the accumulation of 5-hydroxytryptophan (5-HTP) and 3,4-dihydroxyphenylalanine (DOPA), respectively, after inhibition of aromatic amino acid decarboxylase. Acute administration of milnacipran decreased the synthesis of both monoamines equipotently in the rat frontoparietal cortex *(19)*, consistent with an increase in both serotonergic and noradrenergic activity. Chronic administration with milnacipran led to a significant increase in the basal synthesis of both 5-HT and NA *(19)*, although both enzymes were still sensitive to increases in extracellular monoamines.

The release of 5-HT and NA from brain tissue is under the feedback control of autoreceptors, which act to limit any increase in neurotransmitter release. Certain antidepressants, such as the SSRI, citalopram, decrease the sensitivity of the 5-HT autoreceptor after repeated administration *(20)*, thus theoretically permitting a greater increase in neurotransmitter release. Studies measuring the release of 5-HT and NA in rat hypothalamic slices after chronic administration of milnacipran showed that neither the serotonergic *(20,21)* nor the noradrenergic autoreceptors *(22)* were modified.

In addition to autoreceptors that modify the release of their own transmitters, $\alpha_2$-heteroreceptors located on 5-HT terminals have been described *(23)*. These receptors, which are stimulated by NA, decrease the release of 5-HT and are thus a key element in the interaction of these two neurotransmitter systems. Chronic administration of milnacipran (60 mg/kg/d for 14 d with 48 h washout) has been found to produce a significant attenuation of the capacity in NA to decrease [$^3$H]5-HT overflow from hippocampal slices *(21)*, indicating desensitization of the presynaptic $\alpha_2$-heteroreceptors located on serotonergic terminals.

### 2.1.2. Venlafaxine

Venlafaxine inhibits the reuptake of both 5-HT and NA in rat brain synaptosomes, with a somewhat greater affinity for 5-HT *(24;* Table 1), and only a weak inhibition of dopamine uptake (IC$_{50}$ 2.8 $\mu M$) *(24,25)*. Radioligand-binding experiments have shown that venlafaxine has no significant affinity for 5-HT$_1$, 5-HT$_2$, histamine H$_1$, muscarinic cholinergic receptors, $\alpha_1$-, $\alpha_2$-, and $\beta$-adrenoceptors, dopamine, or opiate receptors in rat brain *(24,26)*.

Stimulation of 5-HT$_{1A}$ receptors in the raphe and $\alpha_2$-adrenoceptors in the locus ceruleus causes an inhibition of the firing of serotonergic and noradrenergic neurons, respectively. Thus, the acute inhibition of serotonergic activity in the dorsal raphe and noradrenergic neuronal activity in the locus ceruleus by venlafaxine *(27)* is a logical consequence of its inhibition of monoamine reuptake and consequent increase in synaptic levels of 5-HT and NA. Single and repeated dose administration of venlafaxine reduced the isoproterenol-induced increases of cyclic adenosine monophosphate (cAMP) level in pineal gland, an effect mediated by the activation of $\beta$-adrenoceptors. Repeated treatment with venlafaxine for 2 wk (10 mg/kg, ip, twice daily) reduced the cAMP response *(28)* without decreasing the density of $\beta$-adrenoceptors in rat cortex. The speed of downregulation of $\beta$-adrenoceptor-mediated responses has been suggested to be related to the onset of antidepressant action *(28)*, but the fact that a number of more recently introduced antidepressants (including milnacipran, *see above*) do not modify the function of $\beta$-adrenoceptors after repeated administration *(16)* makes the link between $\beta$-adrenoceptor desensitization and antidepressant response rather tenuous.

## 2.2. Behavioral Effects in Animals

### 2.2.1. Milnacipran

Milnacipran was active in behavioral tests involving the serotonergic system, such as the potentiation of l-tryptophan-induced behavior, as well as those involving the noradrenergic system, such as the antagonism of tetrabenazine-induced hypothermia *(29)*. Milnacipran has neither stimulant nor sedative effects. As expected from in vitro binding data, there were no anticholinergic effects. The slight mydriasis produced by milnacipran was completely antagonized by the $\alpha_1$-adrenoceptor antagonist, prazosin, showing it to be the result of sympathetic noradrenergic stimulation and not parasympathetic cholinergic inhibition *(29)*. Milnacipran decreases rapid eye movement (REM) sleep latency, but has no effect on the duration of REM or slow-wave sleep in rats (J. Laval, personal communication). Milnacipran is active in three widely used models of antidepressant activity: the behavioral despair (Porsolt) test *(30)*, the learned helplessness test *(31)*, and the bulbectomized rat model *(32)*.

### 2.2.2. Venlafaxine

Consistent with its in vitro capacity to inhibit the reuptake of NA, venlafaxine reversed reserpine-induced hypothermia in mice *(27)*. Venlafaxine is devoid of stimulant, sedative, and proconvulsant properties *(27)*. The only effects on sleep were to reduce the latency of the onset and the duration of REM sleep. The antidepressant potential of the drug was suggested by its activity in behavioral models in rats, including the learned helplessness and behavioral despair paradigms *(33)*. In the resident–intruder social interaction paradigm, the opposing effects of acute vs long-term venlafaxine administration in rats were similar to those of many antidepressants *(34)*.

## 3. HUMAN PHARMACOKINETICS

### *3.1. Absorption and Distribution*

#### *3.1.1. Milnacipran*

Peak plasma concentrations of milnacipran are obtained within 0.5–4 h (for review, *see* ref. *35*). The pharmacokinetic plasma profiles of a 50-mg dose of milnacipran, given either as an infusion or as a capsule, are very similar, and indistinguishable 2 h after administration. The bioavailability of milnacipran is thus high, with very little interindividual variability *(36)*. The pharmacokinetics of increasing single (25–200 mg), as well as multiple, twice-daily doses (25–100 mg) of milnacipran show a linear relationship between the dose and the plasma concentration. Plasma-protein binding of milnacipran is low (13%) and nonsaturable. There is thus a low risk of displacement of other drugs from plasma-protein binding sites. The large volume of distribution (5.3 L/kg) indicates extensive tissue penetration.

#### *3.1.2. Venlafaxine*

Studies in healthy volunteers have shown that, following oral absorption, peak plasma concentrations of venlafaxine are achieved after about 2 h *(37)*. The large volume of distribution of venlafaxine (6.8 L/kg after administration of 75 mg, 3 times daily for 3 d), and its low plasma–protein binding (30%), suggest a wide tissue distribution *(37)*.

### *3.2. Metabolism and Elimination*

#### *3.2.1. Milnacipran*

Following oral administration of milnacipran, more than 90% is recovered in urine over 96 h, while the fecal elimination represents less than 5% *(35)*. The elimination is rapid, with approx 85% of the initial dose eliminated in the first 24 h. Milnacipran is excreted essentially as the parent compound and the inactive glucuronic acid conjugate. In urine, 50–60% of the dose is recovered as parent drug, and approx 20% as glucuronide conjugate of milnacipran. The remainder is mainly the pharmacologically inactive *N*-dealkyl-milnacipran and its glucuronide conjugate. Only an insignificant proportion of the dose is excreted as other metabolites. No active metabolites have been detected in plasma in clinically significant amounts. The liver and kidney are similarly involved in the elimination of milnacipran, as illustrated by similar renal and nonrenal clearances values *(36)*. No saturation of the liver capacity to metabolize this drug is expected in case of overdosage *(38)*. The elimination half-life of milnacipran is approx 8 h *(36)*. With a twice-daily dose regimen, the steady-state plasma concentration is reached within 2–3 d, with an increase in peak concentrations of only 50–60%, compared to the single administration. The elimination rate was slightly decreased in the elderly, but the differences in pharmacokinetics parameters were minor and no dose adjustment is required in the elderly, except in those with medium to severe renal impairment.

### 3.2.2. Venlafaxine

Venlafaxine undergoes extensive first-pass metabolism and less than 5% of the parent drug is excreted in the urine. The majority is metabolized in the liver to a major active metabolite, *O*-demethylvenlafaxine, and two minor, less active metabolites, *N*-demethylvenlafaxine and *N,O*-didemethylvenlafaxine (39). *O*-demethylvenlafaxine has a similar potency for the inhibition of the reuptake of 5-HT as the parent compound (Table 1), but it is less active on the reuptake of NA. Because of its potency and circulating levels, it certainly plays a major role in the clinical action of venlafaxine. The two minor active metabolites are less potent and probably do not play a role in the clinical effect.

Excretion of venlafaxine and its metabolites is primarily renal, with 92% of an administered dose recovered in the urine, and less than 2% in the feces of healthy volunteers following administration of $^{14}$C-labeled venlafaxine 50 mg. The elimination half-life is about 4 h for venlafaxine and 10 h for *O*-demethylvenlafaxine *(37)*. Clinical trials have used twice- or three-times-daily administration regimens, which were found to give similar steady-state plasma levels. A slow-release form of venlafaxine is being developed. Clearance of venlafaxine and *O*-demethylvenlafaxine was severely reduced in patients with renal or hepatic impairment, so that dose adjustment is required in these patients *(40,41)*. There is a minor reduction in clearance in elderly patients, but it does not appear to be clinically significant and adjustment appears to be unnecessary *(37)*.

## 3.3. Drug Interactions

Depressed patients, particularly the elderly, often require concomitant medication with an antidepressant. The possibility of additional adverse effects, as a consequence of drug interactions, is therefore a major consideration.

### 3.3.1. Milnacipran

Although detailed studies are not yet available, it seems unlikely that milnacipran is metabolized by, or inhibits, cytochrome P450 2D6 (CYP2D6), an isoenzyme commonly associated with the metabolism of psychotropic drugs *(42)*. In view of its lack of interaction with the liver cytochrome P450 system and its low and nonsaturable plasma-protein binding, the risk of drug–drug interactions with milnacipran is low. Specific studies have shown no significant interaction with psychotropic drugs, such as lithium, benzodiazepines, levomepromazine, and carbamazepine *(35)*; no dose adjustment is required. Although there have been no specific studies of the interaction of milnacipran with monoamine oxidase inhibitors (MAOI), the potential for pharmacological potentiation has led to the recommendation not to associate these compounds *(35)*.

### 3.3.2. Venlafaxine

Venlafaxine is metabolized by the CYP2D6 isoenzyme of cytochrome P450. It also inhibits the CYP2D6 isoenzyme, but less potently than the SSRIs *(43)*, suggesting that the risk of clinically significant interactions between venlafaxine and other drugs that interact with the CYP2D6 isoenzyme is less with venlafaxine than with this class of compounds *(44)*.

Neither lithium *(45)* nor diazepam *(46)* had any significant effect on the pharmacokinetics of venlafaxine when coadministered acutely to volunteers. Evidence from animal studies indicates that serious interactions may occur between 5-HT reuptake inhibitors and nonselective MAOIs *(47)*. Therefore, venlafaxine should not be administered with a MAOI *(48)*.

## 4. CLINICAL STUDIES

### 4.1. Antidepressant Efficacy

#### 4.1.1. Milnacipran

Early trials showed that antidepressant activity comparable to that of 150 mg/d amitriptyline was obtained with 100 mg/d and 200 mg/d milnacipran (administered as a twice-daily regimen) *(49,50)*, while 50 mg/d was ineffective. A placebo-controlled dose-ranging study *(51)* demonstrated, however, that the efficacy with milnacipran 200 mg/d was not superior to that obtained with 100 mg/d; the higher dose was associated with an increased frequency of side effects *(52)*. One hundred mg/d (administered as 50 mg twice daily) has thus been defined as the optimal dose. When given repeatedly at this dose, the plasma levels are situated between 40 and 200 ng/mL, concentrations theoretically sufficient to inhibit the uptake of NA and 5-HT by 80–95% *(51)*. Subgroup-analysis of hospitalized, severely depressed patients vs placebo in controlled trials (Table 2) showed milnacipran to be at least as effective in this group as in the general depressed population. Examination of the qualitative nature of the antidepressant efficacy of milnacipran shows that it acts consistently on all of the core symptoms of depression (anxiety, memory, sleep disorders, and retardation) without producing either sedation or emergence of suicidal thoughts *(51)*.

Meta-analysis of seven double-blind trials comparing 100 mg/d (50 mg twice daily) milnacipran with tricyclic antidepressants showed that the response rate with milnacipran was similar to that with the TCAs (Table 3) but with a far more benign side effect profile *(53)*. The efficacy of milnacipran appears to be superior to that of the SSRIs (Table 3) *(54)*. A meta-analysis of studies comparing milnacipran, 100 mg/d, and the SSRIs, fluoxetine and fluvoxamine, showed that milnacipran therapy was associated with larger changes in symptom scores, and higher response and remission rates, than SSRIs (Table 3), with similar tolerability.

Despite initially encouraging results *(55)* suggesting a rapid onset of action, a meta-analysis of the major comparative studies has shown no difference in onset of action of milnacipran compared to TCAs or SSRIs. The mean time to 50% response rate for milnacipran is in the range of 15–20 d, depending on the study *(52)*.

#### 4.1.2. Venlafaxine

Early studies indicated that venlafaxine was efficacious at doses ranging from 30 to 450 mg/d (administered three times daily) *(41)*. In a pooled analysis of four placebo-controlled studies, Preskorn *(56)* has shown that the efficacy of venlafaxine increases with dose from 25 mg/d to about 200 mg/d (Fig 2). At higher doses, the efficacy is

**Table 2**
**Meta-Analysis of Double-Blind Studies Comparing Milnacipran (50 mg twice daily) with Placebo in the General Depressed Population and in Patients Hospitalized for Depression**

| | HDRS | | | MADRS | | |
|---|---|---|---|---|---|---|
| | Baseline | Δ End point | Responders | Baseline | Δ End point | Responders |
| All depressed patients | | | | | | |
| Milnacipran n = 227 | 26.2 | −12.5[a] | 54.6%[b] | 31.7 | −15.5[a] | 52.4%[a] |
| Placebo n = 211 | 25.7 | −9.6 | 40.3% | 31.4 | −11.5 | 39.8% |
| Hospitalized patients | | | | | | |
| Milnacipran n = 97 | 28.4 | −15.1[a] | 63.9%[a] | 34.3 | −19.2[a] | 62.9%[a] |
| Placebo n = 78 | 27.4 | −8.6 | 39.7% | 33.8 | −11.6 | 41.0% |

[a]$p < 0.01$; [b]$p < 0.001$ milnacipran superior to placebo. Responders are patients with a reduction of HDRS or MADRS greater than 50%.

**Table 3**
**Meta-Analysis of Major Double-Blind Studies Comparing Milnacipran (50 mg twice daily) with TCAs and SSRIs**

| | HDRS | | | MADRS | | | |
|---|---|---|---|---|---|---|---|
| | Baseline | Δ End point | Responders | Baseline | Δ End point | Responders | CGI-3 |
| Milnacipran | 25.9 n = 380 | −14.2 | 64% | 35.0 n = 356 | −19.5 | 63% | 1.98[a] n = 410 |
| TCA | 25.9 n = 398 | −15.2 | 67% | 34.6 n = 373 | −20.9 | 68% | 1.84 n = 432 |
| Milnacipran n = 150 | 27.0 | −15.1[a] | 64%[b] | 33.5 | −19.3[b] | 67%[b] | |
| SSRI n = 156 | 26.5 | −12.2 | 50% | 32.8 | −14.9 | 51% | |

[a]$p < 0.05$; [b]$p < 0.01$ milnacipran superior to TCA or SSRI. Responders are patients with a reduction of HDRS or MADRS greater than 50%.

reduced, probably because of the greater frequency of adverse events, which increase linearly over the whole dose range (Fig 2). In a 6-wk placebo-controlled trial, venlafaxine, at 225 mg/d and 375 mg/d, but not 75 mg/d, improved HDRS anxiety-somatic, cognitive-disturbances, and retardation subscores, but had no effect on the sleep disturbances subscore *(57)*.

A variable-dose, 6-wk study comparing venlafaxine (mean dose administered 182 mg/d) and imipramine (mean dose 176 mg/d) concluded that the two treatments were

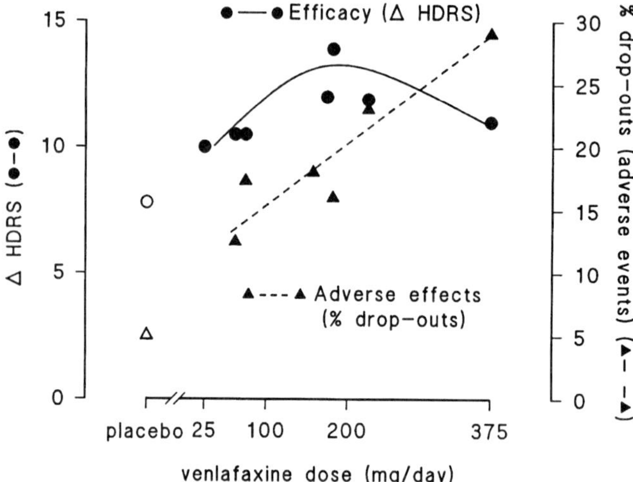



**Fig. 2.** Dose relationship of efficacy (as measured by decrease at end point in HDRS score from baseline) and tolerability (as measured by % drop-out for adverse events). Drawn from ref. *56*.

equivalent, although on certain outcome measures venlafaxine was superior to imipramine *(58)*. When studied over a 12-mo period, venlafaxine (154–176 mg/d) also showed equivalent efficacy with imipramine (138–160 mg/d) in a double-blind study in outpatients with major depression *(59)*. In both studies, venlafaxine was better tolerated.

Comparisons with SSRIs have been made in both hospitalized patients and in depressed outpatients. In a 6-wk study in hospitalized depressed patients with melancholia, venlafaxine (200 mg/d) showed significantly superior efficacy, compared to fluoxetine (40 mg/d) at wk 4 and 6 *(60)* (Fig. 3). In an outpatient population, 75 mg/d venlafaxine was found to have efficacy similar to 20 mg/d fluoxetine, but when the dose of venlafaxine was increased to 150 mg/d, the efficacy was significantly superior to fluoxetine *(61)*. Data that have not as yet been published in detail (reviewed in ref. *41*) suggest that the efficacy of venlafaxine is maintained over 12 mo. In addition, significantly fewer patients receiving venlafaxine relapsed during a 12-mo continuation study than those receiving placebo (11 vs 23%). Although there are no controlled studies with venlafaxine in resistant depression, a 12-wk open study showed that a full or partial response was achieved in one-third of a population of 70 patients resistant to at least two different antidepressants and/or electroconvulsive therapy *(62)*.

Evidence from an early open study suggested that venlafaxine may have an onset of therapeutic action of about 2 wk *(63)*. Double-blind trials with 375 mg/d have shown significant improvement compared to placebo after 2 wk *(57,64)*. There is one report of significant improvement in hospitalized patients with depression and melancholia after only 4 d, following rapid dose titration *(65)*. Response rates (percent of patients with a decrease in HDRS total score of >50%) significantly higher than placebo have also been reported after 1 wk with venlafaxine 100–375 mg/d *(64)*. Further studies are, however, required to confirm the potential rapid onset of action of venlafaxine.

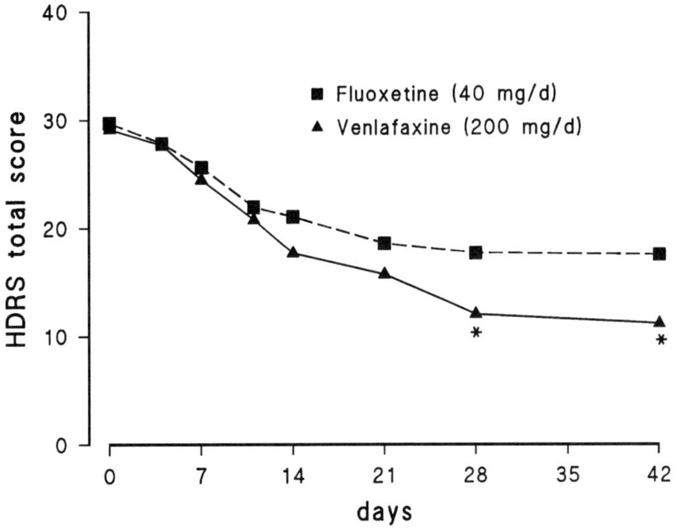


**Fig. 3.** Total mean HDRS values for patients treated with venlafaxine 200 mg/d or fluoxetine 40 mg/d. *$p < 0.05$, venlafaxine significantly superior to fluoxetine. Drawn from ref. *60*.

### 4.2. Tolerability

#### 4.2.1. Milnacipran

Most adverse events reported with milnacipran were of mild to moderate severity, occurring early in treatment, and tending to either diminish or resolve with continuing therapy. An analysis of comparative trials with milnacipran (at 100 mg/d) and TCAs has shown that TCAs were associated with a higher frequency of clinically significant adverse events compared with milnacipran *(52)*. Approximately 7.6% of patients (with milnacipran) withdrew from treatment because of adverse events, compared with 14.8% of patients receiving TCAs. In particular, anticholinergic side effects, such as dry mouth (tricyclics 37%; milnacipran 8%), constipation (15 vs 7%), and tremor (13 vs 3%), were greatly reduced. Only dysuria occurred more than twice as frequently with milnacipran (0.6 vs 2.1%) *(52)*. This may be related to increased noradrenergic stimulation of the urinary tract. Indeed, when modelled in animals, the effect can be antagonized by the $\alpha_1$-adrenoceptor antagonist prazosin (A. Solles, personal communication).

Cardiovascular adverse events were significantly less common in milnacipran-treated patients than in those receiving TCAs. In particular, as expected from its lack of $\alpha_1$-adrenergic antagonistic activity, the incidence of orthostatic hypotension (defined as a decrease of >20 mmHg) was considerably less with milnacipran than with TCAs (21 vs 34%) *(53)*. TCAs produced significant changes in the electrocardiogram (ECG), with prolongation of the PR interval, the QRS complex, and the corrected QT interval. By contrast, milnacipran had no effect on the ECG.

The tolerability of milnacipran was comparable with that of the SSRIs *(52)*. In the total database, 7.6% of patients discontinued treatment with milnacipran, compared with 7.8% of patients receiving SSRIs. Patients treated with SSRIs reported more nausea (SSRIs 20%; milnacipran 11%) and diarrhea (4 vs 2%) than those treated with milnacipran, but less headache (4 vs 8%), dry mouth (4 vs 8%), and dysuria (0.3 vs 2.1%).

### 4.2.2. Venlafaxine

Most adverse events with venlafaxine have been reported early in treatment. Their severity was mild to moderate and tended to decrease or disappear with continuing treatment. The most common side effects reported were nausea, dry mouth, headache, somnolence, and constipation. The occurrence of side effects has been reported as non-dose-related *(41)*. Their severity, however, seems to be dose-related; the number of patients withdrawing because of adverse events was clearly dose-related from 75 mg/d to 375 mg/d (Fig. 2) *(56,64)*. At high doses, nausea is particularly common, experienced by more than 50% of patients receiving 375 mg/d, compared to 32% of patients receiving 75 mg/kg *(66)*.

In a 12-mo study comparing venlafaxine with imipramine, there were far fewer anticholinergic effects, such as dry mouth, associated with treatment with venlafaxine (imipramine, 59%; venlafaxine, 28%). Nausea, however, was more common with venlafaxine (imipramine, 29%; venlafaxine, 50%) *(59)*. Over the 12-mo period, 34% of the patients in the imipramine group withdrew because of adverse events, but only 28% of those treated with venlafaxine withdrew. The tolerability of venlafaxine and the SSRIs appears to be similar. In a study comparing venlafaxine (200 mg/d) to the SSRI, fluoxetine (40 mg/d), 9% of the patients treated with venlafaxine reported nausea; there was 12% nausea in the fluoxetine group. Dry mouth was, however, more frequent with venlafaxine (12%) than with fluoxetine (3%).

Venlafaxine causes a moderate increase in blood pressure *(41)*. A pooled analysis of the drug's database showed that increases in diastolic blood pressure of greater than 10 mmHg are dose-dependent. The frequency ranged from 3% for patients receiving less than 100 mg/d to 13% for patients receiving more than 300 mg/d *(67)*. Discontinuation rate because of hypertension is, however, less than 1%.

## 5. CONCLUSION

### 5.1. Comparison of Milnacipran and Venlafaxine

#### 5.1.1. Preclinical Studies

The two SNRIs, milnacipran and venlafaxine, have more commonalities than differences. In vitro, the two compounds inhibit almost equally the reuptake of both NA and 5-HT, although milnacipran has a twofold greater affinity for NA reuptake; venlafaxine has a threefold greater selectivity for 5-HT. In addition, the major circulating metabolite of venlafaxine, demethylvenlafaxine, is sixfold more potent on the uptake of 5-HT than NA. Microdialysis studies have shown that milnacipran increases both NA and 5-HT synaptic levels in vivo. In view of the selectivity of venlafaxine itself, and of its major metabolite, one would expect venlafaxine to have an in vivo selectivity considerably in favor of 5-HT. To date no such studies have been published. Both compounds show indirect signs (the induction of feedback mechanisms) of increasing NA and 5-HT neurotransmission (inhibition of synthesis with milnacipran and inhibition of firing with venlafaxine). Both compounds are devoid of receptor interactions. In particular, the absence of affinity for the receptors involved in the main side effects of the TCAs, muscarinic cholinergic receptor, the $\alpha_1$-adrenoceptor, and the histamine $H_1$ receptors has been clearly demonstrated for both drugs.

Chronic studies have shown venlafaxine to downregulate $\beta$-adrenoceptors, but milnacipran modifies neither binding nor function of these receptors. This difference between two clearly active antidepressants with similar mechanisms provides fairly conclusive evidence that the downregulation of $\beta$-adrenoceptors, once thought to be an essential part of antidepressant mechanism, is, in fact, an unrelated phenomenon. The effects of chronic administration of milnacipran on various auto- and hetereoreceptors controlling neurotransmitter release (*see* Section 2.1.1.) are of interest, but need to be further investigated before their importance can be clarified. The effects of venlafaxine in similar paradigms would help to indicate whether these effects are common to SNRIs as a class.

In terms of behavioral animal pharmacology, the two compounds are very similar. Both present activities in classical noradrenergic and serotonergic tests, with an absence of anticholinergic effects. The modest mydriasis caused by milnacipran is antagonized by the $\alpha_1$-adrenoceptor agent prazosin, reminding us of the fundamental physiological fact that similar effects can be obtained by parasympathetic (cholinergic) antagonism, or by sympathetic (noradrenergic) stimulation. With this reminder, it is easier to understand occurrence, with both drugs, of a low level of so-called "anticholinergic" (but that are also "pronoradrenergic") clinical side effects, such as dry mouth, constipation and dysuria, despite a total lack of action at the cholinergic receptor.

The SNRIs do not seem to be either stimulant or sedative. Both milnacipran and venlafaxine produce a reduction in the latency for the onset of REM sleep, sometimes thought to be a factor predictive of antidepressant activity. Otherwise, sleep architecture seems to be generally respected by the SNRIs, although venlafaxine does reduce somewhat the duration of REM sleep. Both compounds react positively in the classical behavioral tests indicative of antidepressant action.

### 5.1.2. Clinical Studies

Both drugs are highly bioavailable and rapidly absorbed, well-distributed in the body, and with low plasma–protein binding. In terms of metabolism, however, the compounds are very different. Two-thirds of the absorbed dose of milnacipran is excreted in the urine as unchanged compound or its conjugate, and the rest essentially as a single inactive metabolite. The metabolism is not mediated by, and does not inhibit, the cytochrome P450 isoenzyme 2D6, which metabolizes many psychotropic drugs. Venlafaxine, on the other hand, undergoes extensive first-pass metabolism, so that only 5% is excreted in the urine as the parent compound. Most of the compound circulates as an active metabolite that is more selective for the reuptake of 5-HT than NA by sixfold. The metabolism occurs via cytochrome P450 2D6 isoenzyme, which is inhibited by venlafaxine. Although this inhibition is less than that of the SSRIs, it may still be clinically relevant in certain drug interactions. Milnacipran is eliminated with a terminal half-life of 8 h; the $t_{1/2}$ for venlafaxine is 4 h, and its major metabolite 10 h. Both compounds thus require twice-daily administration. The short half-lives do, however, mean that they achieve steady-state rapidly, within 2 or 3 d.

In terms of efficacy, the minimally active dose with milnacipran is 100 mg/d (50 mg twice daily); 200 mg/d is not systematically more efficacious. At 100 mg/d, the uptake of NA and 5-HT has been calculated to be already very largely inhibited (>80%), thus giving a molecular justification for the absence of improved efficacy at 200 mg/d. With venlafaxine, on the other hand, efficacy is dose-dependent over the range 75 mg/d to at least 200 mg/d (given at least twice daily). Beyond 200 mg/d, the effect may plateau or decrease (Fig. 2), although the high frequency of side effects at these doses is a confounding factor. With venlafaxine, it may be possible to adjust the dose to the severity of the depression, but a heavy cost has to be paid in terms of side effects at the higher doses. Both milnacipran and venlafaxine appear to be at least as efficacious in hospitalized patients as in the general population. There is evidence to support the claims that both milnacipran and venlafaxine have efficacy superior to SSRIs, especially in more severe depression or when melancholia is present. Compared to the SSRIs, both compounds are at least as well-tolerated, and milnacipran and low-dose venlafaxine tend to have a lower incidence of nausea. Both drugs seem to have an efficacy equivalent to TCAs, but with considerably improved tolerability, as expected from their clean biochemical and pharmacological profiles.

There is some evidence that venlafaxine may have a more rapid onset of action, especially at higher doses. However, since the higher doses can only be obtained by dose titration because of poor tolerability, the practical aspects of this possible rapid onset have yet to be demonstrated. The onset of action with milnacipran is similar to other antidepressants. Treatment can, however, be initiated immediately at the therapeutic dose of 100 mg/d, thereby saving a few days of dose buildup, which is required with many antidepressant, especially the tricyclics. In terms of tolerability, the two drugs are generally comparable. At higher doses of venlafaxine, nausea becomes frequent, and its potential effects on blood pressure require regular monitoring. The dysuria seen with milnacipran requires men with prostatic hyperplasia to be treated with caution, and for the drug to be contraindicated where dysuria already exists.

A direct comparison of the two SNRIs can only be achieved by a controlled double-blind comparative trial. This has not yet been carried out, and indeed the design of such a study would not be simple, since venlafaxine can almost be considered to be two drugs: low-dose venlafaxine (75 mg/d) which is well tolerated but submaximally efficacious; and high-dose venlafaxine (>200 mg/d), which has maximal efficacy, but is less well tolerated. A comparison of milnacipran at its optimal dose of 100 mg/d with high- or low-dose venlafaxine would probably give very different results.

### 5.2. The Potential Role of SNRIs in Antidepressant Therapy

The SNRIs, as represented by milnacipran and venlafaxine, are clearly not wonder drugs with the capacity to cure all depression overnight. They do, however, appear to be a useful addition to the psychiatrist's armamentarium, because they have efficacy comparable to the TCAs, with more benign side effect profiles. They appear to have superior efficacy to the SSRIs, especially in more severe depression, with a tolerability that is globally similar. From a scientific point of view, it is encouraging to note that this class of rationally designed drugs, which was predicted to have greater antidepressant efficacy than the more selective compounds, coupled with a benign side effect profile *(14)*, appears, to a large extent, to be fulfilling its promise.

## REFERENCES


1. Schildkraut, J. J. (1965) Catecholamine hypothesis of affective disorder. *Am. J. Psychiatr.* **112,** 509–522.
2. Asnis, G. M., Wetzler, S., Sanderson, W. C., Kahn, R. S., and Van Praag, H. M. (1992) Functional interrelationship of serotonin and norepinephrine—Cortisol responses to mCCP and DMI in patients with panic disorder, patients with depression, and normal control subjects. *Psychiatry Res.* **43,** 65–76.
3. Van Praag, H. M. (1984) Studies in the mechanism of action of serotonin precursors in depression. *Psychopharmacol. Bull.* **20,** 599–602.
4. Richelson, E. (1994) Pharmacology of antidepressants—Characteristics of the ideal drug. *Mayo Clin. Proc.* **69,** 1069–1081.
5. Danish University Antidepressant Group (1986) Citalopram: clinical effect profile in comparison with clomipramine. A controlled multicenter study. *Psychopharmacology* **90,** 131–138.
6. Danish Univeristy Antidepressant Group (1990) Paroxetine—A selective serotonin reuptake inhibitor showing better tolerance, but weaker antidepressant effect than clomipramine in a controlled multicenter study. *J. Affective Disorders* **18,** 289–299.
7. Anderson, I. M. and Tomenson, B. M. (1994) The efficacy of selective serotonin re-uptake inhibitors in depression: a meta-analysis of studies against tricyclic antidepressants. *J. Psychopharmacol.* **8,** 238–249.
8. Sitsen, J. M. A. and Zivkov, M. (1995) Mirtazapine: clinical profile. *CNS Drugs* **4(Suppl 1),** 39–48
9. Potter, W. Z. and Manji, H. K. (1994) Catecholamines in depression: an update. *Clin. Chemistry* **40,** 279–287.
10. Thomas, D. N., Nutt, D. J., and Holman, R. B. (1992) Effects of acute and chronic electroconvulsive shock on noradrenaline release in the rat hippocampus and frontal cortex. *Br. J. Pharmacol.* **106,** 430–434.
11. Nelson, J. C., Mazure, C. M., Bowers, M. B., and Jatlow, P. I. (1991) A preliminary, open study of the combination of fluoxetine and desipramine for rapid treatment of major depression. *Arch. Gen. Psychiatry* **48,** 303–307.
12. Seth, R., Jennings, A. L., Bindman, J., Phillips, J., and Bergmann, K. (1992) Combination treatment with noradrenaline and serotonin reuptake inhibitors in resistant depression. *Br. J. Psychiatry* **161,** 562–565.
13. Wong, D. T., Bymaster, F. P., Mayle, D. A., Reid, L. R., Krushinski, J. H., Robertson, D. W. (1993) LY248686, a new inhibitor of serotonin and norepinephrine uptake. *Neuropsychopharmacology* **8,** 23–33.
14. Moret, C., Charveron, M., Finberg, J. P., Couzinier, J. P., and Briley, M. (1985) Biochemical profile of midalcipran (F 2207), 1-phenyl-1-diethyl-aminocarbonyl-2-aminomethyl-cyclopropane (Z) hydrochloride, a potential fourth generation antidepressant drug. *Neuropharmacology* **24,** 1211–1219.
15. Moret, C. and Briley, M. (1996) In vivo study of monoamines neurotransmission by milnacipran a double noradrenaline and serotonin reuptake inhibiting antidepressant. *Brain Research Association Meeting, Newcastle* (abstract 5.01).
16. Neliat, G., Bodinier, M.-C., Panconi, E., and Briley, M. (1996) Lack of effect of milnacipran, a double noradrenaline and serotonin reuptake inhibitor, on the β-adrenoceptor-linked adenylate cyclase system in the rat cerebral cortex. *Neuropharmacology* **35,** 589–593.
17. Assie, M. B., Charveron, M., Palmier, C., Puozzo, C., Moret, C., and Briley, M. (1992) Effects of prolonged administration of milnacipran, a new antidepressant, on receptors and monoamine uptake in the brain of the rat. *Neuropharmacology* **31,** 149–155.

18. Assie, M. B., Le Lann A. D., Stenger A., and Briley M. (1990) Repeated administration of milnacipran, a new antidepressant, has no effect on a functional beta-adrenergic response in the rat brain. *Eur. J. Pharmacol.* **183,** 741–742.
19. Moret, C. and Briley, M. (1992) Effect of antidepressant drugs on monoamine synthesis in brain in vivo. *Neuropharmacology* **31,** 679–684.
20. Moret, C. and Briley, M. (1990) Serotonin autoreceptor subsensitivity and antidepressant activity. *Europ. J. Pharmacol.* **180,** 351–356.
21. Blier, P., Weiss, M., and De Montigny, C. (1992) Effects of sustained administration of milnacipran on serotonin and noradrenaline neurotransmissions in rat hippocampus. Abstract—Role of Serotonin in Psychiatric Disorders, Castres, P9.
22. Moret, C. Briley, M. (1994) Effect of milnacipran and desipramine on noradrenergic alpha 2-autoreceptor sensitivity. *Prog. Neuropsychopharmacol. Biol. Psychiatry* **18,** 1063–1072.
23. Göthert, M., Huth, H., and Schlicker, E. (1981) Characterization of the receptor subtype involved in alpha-adrenoceptor-mediated modulation of serotonin release from rat brain cortex slices. *Naunyn-Schmiedeberg's Arch. Pharmacol.* **317,** 199–203.
24. Muth, E. A., Haskins, J. T., Moyer, J. A., Husbands, G. E. M., Nielsen, S. T. and Sigg, E. B. (1986) Antidepressant biochemical profile of the novel bicyclic compound WY-45,030 an ethyl cycolhexanol derivative. *Biochem. Pharmacol.* **35,** 4493–4497.
25. Bolden-Watson, C. and Richelson, E. (1993) Blockade by newly-developed antidepressants of biogenic amine uptake into rat brain synaptosomes. *Life Sci* **52,** 1023–1029.
26. Cusack, B., Nelson, A., and Richelson, E. (1994) Binding of antidepressants to human brain receptors: focus on newer generation compounds. *Psychopharmacology* **114,** 559–565.
27. Muth, E. A., Moyer, J. A., Haskins, J. T., Andree, T. H., and Husbands, G. E. M. (1991) Biochemical neurophysiological and behavioral effects of WY-45233 and other identified metabolites of the antidepressant venlafaxine. *Drug. Dev. Res.* **23,** 191–199.
28. Yardley, J. P., Husbands, G. E. M., Stack, G., et al. (1990) 2-Phenyl-2-(1-hydroxycycloalkyl)ethylamine derivatives—Synthesis and antidepressant activity. *J. Med. Chem.* **33,** 2899–2905.
29. Stenger, A., Couzinier, J. P., and Briley, M. (1987) Psychopharmacology of midalcipran, 1-phenyl-1-diethyl-amino-carbonyl-2-aminomethylcyclopropane hydrochloride (F 2207), a new potential antidepressant. *Psychopharmacology* **91,** 147–153.
30. Briley, M., Prost, J. F. and Moret, C. (1996) Preclinical pharmacology of milnacipran. *Int. Clin. Psychopharmacol.* **11(Suppl 4),** 10–14.
31. Lacroix, P., Rocher, N., Gandon, J. M., and Panconi, E. (1995) Antidepressant effects of milnacipran in the learned helplessness test in rats. 8th European College of Neuropharmacology *ECNP Meeting Venice* (abstract).
32. Redmond, A. M., Kelly, J. P., and Leonard, B. E. (1995) The behavioural effects of milnacipran in the olfactory bulbectomised rat model of depression. *Med. Sci. Res.* **23,** 533–534.
33. Lloyd, K. and Mitchell, P. (1992) Preclinical evaluation of venlafaxine, a novel antidepressant drug, in behavioural models of antidepressant activity. *21st Annual Meeting of the American College of Neuropsychopharmacology, San Juan* (abstract 190).
34. Mitchell, P. J. and Fletcher, A. (1993) Venlafaxine exhibits pre-clinical antidepressant activity in the resident-intruder social interaction paradigm. *Neuropharmacology* **32,** 1001–1009.
35. Puozzo, C. and Leonard, B. E. (1996) Pharmacokinetics of milnacipran in comparison with other antidepressants. *Int. Clin. Psychopharmacol.* **11(Suppl 4),** 15–27.
36. Puozzo, C., Rostin, M., Montastruc, J. L., and Houin, G. (1987) Absolute bioavailability study of midalcipran (F 2207) in volunteers, in: *Proc. Eur. Congr. Biopharm. Pharmacokinet.* Aiache, J. M. and Hirtz, J., (eds.) Université Clermont-Ferrand, Clermont-Ferrand, pp. 59–68.

37. Klamerus, K. J., Maloney, K., Rudolph, R. L., Sisenwine, S. F., Jusko, W. J., and Chiang, S. T. (1992) Introduction of a composite parameter to the pharmacokinetics of venlafaxine and its active O-desmethyl metabolite. *J. Clin. Pharmacol.* **32,** 716–724.
38. Puozzo, C., Filaquier, C., and Briley, M. (1985) Plasma levels of F-2207, a novel antidepressant, after a single oral administration in volunteers. *Brit. J. Clin. Pharmacol.* **20,** 291.
39. Howell, S. R., Husbands, G. E. M., Scatina, J. A. and Sisenwine, S. F. (1993) Metabolic disposition of carbon-14 venlafaxine in mouse, rat, dog, rhesus monkey, and man. *Xenobiotica* **23,** 349–359.
40. Troy, S. M., Schultz, R. W., Parker, V. D., Chiang, S. T. and Blum, R. A. (1994) The effect of renal disease on the disposition of venlafaxine. *Clin. Pharmacol. Ther.* **56,** 14–21.
41. Holliday, S. M. and Benfield, P. (1995) Venlafaxine: a review of its pharmacology and therapeutic potential in depression. *Drugs* 49, 280–294.
42. Nemeroff, C. B., DeVane, C. L., and Pollack, B. G. (1996) Newer antidepressants and the cytochrome P450 system. *Am. J. Psychiatry* **153,** 311–320.
43. Otton, S. V., Ball, S. E., Cheung, S. W., Inaba, T., Rudolph, R. L. and Sellers, E. M. (1996) Venlafaxine oxidation *in vitro* catalysed by CYP2D6. *Brit. J. Clin. Pharmacol.* **41,** 149–156.
44. Ereshefsky, L. (1996) Drug-drug interactions involving antidepressants: focus on venlafaxine. *J. Clin. Psychopharmacol.* **16**(Suppl. 3), 3 SU50S.
45. Troy, S. M., Parker, V. D., Hicks, D. R, Boudino, D., and Chiang, S. T. (1996) Pharmacokinetic interaction between multiple-dose venlafaxine and single-dose lithium. *J. Clin. Pharmacol.* **36,** 175–181.
46. Troy, S. M., Lucki, I., Peirgies, A. A., Parker, V. D., Klockowski, P. M., and Chiang, S. T. (1995) Pharmacokinetic and pharmacodynamic evaluation of the potential drug interaction between venlafaxine and diazepam. *J. Clin. Pharmacol.* **35,** 410–419.
47. Marley, E. and Wozniak, K. M. (1983) Clinical and experimental aspects of interactions between amine oxidase inhibitors and amine reuptake inhibitors. *Psychol. Med.* **13,** 735–749.
48. Phillips, S. D. and Ringo, P. (1995) Phenelzine and venlafaxine interaction. *Am. J. Psychiatry* **152,** 1400–1401.
49. Ansseau, M., Von Frenckell, R., Mertens, C., et al (1989) Controlled comparison of two doses of milnacipran (F 2207) and amitriptyline in major depressive inpatients. *Psychopharmacology* **98,** 163–168.
50. Ansseau, M., Von Frenckell, R., Papart, P., et al (1989) Controlled comparison of milnacipran (F2207) 200 mg and amitriptyline in endogenous depressive inpatients. *Hum. Psychopharmacol.* **4,** 221–227.
51. Lecrubier, Y., Pletan, Y., Solles, A., Tournoux, A., and Magne, V. (1996) Clinical efficacy of milnacipran. Placebo-controlled trials. *Int. Clin. Psychopharmacol.* **11(Suppl 4),** 29–33.
52. Montgomery, S. A., Prost, J. F., Solles, A., and Briley, M. (1996) Efficacy and tolerability of milnacipran: an overview. *Int. Clin. Psychopharmacol.* **11(Suppl 4),** 47–57.
53. Kasper, S., Pletan, Y., Solles, A., and Tournoux, A. (1996) Comparative studies with milnacipran and tricyclic antidepressants in the treatment of patients with major depression: a summary of clinical trial results. *Int. Clin. Psychopharmacol.* **11(Suppl 4),** 35–39.
54. Lopez-Ibor, J., Pletan, Y., Solles, A., Tournoux, A. and Prost, J. F. (1996) Milnacipran and selective serotonin reuptake inhibitors in major depression. *Int. Clin. Psychopharmacol.* **11 (Suppl 4),** 41–46.
55. Serre, C., Clerc, G., and Escande, M. (1986) An early clinical trial of midalcipran, 1-phenyl-1-diethyl aminocarbonyl 2-aminomethyl cyclopropane (Z) hydrochloride, a potential fourth generation antidepressant. *Curr. Ther. Res.* **39,** 156–164.
56. Preskorn, S. H. (1994) Antidepressant drug selection: criteria and options. *J. Clin. Psychiatry* **55(Suppl A),** 6–22.

57. Schweizer, E., Weise, C., Clary, C., Fox, I., and Rickels, K. (1991) Placebo-controlled trial of venlafaxine for the treatment of major depression. *J. Clin. Psychopharmacol.* **11,** 233–236.

58. Schweizer, E., Feighner, J., Mandos, L. A., and Rickels, K. (1994) Comparison of venlafaxine and imipramine in the acute treatment of major depression in outpatients. *J. Clin. Psychiatry* **55,** 104–108.

59. Shrivastava, R. K., Cohn, C., Crowder, J., et al (1994) Long-term safety and clinical acceptability of venlafaxine and imipramine in outpatients with major depression. *J. Clin. Psychopharmacol.* **14,** 322–329.

60. Clerc, G. E., Ruimy, P., and Verdeau Pailles, J. (1994) A double-blind comparison of venlafaxine and fluoxetine in patients hospitalized for major depression and melancholia. *Int. Clin. Psychopharmacol.* **9,** 139–143.

61. Dierick, M., Ravizza, L., Realini, R., and Martin, A. (1996) A double-blind comparison of venlafaxine and fluoxetine for treatment of major depression in outpatients. *Prog. Neuropsychopharmacol. Biol. Psychiatry* **20,** 57–71.

62. Nierenberg, A. A., Feighner, J. P., Rudolph, R., Cole, J. O., and Sullivan, J. (1994) Venlafaxine for treatment-resistant unipolar depression. *J. Clin. Psychopharmacol.* **14,** 419–423.

63. Goldberg, H. L. and Finnerty, R. (1988) An open-label, variable-dose study of WY-45,030 (venlafaxine) in depressed outpatients. *Psychopharmacol. Bull.* **24,** 198–199.

64. Khan, A., Fabre, L. F., and Rudolph, R. (1991) Venlafaxine in depressed outpatients. *Psychopharmacol. Bull.* **27,** 141–144.

65. Guelfi, J. D., White, C., Hackett, D., Guichoux, J. Y. and Magni, G. (1995) Effectiveness of venlafaxine in patients hospitalized for major depression and melancholia. *J. Clin. Psychiatry* **56,** 450–458.

66. Artigas, F. (1995) Selective serotonin/noradrenaline reuptake inhibitors (SNRIs). Pharmacology and therapeutic potential in the treatment of depressive disorders. *CNS Drugs* **4,** 79–89.

67. *Physicians Gen Rx.* (1995) Mosby-Year Book, St Louis, MO, pp. 1959–1962.

**3**

# 5-HT-Moduline

*Novel Therapeutic Strategy for Antidepressant Action*

**Gilles Fillion, Laure Seguin, Olivier Massot, Jean-Claude Rousselle, Marie-Paule Fillion, Isabelle Cloëz-Tayarani, Brigitte Grimaldi, Jean-Christophe Seznec, and Nicole Prudhomme**

## 1. INTRODUCTION

The serotoninergic system (serotonin = 5-hydroxytryptamine = 5-HT) has been implicated in a vast number of physiological and pathological events in vertebrates *(12)*. Surprisingly, serotoninergic activity does not appear essential for any of the physiological processes in which it has been implicated. This characteristic is consistent with the hypothesis that the main role of the 5-HT system is to exert a modulatory control on such physiological functions.

Although species differences indicate that the serotoninergic system is not identical in all vertebrates *(1)*, the general structure of the system is in accordance with its functional modulatory role. Indeed, on the one hand, it is very centralized; all cellular bodies of the serotoninergic cells are located in a single area (raphe); on the other hand, axonal projections are present in almost all areas of the brain. Moreover, a significant proportion of 5-HT neurons develop additional varicosities (or neuron terminal equivalents) along the axons, which, in addition, are highly arborescent. This structure confers to the 5-HT neuronal system an enormous increase in its capacity to interact with other neurons. Thus, it has been calculated that a single axon projecting from the raphe to the cortex in rat brain could possess up to 500,000 terminals (varicosities) *(3)*. Although it cannot be verified that all these terminals are functional, reciprocally, it has not been demonstrated that the reverse is true. Therefore, it appears very likely that the serotoninergic system is adequately built up to exert a control on various other neurotransmissions and, thus, to efficiently participate in the homeostasis of the brain.

Serotoninergic activity is like that of an oscillator. Such a device delivers a signal of a variable amplitude with a variable frequency, and is used in the domain of physics to control diverse functions. The serotoninergic system behaves like an oscillator, since the signal, 5-HT, is released in variable amounts at various frequencies. The frequency of

*From:* Antidepressants: *New Pharmacological Strategies*
*Edited by: P. Skolnick, Humana Press Inc., Totowa, NJ*

53

the oscillator corresponds to that of the discharges of the neuron (several action potentials per second); it is regulated, among others, by 5-HT$_{1A}$ autoreceptors located on 5-HT cell bodies in the raphe. The amplitude of the signal corresponds to the amount of 5-HT released by the serotoninergic varicosities, and is controlled by 5-HT$_{1B/D}$ autoreceptors located on 5-HT terminals.

These receptors, previously called 5-HT$_{1B}$ in rodents and 5-HT$_{1D\beta}$ in other species, including human, are now called r5-HT$_{1B}$ and h5-HT$_{1B}$, respectively *(4)*. The role of 5-HT$_{1B}$ receptors in the release of 5-HT leads to important functional consequences, because the latter receptors are able to finely modulate the tonic serotoninergic activity, which controls other neurotransmissions via the high number of varicosities able to release 5-HT. Therefore, any mechanism affecting the efficacy of 5-HT$_{1B}$ receptors may be of importance in physiological as well as pathological situations, and may also represent an interesting direction in the study of new therapeutic agents.

## 2. REGULATION OF 5-HT$_{1B}$ RECEPTOR ACTIVITY

The working hypothesis that we postulated was that 5-HT$_{1B}$ receptors may change their functional activity in certain physiological situations. To test this hypothesis, the sensitivity of 5-HT$_{1B}$ receptors was measured ex vivo in animal models submitted to stressful situations.

Rats (male Wistar, 220–250 g) were submitted to an acute stress (restraint stress in a glass tube for 40 min), immediately sacrificed by decapitation, and brains dissected rapidly on ice. Substantia nigra, an area rich in 5-HT$_{1B}$ receptors *(5,6)*, was then homogenized and washed by centrifugation. Aliquots of the resuspended material were prepared to test the activity of a 5-HT$_{1B}$ receptor agonist (CP 93129) on adenylyl cyclase activity. It was shown that rats subjected to a single stress session exhibited a marked decrease of the 5-HT$_{1B}$ receptor activity in substantia nigra *(7)* (Seguin et al., in preparation). Indeed, the dose–response curve illustrating the effect of a 5-HT$_{1B}$ agonist on cAMP production was shifted to the right. A similar change was observed when measuring the cellular functional activity of 5-HT$_{1B}$ receptors, i.e., the inhibitory effect of CP 93129 on K$^+$-evoked release of 5-HT from rat brain synaptosomes *(7)*. Therefore, these results indicated that an acute stress (immobilization) induced the desensitization of 5-HT$_{1B}$ receptors. The effect was observed very early after stress. Potential changes in the sensitivity of 5-HT$_{1A}$ receptors were tested in parallel assays measuring their effect on adenylyl cyclase activity; it was shown that 5-HT$_{1A}$ receptor sensitivity was not affected during the first hours after stress and was altered only 24 h later *(8)*, suggesting that the mechanisms involved in the regulation of 5-HT$_{1A}$ receptors were complex and composed of different steps. Therefore, a comparison between the functional changes affecting 5-HT$_{1A}$ and 5-HT$_{1B}$ receptors suggests that the mechanisms that lead to the alteration of 5-HT$_{1B}$ receptor activity are precocious and possibly directly targeted at these receptors.

The observed desensitization of 5-HT$_{1B}$ receptors should affect serotoninergic activity. Indeed, it is known that a number of 5-HT$_{1B}$ receptors are located on 5-HT terminals, where they control the release of the amine via a negative feedback mechanism (i.e., the electrical activity of the neuron occurring with a certain frequency will lead to an increase of the 5-HT concentration in the synaptic cleft). This increase is regulated via the inhibitory action of 5-HT$_{1B}$ autoreceptors stimulated by the released 5-HT. In a steady state, the result of this regulatory mechanism is to maintain a constant concentration of 5-HT in the synaptic cleft. The desensitization of 5-HT$_{1B}$ receptors decreases the efficacy of the inhibitory control of the receptor on amine release, and accordingly will increase the 5-HT concentration in the synaptic cleft. This mechanism is supported by microdialysis experiments that demonstrate enhancement of the release of 5-HT in various areas of the brain after acute psychological or restraint stress *(9–11)*.

A second series of experiments was undertaken to test whether a chronic situation would also affect 5-HT$_{1B}$ receptor activity. Briefly, three groups of animals were compared: Group 1, naïve unmanipulated rats; Group 2, rats trained for 5 wk to run at a moderate rate; and Group 3, animals that had to run according to an intensive training program (Seguin et al., in preparation). After 5 wk of moderate training, a significant decrease in 5-HT$_{1B}$ receptor sensitivity was observed in Group 2, manifested by a significant decrease in the apparent affinity of a 5-HT$_{1B}$ specific agonist (CP 93129) to inhibit adenylyl cyclase activity (Fig. 1). In parallel, a similar desensitization of 5-HT$_{1B}$ receptors was also observed in Group 3. These observations suggest that serotoninergic activity significantly increased in the two latter groups of animals, enhancing the tonic control exerted on various other neurotransmitters. However, when the animals of these two groups were further submitted to an acute immobilization stress, different effects were observed on 5-HT$_{1B}$ receptors. Indeed, no significant additional desensitization of 5-HT$_{1B}$ receptor was observed in the animals of Group 2, but a further desensitization occurred in the animals of Group 3 (Fig. 1). This observation indicates that, in the latter group of animals, an additional release of 5-HT was induced by the acute stress, presumably leading to behavioral changes.

The results of this series of experiments indicate that 5-HT$_{1B}$ receptors are regulated by physiological mechanisms that occur very soon after delivering the stimulus responsible for the observed change. Several hypotheses could be presented to tentatively explain the mechanisms involved in the desensitization of the receptor. Among them, the existence of an endogenous compound able to regulate the sensitivity of the receptor appears very promising.

## 3. ISOLATION AND PURIFICATION OF AN ENDOGENOUS COMPOUND MODULATING THE ACTIVITY OF 5-HT$_{1B}$ RECEPTORS

The existence of an endogenous compound able to specifically interact with 5-HT$_1$ receptors was hypothesized long ago *(12)*, but the fact that multiple 5-HT$_1$ receptor subtypes were discovered during the 1980s did not favor studies on such a compound *(2,13)*. Nonetheless, the purification of such a potential compound from brain was undertaken on the basis of its capacity to interact with the binding of [$^3$H]5-HT to 5-HT$_1$ receptors. Rat and bovine brains were extracted using classical protocols involving acidic and organic extractions. The extracts were then submitted to sequential HPLC analysis using various combinations of solid-phase matrix and mobile phases *(14)*. A seven-step chromatography protocol resulted in the purification to homogeneity of a fraction containing an endogenous compound. The corresponding index of purification was greater than 1 million-fold (Fig. 2). An analytic procedure involving

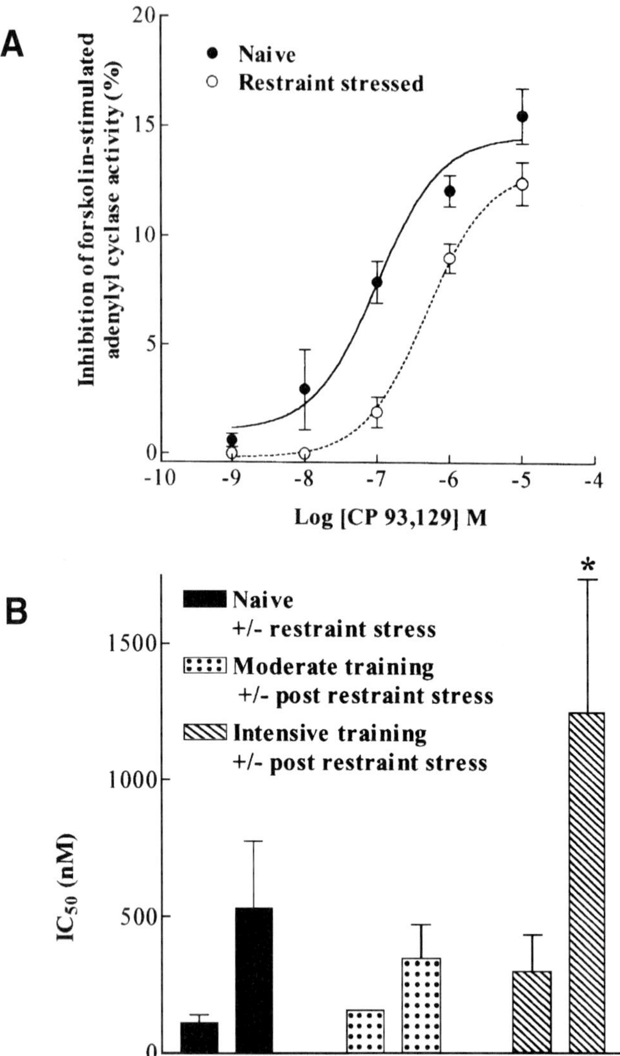

**Fig. 1. (A)** Comparison of the concentration-dependent inhibitory effect of CP 93,129 on forskolin-stimulated adenylyl cyclase activity in the substantia nigra homogenates from naïve and restraint stressed rats. Data were analyzed by nonlinear regression analysis (Prism 2.0, Graph Pad) and expressed as % of forskoline-stimulated adenylyl cyclase activity in the absence of CP 93,129. Each point represents mean ± SEM of three independent experiments conducted in duplicates. **(B)** Half-maximal inhibitory effects ($IC_{50}$) expressed in $nM$ (mean ± SEM of three independent experiments) of CP 93,129 on adenylyl cyclase activity in the naïve and intensive trained rats submitted or not to restraint stress (upper panel). $*p > 0.05$.

NMR analysis, amino acid analysis, and protein microsequencing resulted in the characterization of the isolated compound as a short peptide consisting of four amino acids: Leu—Ser—Ala—Leu. An identical sequence was obtained from both rat and bovine brain.

The principal property of this peptide was its capacity to antagonize the binding of [$^3$H]5-HT to 5-HT$_{1B}$ receptors. A careful study of various derivatives of the peptide

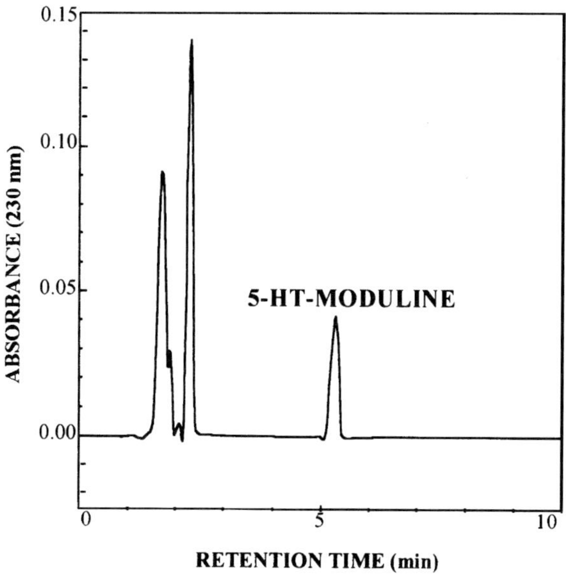


**Fig. 2.** Endogenous 5-HT-Moduline was purified by a seven-step chromatographic protocol. The chromatographic pattern represented herein is the last step of the purification on a $C_{18}$ reverse phase column *(14)*. The retention time of 5-HT-Moduline was 5.50 min in the experimental conditions used.

demonstrated that the observed activity was restricted to the native peptide; alterations at either the N-terminal or sequence drastically decreased its activity *(15)*.

Thus, these studies demonstrated the existence of an endogenous peptide contained within the brain of mammals, capable of interacting with serotonin receptors. Accordingly, the peptide was called 5-HT-Moduline.

### 3.1. Properties of 5-HT-Moduline

#### 3.1.1. Binding Assays

Binding studies were undertaken to more closely study the interaction of 5-HT-Moduline with $5-HT_1$ receptors. It was first demonstrated that the peptide affected the binding of $[^3H]5$-HT to $5-HT_{1B}$ receptors. Rat brain membranes incubated in the presence of increasing concentrations of $[^3H]5$-HT and 0.1 $\mu M$ 8-OH-dipropylaminotetraline (8-OH-DPAT) exhibited a high-affinity binding termed $5-HT_{1nonA}$ binding. This represents binding mainly to $5-HT_{1B}$ and, to a lesser extent, $5-HT_{1E}$ and $5-HT_{1F}$ receptors. It was demonstrated that 5-HT-Moduline antagonized $5-HT_{1nonA}$ binding in a noncompetitive manner. Similar results were obtained when binding to $5-HT_{1B}$ receptors was studied using $[^{125}I]$cyanopindolol (a $5-HT_{1B}$ antagonist). These observations indicated that the site that recognized the amine was not identical to that which binds the peptide (Fig. 3); however, the two sites were interacting and, therefore, were presumably located on the same protein, i.e., the $5-HT_{1B}$ receptor. The apparent affinity of the peptide for its site on $5-HT_{1B}$ receptor was very high, since the $EC_{50}$ was close to $10^{-10}M$. The demonstration that 5-HT-Moduline directly interacted with $5-HT_{1B}$ receptor protein was shown using cultured cell lines transfected with the $5-HT_{1B}$ receptor

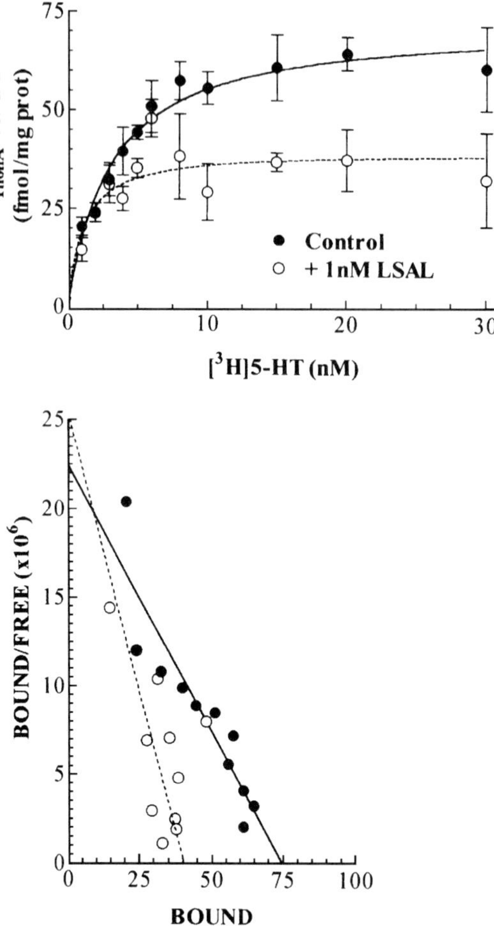

**Fig. 3.** Binding interactions of 5-HT-Moduline on 5-HT-$_\text{1nonA}$ receptors. Rat membranes were incubated for 30 min at 25°C with increasing concentrations (1–30 n$M$) of [$^3$H]5-HT 0,1 µ$M$ of 8-OH-DPAT, in the absence (•) or in the presence (○) of 5-HT-Moduline (1 n$M$). Nonspecific binding were determined in the presence of 10 µ$M$ 5-CT. Each value represents the mean ± SEM of triplicate determinations. The presented data illustrate typical experiment with the corresponding Scatchard plot. Binding parameters were $K_d$ = 3.2 n$M$ and $B_\text{max}$ = 73 fmol/mg of protein for control; and $K_d$ = 1.1 n$M$ and $B_\text{max}$ = 40 fmol/mg of protein in the presence of 1 n$M$ 5-HT-Moduline.

gene expressing the receptor protein (Fig. 4). The peptide was shown to interact with the binding of [$^3$H]5-HT to these receptors with parameters very similar to those observed in rat brain membrane preparations.

The property of 5-HT-Moduline to interact with 5-HT$_\text{1B}$ receptors in rat (r5-HT$_\text{1B}$) could be extended to h5-HT$_\text{1B}$. Indeed, the inhibitory activity of 5-HT-Moduline was shown in both guinea pig brain membrane preparations containing h5-HT$_\text{1B}$ receptors and a cell line transfected with the h5-HT$_\text{1B}$ receptor gene (Fig. 5). The parameters corresponding to the biochemical interaction are very similar to those observed in the

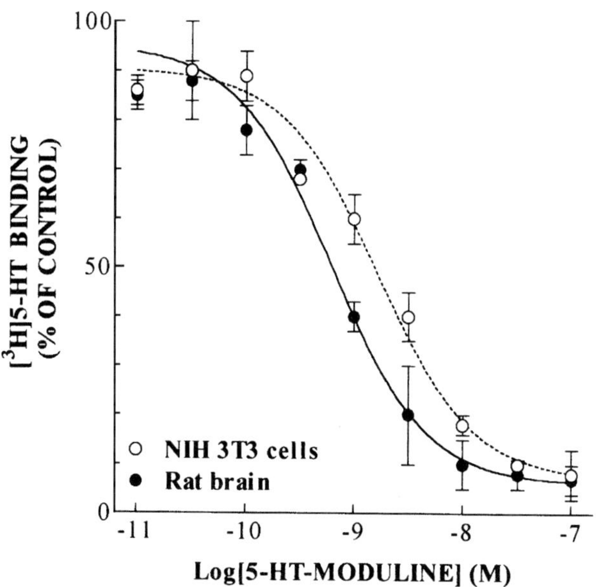



**Fig. 4.** Dose–response curves of 5-HT-Moduline on [3H]5-HT binding. Rat cortical membranes (•) or NIH 3T3 transfected cell membranes (o) were incubated for 30 min at 25°C with [3H]5-HT (20 n$M$) and increasing concentrations of 5-HT-Moduline ($10^{-11}$–$10^{-7}M$). In rat, binding experiments were carried out in the presence of 0.1 μ$M$ 8-OH-DPAT; nonspecific and 5-HT$_{1E/1F}$ bindings were determined in the presence of 20 n$M$ 5-CT. Specific r5-HT$_{1B}$ binding was calculated as the difference between 5-HT$_{1nonA}$ binding and residual binding measured in the presence of 5-CT. Each point is the mean ± SEM of triplicate determinations. The corresponding IC$_{50}$ 5-HT-Moduline were 0.6 and 1.2 n$M$ in rat and NIH 3T3 transfected cell, respectively.

case of r5-HT$_{1B}$ receptors. This result is not surprising, since h5-HT$_{1B}$ and r5-HT$_{1B}$ receptor genes exhibit a high homology, with a 93–95% identity of their amino acid sequences *(16,17)*. Moreover, they are clearly functionally equivalent in rodents and humans, as initially proposed by Hoyer et al. *(18)*.

Nevertheless, this property of 5-HT-Moduline was further confirmed by the use of radiolabeled 5-HT-Moduline. Indeed, [3H]5-HT-Moduline bound to rat brain membrane preparations with an affinity constant very similar to the apparent affinity constant observed in [3H]5-HT competition experiments ($K_d$ in the $10^{-10}M$ range); the binding corresponded to the existence of a single population of sites. Similar binding properties were observed not only in guinea pig brain membrane preparations, but also in horse, bovine, and human brain membranes, indicating that the existence of the site recognizing 5-HT-Moduline was maintained throughout evolution. Accordingly, since the peptide was found not only in rodent, but also in horse and bovine brain, it is very likely that the peptide has been conserved all along the course of evolution; this constitutes a serious argument in favor of its functional importance. A final confirmation that 5-HT-Moduline interacts with the 5-HT$_{1B}$ receptor protein is that the radiolabeled peptide has been shown to bind with a similar affinity to cells

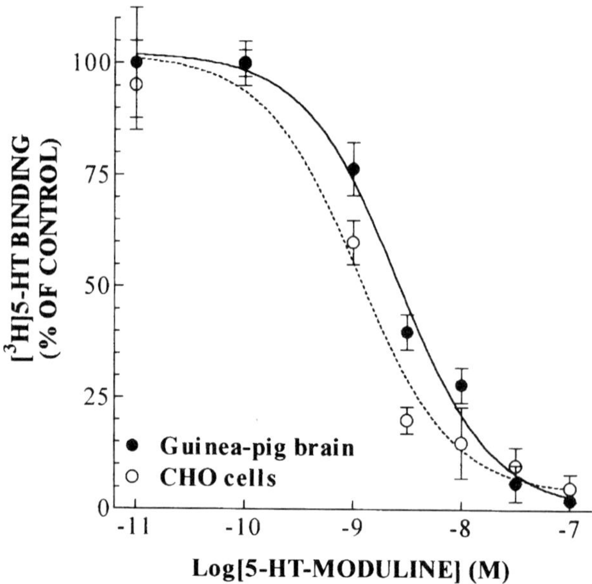



**Fig. 5.** Dose–response curves of 5-HT-Moduline on [³H[5-HT binding. Guinea pig cortical membranes (•) or CHO-transfected cell membranes (○) were incubated for 30 min at 25°C with [³H]5-HT-Moduline (20 n$M$) and increasing concentrations of 5-HT-Moduline ($10^{-11}$–$10^{-7}M$). In guinea pig, binding experiments were carried out in the presence of 0.1 µ$M$ 8-OH-DPAT; nonspecific and 5-HT$_{1E/1F}$ binding were determined in the presence of 20 n$M$ 5-CT. Specific h5-HT$_{1B}$ binding was calculated as the difference between 5-HT$_{1nonA}$ binding and residual binding measured in the presence of 5-CT. Each point is the mean ± SEM of triplicate determinations. The corresponding IC$_{50}$ of 5-HT-Moduline were 1.5 and 1 n$M$ in guinea pig and CHO-transfected cell, respectively.

expressing r5-HT$_{1B}$ and cells expressing h5-HT$_{1B}$, but the wild-type cells are not labeled (Fig. 6).

Autoradiographic studies carried out on rat and guinea pig also demonstrated the binding of [³H]5-HT-Moduline in brain sections *(15)*. Moreover, these experiments indicated that the binding was heterogeneously distributed within the brain and suggested that it had a distribution similar to that of 5-HT$_{1B}$ receptors, as expected.

5-HT-Moduline was shown to bind in a noncompetitive manner and with a high apparent affinity to r5-HT$_{1B}$ and h5-HT$_{1B}$ receptor protein. Thus, the question that had to be answered was related to the specificity of the interaction. Binding of a series of radiolabeled ligands specific for various receptors to other neurotransmitters were studied and the effect of 5-HT-Moduline examined on these receptors. It was shown that none of the studied receptors were affected by the peptide, i.e., noradrenergic, adrenergic, dopaminergic, muscarinic, histaminergic, benzodiazepine, and opiate receptors (*see* Table 1). This result strongly suggests that 5-HT-Moduline does not interact with any other neurotransmitter receptors; however, these experiments merit to be extended to other neurotransmitters. Furthermore, it was interesting to note that 5-HT-Moduline appeared to be devoid of any activity on other 5-HT receptors, i.e., 5-HT$_{1A}$, 5-HT$_{1E}$,

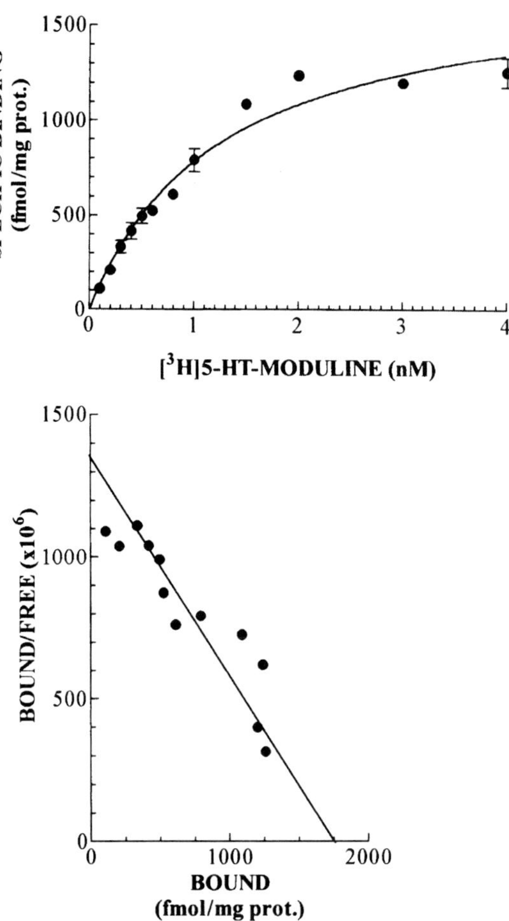

**Fig. 6.** [3H]5-HT-Moduline binding in CHO cells transfected with the gene coding for the h5-HT$_{1B}$ receptor. [3H]5-HT-Moduline binding: Cell membranes (200 μg of protein in a total volume of 10 mL) were incubated for 60 min at 25°C in a 80 m$M$ HEPES buffer, pH 7.4. Nonspecific binding was determined in the presence of 1 μ$M$ of nonlabeled 5-HT-Moduline and represented about 10% of the total binding at 4 n$M$ of [3H]5-HT-Moduline. Each point is the mean ± SEM of triplicate determinations. Binding parameters: $K_d$ = 1.2 n$M$ and $B_{max}$ = 1767 fmol/mg of protein.

5-HT$_{1F}$, 5-HT$_3$, 5-HT$_4$, and 5-ht$_6$ receptor subtypes (*see* Table 1). Therefore, these results strongly suggest that the functional role of 5-HT-Moduline would be restricted to an alteration of 5-HT$_{1B}$ receptor function.

### 3.1.2. Functional Assays

Since it was demonstrated that 5-HT-Moduline exerted a specific interaction at 5-HT$_{1B}$ receptors, it was of interest to determine whether this effect would have any functional consequence. The inhibitory effect of 5-HT on its own neuronal release was determined in superfusion experiments carried out using synaptosomes loaded with [3H]5-HT. The release of radiolabeled amine was evoked by a K+ shock; the stimula-

**Table 1**
**Pharmacological Specificity of 5-HT-Moduline**

| Receptors | 5-HT-MODULINE $IC_{50}$, n$M$ |
|---|---|
| 5-HT Receptors | |
|   5-HT$_{1A}$ | 1000 ± 15 |
|   5-HT$_{1B}$ | 0.5 ± 0.2 |
|   5-HT$_{1E/1F}$ | >1000 |
|   5-HT$_{2A}$ | >10,000 |
|   5-HT$_3$ | >10,000 |
|   5-ht$_6$ | >10,000 |
| Other receptors | |
|   $\alpha_1$-Adrenergic | >10,000 |
|   $\beta_1$-Adrenergic | >10,000 |
|   Dopaminergic D$_2$ | >10,000 |
|   Muscarinic cholinergic | >10,000 |
|   Opiate | >10,000 |
|   Histaminergic H$_1$ | >10,000 |
|   Benzodiazepine | >10,000 |

The efficacy of 5-HT-Moduline on serotoninergic and other receptors was determined in dose–response curves in rat brain cortical membranes using specific radiolabeled ligands for each receptor. Binding assays were performed as described previously (14). The IC50 were determined using a nonlinear regression fit (Prism 2.0, Graph Pad software). Data are given as the mean ± SEM of three independent determinations.

tion of 5-HT$_{1B}$ autoreceptors by 5-HT (or a r5-HT$_{1B}$ specific agonist) dose-dependently decreased the evoked release of the amine. In the presence of the peptide, the inhibitory effect of 5-HT or CGS 12066 B linked to the activation of the autoreceptor was reversed (Fig. 7). This result was shown to be concentration dependent and confirmed the antagonistic activity of the peptide on 5-HT$_{1B}$ receptor function. This effect was observed at low concentrations of the peptide, in agreement with its apparent affinity constant in binding experiments.

A further demonstration of the antagonistic activity of 5-HT-Moduline on the 5-HT$_{1B}$ receptor was obtained in vivo. The animal model chosen to examine this aspect was described by Francès et al. *(19)*, since it was demonstrated by these authors that 5-HT$_{1B}$ receptors played a key role in this model. Briefly, mice are isolated singly in a cage for 1 wk and the behavior of these animals is compared to that of mice kept in a group of five animals. The test consists of measuring the exploratory behavior and escape attempts of isolated mice vs those of grouped mice, when placed under a beaker in the presence of another normal mouse used as an "object." The isolated mice exhibit a marked deficit of their behavior, which is totally reversed after administration of a drug acting at 5-HT$_{1B}$ receptors *(15)*. The peptide administered intracerebroventricularly (compared to NaCl administration) has no effect of its own on the behavior of the mouse, but dose-dependently reverses the effect of a 5-HT$_{1B}$ stimulation (Table 2). Here again, the peptide has antagonistic properties. Current experiments show that the

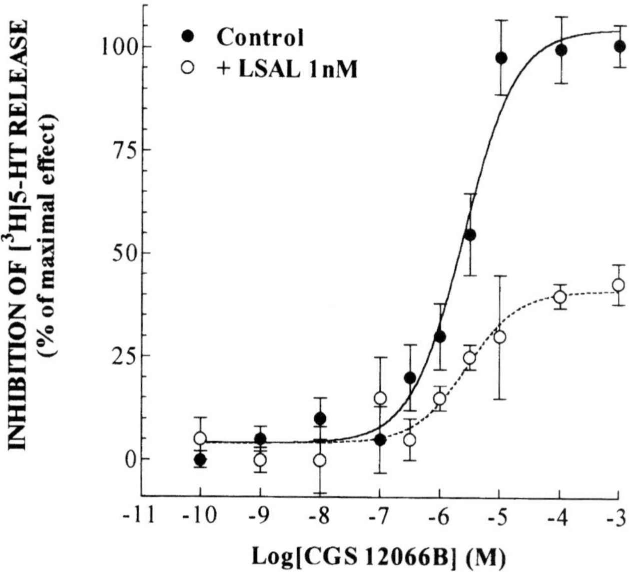


**Fig. 7.** Effects of 5-HT-Moduline on the synaptosomal release of 5-HT. Hippocampal synaptosomes, loaded with 30 n$M$ of [³H]5-HT, were incubated for 3 min with increasing concentrations of CGS 12066B $10^{-10}$–$10^{-3}M$, in the absence (•) or in the presence of 1 n$M$ 5-HT-Moduline (○). At the end of the incubation period, a 5 min $K^+$ stimulation (15 m$M$) was applied and the released radioactivity was measured by liquid scintillation counting.

effects of the peptide may be observed using other in vivo models, i.e., in food intake mechanisms (Seznec et al., in preparation).

### 3.1.3. Release of 5-HT-Moduline from Synaptosomal Preparation

An important question arises regarding the origin of 5-HT-Moduline. The antibodies raised against the peptide (Grimaldi et al., in preparation) will allow a precise examination of this point. However, a series of experiments was carried out to examine whether the peptide could be released according to mechanisms classically accepted for a neurotransmitter/neuromodulator. Thus, synaptosomal preparations were obtained from rat brain and submitted to a $K^+$ shock (30 m$M$) in the presence or absence of $Ca^{2+}$. The presence of 5-HT-Moduline, potentially released in the extrasynaptosomal fluid, was determined in the supernatant after centrifugation of the synaptosomal preparation. The released peptide was isolated, identified by HPLC analysis, and quantified by radioreceptor assay. The results demonstrate that 5-HT-Moduline was released in a $K^+$, $Ca^{2+}$-dependent manner, suggesting that it behaves similarly to other neurotransmitters/neuromodulators.

## 4. 5-HT-MODULINE: A NEW STRATEGY FOR MODULATING 5-HT ACTIVITY

It is now clearly demonstrated that a small tetrapeptide (5 HT-Moduline), specifically interacting with 5-HT$_{1B}$ receptors, is present in mammalian brain. It is also shown that the corresponding interaction leads to an antagonistic effect on the 5-HT$_{1B}$ function. To what extent does this mechanism play a significant role in regulating 5-HT activity? On

**Table 2**
**Behavioral Effect of 5-HT-Moduline**

| Animal treatment | Effect, % |
| --- | --- |
| Grouped mice | |
|   C1Na | $100 \pm 5$ |
| Isolated mice | |
|   C1Na | $50 \pm 10$ |
|   5-HT-Moduline (50 µg icv) | $60 \pm 15$ |
|   Scrambled peptide (50 µg icv) | $45 \pm 20$ |
|   RU 24969 (4 mg/kg ip) | $100 \pm 20$ |
|   RU 24969 + 5-HT-Moduline | $50 \pm 5$** |
|   RU 24969 + Scrambled peptide | $100 \pm 5$ |

Mice were tested in the social interaction test. RU 24969 was injected 4 mg/kg ip 30 min before the test; 5-HT-Moduline (LSAL) or a scrambled peptide (ALLS) were administered icv at a dose of 50 µg 45 min before the test. Results are expressed in percentage ± SEM of escape attempts of grouped mice. This experiment was repeated three times. Five mice were tested in each group. **$p < 0.01$.

a theoretical basis, the antagonistic effect of 5-HT-Moduline on the 5-HT autoreceptor mediating the negative feedback regulatory mechanism occurring at 5-HT terminal will lead to an increase of the 5-HT activity, resulting from an increased release of the amine. Accordingly, the presence/absence of 5-HT-Moduline represents an efficient way of regulating 5-HT release and thus serotoninergic activity.

Experimental results obtained in animal model after stress have clearly shown that a single acute stress reduces the efficacy of $5\text{-HT}_{1B}$ receptors, and therefore likely results in an increase in 5-HT release. The latter proposal is supported by numerous experimental findings (using a microdialysis technique) that acute stress induces a rapid release of 5-HT in cortex, hippocampus, or hypothalamus of the animal *(9–11)*. These results suggest that 5-HT-Moduline might be, at least in part, responsible for an increase in 5-HT activity. To further test this hypothesis, naïve rats received (via icv administration) either NaCl or 5-HT-Moduline (30–50–100–500 µg). Animals were sacrificed 15 min later and the sensitivity of $5\text{-HT}_{1B}$ receptors was immediately tested, as previously described (*see above*). It was shown that the peptide had a very similar effect on the autoreceptor as that described after a stressful situation. This observation suggests that acute stress may induce the release of 5-HT-Moduline, which will desensitize the receptor leading to an increase in 5-HT activity. To validate this hypothesis, we are currently examining the release of 5-HT-Moduline under these conditions, and determining the kinetics of peptide release following a stressful stimulus.

Various questions are still open and need to be answered in order to better understand the role of 5-HT-Moduline in the regulation of 5-HT activity. The molecular mechanism involved in the interaction of the peptide with the receptor should be precisely defined, although a number of experimental observations tend to suggest that it corresponds to an allosteric interaction. The metabolism of the peptide is not yet precisely defined, although it is known that a rapid degradation occurs in cerebral tissue, as generally observed for other neurotransmitters. An interesting point to be clarified is whether 5-HT-Moduline is also released at $5\text{-HT}_{1B}$ receptors located on nonserotoninergic terminals. Indeed, 5-HT$_{1B}$ receptors are not only presynaptic autoreceptors located on 5-HT terminals, where they control the release of 5-HT; they also are presynaptic heteroreceptors located on nonserotoninergic terminals (e.g., cholinergic, glutamatergic, noradrenergic, and GABA-ergic), where they modulate the release of the corresponding neurotransmitter *(20–23)*. Exogenous 5-HT-Moduline is also able to affect the function of the heterologous 5-HT$_{1B}$ receptors (unpublished experiments); however, the potential regulation of these receptors by endogenously released 5-HT-Moduline has still to be examined. This latter aspect represents an important point in the study of the properties of 5-HT-Moduline in regulating the 5-HT control on other neurotransmissions.

## 5. 5-HT-MODULINE: STRATEGY FOR A NOVEL ANTIDEPRESSANT?

From both a theoretical and experimental standpoint, 5-HT-Moduline appears to be a potential regulator of 5-HT activity, particularly in controlling the release of this amine. The main characteristic is that the interaction of the peptide with the 5-HT$_{1B}$ receptor leads to an increase in 5-HT activity. This effect is reminiscent of a very common effect of various antidepressants, in particular SSRIs, which specifically increase 5-HT activity. Therefore, the proposed hypothesis is that 5-HT-Moduline may correspond to an important endogenous factor whose functional role is a rapid adaptation of the 5-HT control to favor the elaboration of the pertinent response of the CNS to a stimulus. Thus, it is expected that this peptide may play an important role in the regulation of affect and also in many other functions that are modulated by the serotoninergic system. If it is assumed that 5-HT-Moduline plays a significant role in the control of affect, it is not excluded that it may be responsible, at least in part, for the placebo effect observed in therapeutic treatments of depression. Therefore, drugs which could interfere with that mechanism may have interesting therapeutic properties.

Finally, it should be emphasized that 5-HT-Moduline is specific to 5-HT$_{1B}$ receptors. As such, it represents the first identified endogenous allosteric modulator that interferes with G protein-related receptor. It is likely that, among the large multigenic family of seven transmembrane domain receptors, other members of this family are sensitive to various modulines analogous to 5-HT-Moduline and specific for a single receptor. Studies of these other potential modulines may bring further important progress in the knowledge of depression and new antidepressant tools.

## REFERENCES


1. Jacobs, B. L. and Azmitia, E. C. (1992) Structure and function of the brain serotonin system. *Physiol. Rev.* **72,** 165–228.
2. Zifa, E. and Fillion, G. (1992) 5-hydroxytryptamine receptors. *Pharmacol. Rev.* **44,** 401–458.
3. Audet, M. A., Descarries, L., and Doucet, G. (1989) Quantified regional and laminar distribution of the serotonin innervation in the anterior half of adult rat cerebral cortex. *J. Chem. Neuroanat.* **2,** 29–44.
4. Hartig, P. R., Hoyer, D., Humphrey, P. P. A., and Martin, G. R. (1996) Alignment of receptor nomenclature with the human genome—Classification of 5-HT$_{1B}$ and 5-HT$_{1D}$ receptor subtypes. *Trends Pharmacol. Sci.* **17,** 103–105.

5. Hoyer, D., Engel, G., and Kalkman, H. O. (1985) Characterisation of the 5-HT$_{1B}$ recognition site in rat brain: binding studies with [$^{125}$I]iodocyanopindolol. *Eur. J. Pharmacol.* **118,** 1–12.

6. Bruinvels, A. T., Palacios, J. M., and Hoyer, D. (1993) Autoradiographic characterization and localization of 5-HT$_{1D}$ compared to 5-HT$_{1}$B binding sites in rat brain. *Naunyn Schmiedeberg's Arch. Pharmacol.* **347,** 569–582.

7. Bolaños-Jiménez, F., Manhães de Castro, R., Seguin, L., Cloëz-Tayarani, I., Monneret, V., Drieu, K., and Fillion, G. (1995) Effects of stress on the functional properties of pre- and postsynaptic 5-HT$_{1B}$ receptors in the rat brain. *Eur. J. Pharmacol.* **294,** 531–540.

8. Manhães de Castro, R., Bolaños-Jiménez, F., Seguin, L., Sarhan, H., Drieu, K., and Fillion, G. (1996) Sub-chronic cold stress reduced 5-HT$_{1A}$ receptor responsiveness in the old but not in the young rat. *Neurosci. Lett.* **203,** 21–24.

9. Shimizu, N., Take, S., Hori, T., and Omura, Y. (1992) In vivo measurement of hypothalamic serotonin release by intracerebral microdialysis: significant enhancement by immobilization stress in rats. *Brain Res. Bull.* **28,** 727.

10. Kawahara, H., Yoshida, M., Yokoo, H., Nishi, M., and Tanaka, M. (1993) Psychological stress increases serotonin release in the rat amygdala and prefrontal cortex assessed by in vivo microdialysis. *Neurosci. Lett.* **162,** 81.

11. Vahabzadeh, A. and Fillenz, M. (1994) Comparison of stress-induced changes in noradrenergic and serotoninergic neurons in the rat hippocampus using microdialysis. *Eur. J. Neurosc.* **6,** 1205.

12. Fillion, G. and Fillion, M. P. (1981) Modulation of affinity of postsynaptic serotonin receptors by antidepressant drugs. *Nature* **292,** 349–351.

13. Hoyer, D., Clarke, D. E., Fozard, J. R., Hartig, P. R., Martin, G. R., Mylecharane, E. J., Saxena, P. R., and Humphrey, P. P. A. (1994) International union of pharmacology. Classification of receptors for 5-hydroxytryptamine (serotonin). *Pharmacol. Rev.* **46,** 157–203.

14. Rousselle, J.-C., Massot, O., Delepierre, M., Zifa, E., and Fillion, G. (1996) Isolation and characterization of an endogenous peptide from rat brain interacting specifically with the serotoninergic$_{1B}$ receptor subtypes. *J. Biol. Chem.* **271,** 2, 726–735.

15. Massot, O., Rousselle, J.-C., Fillion, M.-P., Grimaldi, B., Cloëz-Tayarani, I., Fugelli, A., Prudhomme, N., Rousseau, B., Seguin, L., Hen, R., and Fillion, G. (1996) 5-HT-Moduline, a new endogenous cerebral peptide controls the serotoninergic activity via its specific interaction with 5-HT$_{1B/1D}$ receptors. *Mol. Pharmacol.,* in press.

16. Peroutka, S. J. (1994) Molecular biology of serotonin (5-HT) receptors. *Synapse* **18,** 241–260.

17. Saudou, F. and Hen, R. (1994) 5-Hydroxytryptamine receptor subtypes: molecular and functional diversity. *Adv. Pharmacol.* **30,** 327.

18. Hoyer, D. and Middlemiss, D. N. (1989) Species differences in the pharmacology of terminal 5-HT autoreceptors in mammalian brain. *TIPS* **10,** 130–132.

19. Francès, H. (1988) New animal model of social behavioural deficit: reversal by drugs. *Pharmacol. Biochem. Behav.* **29,** 467–470.

20. Bobker, D. H. and Williams, J. T. (1989) Serotonin agonists inhibit synaptic potentials in the rat locus ceruleus *in vitro* via 5-hydroxytryptamine$_{1A}$ and 5-hydroxytryptamine$_{1B}$ receptors. *J. Pharmacol. Exp. Ther.* **250,** 37–43.

21. Boeijinga, P. H. and Boddeke, H. W. G. M. (1993) Serotoninergic modulation of neurotransmission in the rat subicular cortex in vitro: a role for 5-HT$_{1B}$ receptors. *Naunyn-Schmiedeberg's Arch. Pharmacol.* **348,** 553–557.

22. Johnson, W. S., Mercuri, N. B., and North, R. A. (1992) 5-hydroxytryptamine$_{1B}$ receptors block the GABA$_B$ synaptic potential in rat dopamine neurons. *J. Neurosci.* **12,** 2000–2006.

23. Maura, G. And Raiteri, M. (1986) Cholinergic terminals in rat hippocampus possess 5-HT$_{1B}$ receptors mediating inhibition of acetylcholine release. *Eur. J. Pharmacol.* **129,** 333–337.
24. Molderings, G. J., Fink, K., Schliker, E., and Göthert, M. (1987) Inhibition of noradrenaline release via presynaptic 5-HT$_{1B}$ receptors of the rat vena cava. *Naunyn-Schmiedeberg's Arch. Pharmacol.* **336,** 245–250.

**4**

# Reversible Inhibitors of Monoamine Oxidase A (RIMAs)

## *Where Can We Go from Here?*

**Vivette Glover**

## 1. INTRODUCTION

The newest class of antidepressant monoamine oxidase inhibitors in the clinic is the selective reversible inhibitors of monoamine oxidase A (RIMAs). These are considerably safer than the older, nonselective irreversible monoamine oxidase inhibitors (MAOIs), in that they have much less interaction with tyramine and certain drugs. Many trials have shown them to be of benefit in depression, and there is room for more variants of this type of drug to be developed. However, RIMAs also have less effect on both 5-hydroxytryptamine and dopamine oxidation than the older drugs, and it is possible that they are also less effective as antidepressants. In this chapter, it is suggested that nonselective but reversible monoamine oxidase inhibitors, a class of drug not yet tried in the clinic, might combine the efficacy of the old drugs and retain a greater safety margin.

## 2. MONOAMINE OXIDASE

Monoamine oxidase (MAO) is an important enzyme in the metabolism of a wide range of monoamine neurotransmitters, including noradrenaline, dopamine, and 5-hydroxytryptamine (5-HT), of trace amines, such as phenylethylamine (PEA), and of dietary amines, such as tyramine. Its function is both to remove exogenous amines, which could interfere with the functioning of monoamine neurotransmitters, and to deaminate the endogenous monoamines. MAO exists in two forms, A and B, with different substrate and inhibitor sensitivities (Table 1), and also with certain differences in its properties in different species. Both forms are located in the mitochondrial outer membrane, and both are widely distributed in tissues *(1)*. MAO A and B have now been cloned and located to adjacent regions of the X chromosome in the p11.3–11.4 region *(2,3)*.

Much has been earned recently about the function of these two forms in humans by a detailed examination of the clinical and metabolic characteristics of certain very rare


*From:* Antidepressants: *New Pharmacological Strategies*
*Edited by: P. Skolnick, Humana Press Inc., Totowa, NJ*

**Table 1**
**Substrates and Inhibitors of MAO A and B**

| | MAO A | | MAO B |
|---|---|---|---|
| Substrates | | Mixed | |
| | Noradrenaline | Dopamine | PEA |
| | 5-HT | Tyramine | Tele-methylhistamine |
| Inhibitors | | Nonselective irreversible | |
| First generation | | | |
| | | Iproniazid (Marsilid) | |
| | | Tranylcypromine | |
| | | Phenelzine | |
| Second generation | | Selective irreversible | |
| | Clorgyline | | Deprenyl |
| Third generation | RIMA | Selective reversible | |
| | Moclobemide | | Lazabemide |
| | Brofaramine | | |
| | Toloxatone | | |
| | Befloxatone | | |

individuals who lack either or both forms *(4)*. Some people also have a deletion of the X chromosome, resulting in a combined loss of the genes for both MAO A and B and the adjacent gene for Norrie disease, which causes blindness and impaired hearing *(5)*. Those who lacked MAO B showed no obvious clinical defects, but those who lacked MAO A were characterized by borderline mental retardation and impaired impulse control and aggression. These are not the effects of inhibiting MAO pharmacologically, and suggest that they are the result of the lack of MAO A during neurodevelopment. In terms of neurochemistry, the subjects deficient in MAO B also showed much less abnormality than those lacking MAO A. In those subjects lacking MAO B, the main effects were a loss of platelet MAO B, as could be expected, and an increase in peripheral levels of the trace amine, PEA, its major endogenous substrate. With those subjects lacking MAO A, there was an increase in the platelet content of 5-HT, and a substantial decrease in plasma level of the major catecholamine metabolites. There were corresponding increases in metanephrine and normetanephrine, and particularly in their sulfoconjugates. These findings were even more pronounced in the individuals lacking both MAO A and B. All this suggests that, overall, MAO A is more important than MAO B in the metabolism of the major neurotransmitter monoamines, and, indeed, one may wonder what the necessity is for the B form of the enzyme at all. However, measurement of peripheral effects only gives a global picture, and there is evidence that MAO B is highly localized in specific cells and regions, and that it has a significant role in these particular areas.

## 3. HISTORY OF MAO INHIBITORS (MAOIs)

The discovery, in the 1950s, that drugs that inhibit MAO also have antidepressant properties was one of the key findings that led to the monoamine theory of depression and the whole era of biological psychiatry. The discovery was a mixture of serendipity and scientific acumen *(6)*. After the Second World War, there were large stocks of hydrazine (which had been used for making rocket fuel) left over. This material was cheap and available to the pharmaceutical industry. Fox, at Hoffmann LaRoche, used hydrazine for drug synthesis, and screened what he considered to be some intermediates in a range of animal models, including one for tuberculosis (TB). Iproniazid and isoniazid were shown to work in this model and were tried in TB patients in 1951. They were both found to be remarkably effective, but iproniazid was also reported to have pronounced psychiatric side effects, including inducing a great sense of well-being in many patients. In fact, about 20% of the patients became psychotic and iproniazid was replaced by isoniazid, which did not have these side effects. In 1952, Zeller and colleagues showed that iproniazid, but not isoniazid, was an inhibitor of MAO in vitro *(7)*. In 1956, it was found that, if used before reserpine in animal models, iproniazid caused them to be more alert, rather than sedated. Meanwhile, Nathan Kline had been using reserpine in his psychiatric patients as a tranquilizer, and had listed the desired properties of a drug that should have opposite effects, acting as a "psychic energizer," as he called it, for use in depression. When he read about the action of iproniazid in TB patients, in the test tube, and in animal models, he wondered whether this was the drug he had been searching for *(8)*. He conducted the first open trial in depressed inpatients and outpatients in 1956 and was very encouraged by the results. He and his colleagues reported their findings in 1957 *(9)*. In 1958, iproniazid (Marsilid®) was used in 400,000 people with depression as an antidepressant treatment.

The first generation of MAOIs were the irreversible inhibitors, such as iproniazid, phenelzine, and tranylcypromine. Even though these were, and are, in fact, effective antidepressants, there has been a general reluctance to use them for two reasons. The first was the "cheese effect." This is a hypertensive crisis that can be, on occasion, fatal. Such a crisis can take place when someone taking an MAOI also consumes a tyramine-containing food, such as cheese. Toxic interactions with other drugs that are metabolized by MAO can also occur. The other major reason was the 1965 report by the MRC in Britain that phenelzine was no more effective than placebo *(10)*. This had great influence, although, in fact, the report was very misleading, because the dose of phenelzine used in the trial was too low to be effective.

In the 1960s, it was found that MAO existed in two forms, termed MAO A and MAO B. These enzymes could be delineated with the selective irreversible inhibitors, clorgyline for MAO A, and deprenyl for MAO B, the second generation of MAOI. Deprenyl has been widely used in Parkinson's disease (dopamine in the human striatum is preferentially deaminated by MAO B *[11]*) and has been claimed to increase the lifespan of rodents *(12)*. However, at selective doses, it is not an antidepressant *(13)*, and clorgyline (which is an antidepressant) has almost as pronounced a cheese effect as the nonselective MAOI *(14)*.

## 4. THE DEVELOPMENT OF THE RIMA

The third generation of MAOI consists of the reversible inhibitors of MAO A (RIMA), typified by moclobemide *(15)* (Tables 1 and 2). Moclobemide is an effective MAO A inhibitor in vivo, but not in vitro, and the presumed active metabolite has not

**Table 2**
**Comparison of the Inhibitory Potencies of Some MAOIs**

|                  | Ki MAO A   | Ki ratio (B/A) | Half-life    | Ref. |
|------------------|------------|----------------|--------------|------|
| Befloxatone      | 2.3 n$M$   | 117            | 2 h          | *16* |
| Moclobemide      | 6 μ$M$     | >100           | 16–20 h      | *15* |
| Tranylcypromine  | 0.18 μ$M$  | 4.1            | Irreversible | *15* |

yet been identified. No other pronounced pharmacological activity has yet been attributed to it. Brofaramine was developed for some time, but has now been withdrawn. Befloxatone is currently under development *(16)*. These drugs appear to be free of the cheese effect and other drug interactions. They also appear to be pharmacologically cleaner than previous MAOIs, with less interactions with other systems generally. There is general agreement that the RIMAs are a much safer form of MAOI than those of the first generation *(15)*.

Moclobemide has been widely tested in several thousand patients in many different studies. These studies have been recently reviewed by Paykel (1995) *(17)*. His conclusion is that moclobemide is a safe and effective antidepressant, beneficial in a range of depressive subgroups, but with no evidence for preferential efficacy in any one type. In these reports, moclobemide was comparable with amitryptiline, imipramine, clomipramine, and the selective serotonin uptake inhibitors (SSRI). It has fewer side effects than the tricyclics or SSRIs, and is safe in overdose. There is thus room for the development of other drugs of this type, to allow for different individual responses. Befloxatone is both more potent and has a much shorter half-life than moclobemide *(16)* (Table 2).

There are reports that the older, nonselective MAOIs are effective in a range of disorders in addition to depression, such as panic disorder, social phobia, and bulimia *(18,19)*. It will be of interest to determine whether the RIMA are also effective and safe in this wide range of psychiatric disorders.

At present, moclobemide is certainly not proving as popular for depression in the marketplace as SSRIs, such as fluoxetine and sertraline. This may be partly a result of marketing. However, it might be that patients are finding it less effective, despite the results of extensive clinical trials. Also, many psychiatrists believe that the older antidepressants are the most effective in the most severely ill patients. These issues require further study, but do suggest that there may still be room for yet a new generation of MAOIs, with the efficacy of the first generation, and the safety of the third.

## 5. THE LOCALIZATION OF MAO A AND B IN THE BRAIN

To understand the possible advantages and limitations of the RIMAs, we need to know about both the selective substrates of the two forms of MAO and the selective localization of these two forms in the brain and other tissues *(20)*. There has been extensive characterization of the kinetic properties of the two forms of the enzyme in vitro (e.g., ref. *21*), and a summary of the selective substrates is included in Table 1. The specificity of both substrates and inhibitors is relative, and most compounds can become nonselective at higher concentrations.

**Table 3**
**Localization of MAO A and MAO B in the Brain**

| Ref. | Rat brain MAO A *(22)* | Rat brain MAO B *(22)* | MAOA mRNA rat brain *(24)* | MAO B mRNA rat brain *(24)* | Human brain MAO A *(23)* | Human brain MAO B *(23)* | Human brain MAO A *(25)* | Human brain MAO B *(25)* |
|---|---|---|---|---|---|---|---|---|
| Noradrenaline neurons | | | | | | | | |
| Locus coeruleus | ++++ | – | 10 | 0 | + | – | 12.7 | 4.3 |
| 5-HT neurons | | | | | | | | |
| Dorsal raphe | – | ++++ | 4 | 10 | – | + | 4.3 | 5.6 |
| Dopamine neurons | | | | | | | | |
| Substantia nigra | – | – | 1.5 | 0.5 | – | – | 4.7 | 4.0 |
| Caudate/ putamen | – | – | | | | | 2.4–5.0 | 6.0–13.1 |
| Cerebellum (Bergman cell layer) | nd | nd | 0 | 4 | nd | nd | 0.3–5.7 | 2.7–6.1 |
| Pineal | – | + + | 0.5 | 7.5 | nd | nd | | |
| Area postrema | – | ++<br>+ +<br>++ | 0.5 | 7.0 | | | | |

Symbols and numbers reflect intensity of staining:

*(22)* –, background staining; + to ++++, minimal to maximal staining above background.

*(23)* +, stained neurons observed; –, no neurons stained; nd, not determined.

*(24)* mRNAs. Data are expressed as arbitrary units derived from optical density readings of radioautographic films. Values range from 0–1 (background) to 10 (maximum).

*(25)* Specific binding of [$^3$H]Ro 41-1049 and [$^3$H]lazebemide for quantification of MAO A and MAO B, respectively (specific binding in pmol/mg protein).

The cellular localization of MAO has been studied by various methods, including histochemistry in rat *(22)* and human brain *(23)*, and by tracing the cellular expression of mRNAs for the two forms in rat brain *(24)*. Table 3 shows the results of some of these studies for the localization of the two forms of MAO in the main monoamine cell centers of the brain. The overall picture, using different techniques and both rat and human brain, are generally similar, showing a relative concentration of MAO A in the locus ceruleus, of MAO B in the raphe, and of neither in the substantia nigra. The use of radiolabeled specific ligands for autoradiography *(25)*, has reproduced this overall pattern and also confirms that the human raphe, which contains high levels of MAO B, has notable MAO A also.

MAO B has a very wide distribution in glia throughout the brain, but in the human brain many glia seem to contain some MAO A, also *(23)*. MAO B is also concentrated in other areas, such as the pineal and circumventricular regions. MAO B mRNA has also been shown to be present in histaminergic neurons *(24)*.

A comparison of Tables 1 and 3 indicates that MAO has a different role in the cell bodies of different monoamine neurotransmitters. Noradrenaline is a major substrate for MAO A, and MAO A, not MAO B, is concentrated in the center for noradrenergic neurons, the locus ceruleus. With 5-HT neurons, the position is different; here the cell bodies contain predominantly the form that has only low affinity for the neurotransmitter involved, MAO B. Thus, although inhibition of MAO A will have a profound effect on noradrenaline oxidation, one needs to inhibit both MAO A and MAO B for maximal effects on the metabolism of 5-HT in the raphe. It is not known why there should be this discrepancy between the different systems.

In the substantia nigra, containing the cell bodies of nigrostriatal dopaminergic neurons, there is not an enrichment of either form. In the human, but not the rat, caudate and putamen, there is a particularly high level of MAO B in the glia surrounding the dopaminergic terminals. Dopamine is actually a similar substrate for both MAO A and B, both in terms of $K_m$ and $V_{max}$ *(26)*. The form that it is predominantly metabolized by will depend on the ratio of the two forms in a particular region. In the rat brain, this is MAO A. In the human brain, there is a relatively higher ratio of MAO B/A, particularly in the striatum, and, in postmortem brains from people who have been treated with the selective MAO B inhibitor (−)-deprenyl, dopamine levels were found to be significantly increased *(27)*. However, to obtain maximal effects, one would expect to need to inhibit both MAO A and B *(26)*.

Celada and Artigas *(28)* have shown, using in vivo microdialysis of the rat brain, that when tranylcypromine (a nonselective MAOI) was given intraperitoneally, 4 h after treatment there was a very large increase in 5-HT levels, both in the raphe (63-fold) and in the frontal cortex (11-fold). Clorgyline, or deprenyl alone, had a very modest effect at the raphe, less than a 50% increase in 5-HT levels, and no significant effect in the frontal cortex. However, when both were combined, a dramatic increase was again seen at both sites. Similar results were obtained with brofaramine, and brofaramine plus deprenyl *(29)*. With befloxatone also, it has been shown that, although systemic administration of the drug resulted in a large increase in 5-HT levels in the whole brain postmortem *(16)*, in vivo microdialysis showed no immediate increase in the frontal cortex, but pargyline did. Befloxatone did cause a 100% increase in noradrenaline *(30)*. This is the best evidence to date that both forms of MAO are involved in the functional oxidative deamination of 5-HT in the raphe in vivo, and that both need to be inhibited for pronounced effects on its local levels (*see* Fig. 1). This has only been shown in the rat, but the distribution of the two forms of MAO in and around the raphe seem to be similar in humans.

A further, possibly important, role for MAO B is in the metabolism of PEA. This is a trace amine, present in relatively small amounts in the brain. However, there is some evidence that it acts as a neuromodulator, particularly to activate the dopaminergic system *(31)*, and its levels in the brain are increased enormously after inhibition of MAO B *(20)*.

Taken together, these findings suggest that the inhibition of both MAO B and MAO A may well be important in the development of antidepressants more effective than the RIMA.

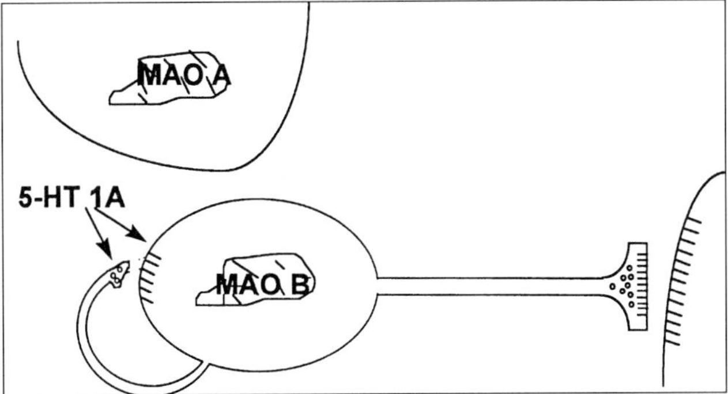



**Fig. 1.** Postulated role for MAO A and MAO B in the metabolism of 5-HT in the cell bodies of the raphe nucleus.

### 5.1. The Cheese Effect

RIMA are relatively free from the most dangerous side effect of the first generation of MAOI, the cheese effect. This is the hypertensive crisis arising from interactions with particular foods or drugs, typified by the high levels of tyramine that are contained in certain cheeses. During treatment with the classical irreversible MAOIs, ingestion of a quantity of tyramine as low as 6 mg can produce a moderate increase in systolic blood pressure, and 20 mg can produce a severe hypertensive reaction *(32)*.

Moclobemide, and all the other RIMAs tested, have a low cheese effect *(33)*. Da Prada and his colleagues have examined the tyramine content of a large number of different cheeses, wines, and other foods, and shown that large meals of high risk foods contained 10–36 mg of tyramine. Oral tyramine tests in humans showed that amounts of tyramine below 100 mg were unlikely to elicit a hypertensive reaction with moclobemide *(33)*.

The localization of MAO A throughout noradrenergic neurons helps to explain the mechanism of the cheese effect. When MAO A is inhibited, tyramine from cheese or other foodstuffs will act to release the amine from its storage vesicles into the cytoplasm, from which it will be transferred in increased concentrations into the extracellular gap (Fig. 2). The reason for the safety of the RIMA is thought to be that because they are competitive and reversible, high levels of tyramine will displace them from the active site and still be metabolized. Because tyramine penetrates the brain poorly, it is possible for it to displace reversible inhibitors from peripheral sites, and inhibition is maintained in the brain.

MAO B, which remains uninhibited, may contribute to the metabolism of tyramine. However, because clorgyline, which is a selective (but irreversible) MAO A inhibitor, has a very strong cheese effect *(14)*, the role of MAO B seems relatively minor, and the reversibility seems likely to be the prime reason for the improved safety. The intravenous pressor response to tyramine in volunteers taking moclobemide, plus the irreversible MAO B inhibitor deprenyl, has recently been shown to be raised about eight-fold *(34)*. However, deprenyl is a drug with several different pharmacological actions. The oral tyramine response in humans with a nonselective reversible inhibitor may well be less pronounced, and needs to be investigated.

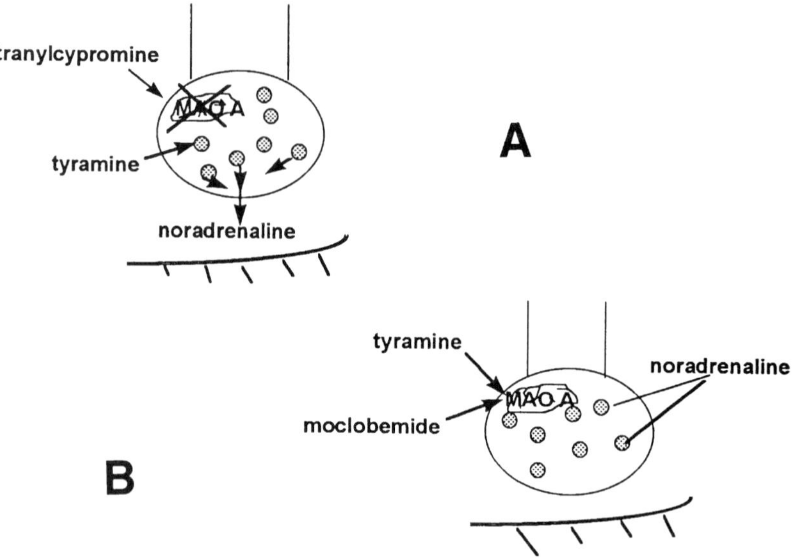


**Fig. 2. (A)** The action of tyramine in the presence of irreversible MAOI releases noradrenaline from storage vesicles into the cytoplasm. **(B)** Tyramine competes with reversible MAOI and is metabolized by mitochondrial MAO A in the noradrenergic terminal.

## 6. Other New Directions

Other possibilities with MAOI development may be to try to make drugs targeted to particular organs, such as the brain, or to particular cells, such as 5-HT neurons. An attempt has been made to make an MAOI targeted at the brain by making a pro-drug that is only activated by decarboxylation, and using it together with a peripheral decarboxylase inhibitor *(18)*.

Both urine and tissues contain endogenous MAOIs, which have been called tribulin, the level of which is increased by a range of conditions of stress or anxiety *(35)*. In understanding the clinical action of MAOIs, it may prove helpful to understand more about any interactions with endogenous factors acting in the same way. Tribulin is of low mol wt and is a nonpeptide. The chemical composition of tribulin is not completely understood, but isatin (2,3-dioxindole) has been identified as one major component *(36)*. It is a somewhat selective inhibitor of MAO B. Tribulin also contains more selective MAO A inhibitory component(s) of unknown identity *(37)*. Effective antidepressants may act by causing an increase in the level of these endogenous MAOIs. Both electroconvulsive therapy *(38)* and the folk herbal remedy Rhazya Stricta have been shown to cause a significant increase in endogenous MAOIs, particularly MAO A inhibitors, in rat brain (unpublished observations).

It is also possible that the changes induced in monoamines by current antidepressants are not the direct cause of the antidepressant action, but in turn act on some further system. Barden et al. *(39)* discuss the possibility that the prime abnormality in depression is a failure of feedback control in the HPA axis, possibly caused by an excess of glucocorticoid receptors in the hippocampus. Many different antidepressant drugs, including moclobemide, have been shown to normalize cortisol levels in depressed subjects, and induce glucocorticoid receptors in animal models. The time-course of this induction is similar to that of the clinical delay of 2–3 wk in the efficacy of antidepressant drugs. If this model is correct, then it may be that antidepressant drugs in the future should be designed to act directly, and more rapidly on the HPA system.

## CONCLUSION

For longer-term future directions, we may need to understand more about both the biological basis of depression, and the mechanism of action of antidepressants. The role of the endocrine system may well be found to be of primary importance. Even though the major focus has been on monoamines and the various monoamine theories of depression, it is also possible that the central defect in depression is in the HPA axis, and that the monoamine abnormalities are secondary *(39)*.

There remains a need for new antidepressants, particularly for a faster-acting antidepressant, without the usual 2-wk delay before onset of action, and for a safe antidepressant that acts in severe or resistant depression. It is the impression of many psychiatrists that the newer, safer, antidepressants, such as the selective 5-HT reuptake inhibitors and RIMAs, are most effective in mild depression. For severe depression, the older drugs often work well, but have adverse side effects.

Thus, any antidepressant that acts in a new way could be of interest for either of these categories. In terms of MAOIs, the differential localization and substrate specificity of MAO A and MAO B may have implications for the antidepressant effects of RIMAs. Theoretically, one would expect these drugs to have as much effect on noradrenaline metabolism as the old irreversible, nonselective, MAO inhibitors, but less effect on 5-HT and dopamine oxidation. If the safety largely lies with reversibility, then it might be that nonselective, but reversible, drugs would be the best type of antidepressant MAOI. A specific (not acting on other enzymes or receptors), reversible (thus eliminating the cheese effect), but nonselective, acting on both MAO A and MAO B, would be of great interest for severe depression, and has not yet been developed.

## REFERENCES


1. Glover, V. and Sandler, M. (1986) Clinical chemistry of monoamine oxidase. *Cell Biochem. Funct.* **4,** 89–97.
2. Zheng-Yi, C., Hotamisligil, G. S., Huang, J.-Q., et al. (1991) Structure of the human gene for monoamine oxidase type A. *Nucleic Acids Res.* **19,** 4537–4541.
3. Tivol, E. A., Shalish, C., Schuback, D. E., et al. (1996) Mutational analysis of the human MAO A gene. *Am. J. Med. Genet.* **67,** 92–97.
4. Lenders, J. W. M., Eishofer, G., Abeling, N. G., et al. (1996) Specific genetic deficiencies of the A and b isoenzymes of monoamine oxidase are characterised by distinct neurochemical and clinical phenotypes. *J. Clin. Invest.* **97,** 1010–1019.
5. Chen, Z.-C., Denney, R. M., and Breakfield, X. O. (1995) Norrie disease and MAO genes: nearest neighbours. *Human Molec Genet.* **4,** 1729–1737.
6. Davis, W. A. (1958) The history of Marsilid *J. Clin. Exp. Psychpath.* **6(2 (Supp1),** 1–10.

7. Zeller, E. A., Barsky, J., Fouts, J. R., et al. (1952) Influence of isonicotinic acid hydrazine (INH) and 1-isonicotinyl-2-isopropyl hydrazide (IIH) on bacterial and mammalian enzymes. *Experientia* **8**, 349–350.

8. Kline, N. S. (1970) Monoamine oxidase inhibitors: an unfinished picaresque tale, in *Discoveries in Biological Psychiatry* Ayd, F.J. and Blackwell, B., ed. Lippincott, Philadelphia, pp. 194–204.

9. Loomer, H. P., Saunders, J. C., and Kline, N. A. (1957) Clinical and pharmacological evaluation of iproniazid as a psychic energizer, in *Research in Affects* Psychiatric Research Reports No 8 of the American Psychiatric Association. Cleghorn, R. A., ed Washington DC, pp. 129–141.

10. Report to the MRC. (1965) Clinical trial of the treatment of depressive illness. *BMJ*, 881.

11. Glover, V., Sandler, M., Owen, F., and Riley, G. J. (1977) Dopamine is a monoamine oxidase B substrate in man. *Nature,* **265,** 80–81.

12. Knoll, J. (1993) The pharamcological basis of the beneficial effects of (–)-deprenyl (selegeline) in Parkinson's and Alzheimer's disease. *J. Neur. Transm.* **40(Suppl),** 69–93.

13. Mendis, N., Pare, C. M. B., Sandler, M., Glover, V., and Stern, G. (1981) Is the failure of (–)-deprenyl, a selective monoamine oxidase B inhibitor, to alleviate depression, related to its freedom from the cheese effect? *Psychopharmacology* **72,** 275–277.

14. Bieck, P. R., Antonin, K.-H., and Schulz, R. (1993) Clinical pharmacology of MAO inhibitors, in *Monoamine Oxidase* Yasuhara, H., ed. VSP. pp. 177–196.

15. Laux, G., Volz, H.-P., and Moller, H.-J. (1995) Newer and older monoamine oxidase inhibitors. *CNS Drugs* **3,** 145–158.

16. Rovei, V., Caille, D., Curet, O., Ego, D., and Jarreau, F.-X. (1994) Biochemical pharmacology of befloxatone (MD370503), a new potent reversible MAO-A inhibitor. *J. Neural Transm.* **41(Suppl),** 339–347.

17. Paykel, E. S. (1995) Clinical efficacy of reversible and selective inhibitors of monoamine oxidase A in depression. *Acta Psychiatr. Scand.* **91(Suppl 386),** 22–27.

18. Nutt, D. and Glue, P. (1989) Monoamine oxidase inhibitors: rehabilitation from recent research? *Brit. J. Psychiatry* **154,** 287–291.

19. Liebowitz, M. R., Hollander, E., Schneier, F., et al. (1990) Reversible and irreversible monoamine oxidase inhibitors in other psychiatric disorders. *Acta Psychiatry Scand.* **360(Suppl),** 29–34.

20. Berry, M. D., Jurio, A. V., and Paterson, I. A. (1994) The functional role of monoamine oxidases A and B in the mammalian central nervous system. *Prog. Neurobiol.* **42,** 375–391.

21. Singer, T. P., Von Korff, R. W., and Murphy, D. L. (1979) *Monoamine Oxidase: Structure, Function and Altered Function.* New York: Academic.

22. Willoughby, J., Glover, V., and Sandler, M. (1988) Histochemical localisation of MAO A and B in rat brain. *J. Neural. Transm.* **74,** 29–42.

23. Konradi, C., Kornhuber, J. K., Froelich, L., et al. (1989) Demonstration of monoamine oxidase-A and -B in the human brainstem by a histochemical technique. *Neuroscience* **33,** 383–400.

24. Luque, J. M., Kwan, S.-W., Abell, C. W., et al. (1995) Cellular expression of mRNAs encoding monoamine oxidases A and B in the rat central nervous system *J. Comp. Neurol.* **363,** 665–680.

25. Saura, J., Bleuel, Z., Urlich, J., et al. (1996) Molecular neuroanatomy of human monoamine oxidases A and B revealed by quantitative enzyme radiography and in situ hybridization histochemistry. *Neuroscience* **70,** 755–774.

26. Glover, V., Elsworth, J., and Sandler, M. (1980) Dopamine oxidation and its inhibition by (–)-deprenyl in man. *J. Neural. Transm.* **16(Suppl),** 163–172.

27. Riederer, P. and Youdim, M. (1986) Brain monoamine oxidase activity and monoamine metabolism in Parkinson patients treated with l-deprenyl. *J. Neurochem.* **46,** 1349–1356.

28. Celada, P. and Artigas, F. (1993) Monoamine oxidase inhibitors increase preferentially extracellular 5-hydroxytryptamine in the midbrain raphe nuclei. A brain microdialysis study in the awake rat. *Naunyn-Schmiedeberg's Arch. Pharmacol.* **347,** 583–590.

29. Celada, P., Bel, N., and Artigas, F. (1994) The effects of brofaramine, a reversible MAO-A inhibitor, on extracellular serotonin in the raphe nucei and frontal cortex of freely moving rats. *J. Neural. Transm.* **41(Suppl),** 357–363.

30. Curet, O., Damoiseau, G., Labsune, J.-P. et al. (1994) Effects of befloxatone, a new potent reversible MAO A inhibitor, on cortex and striatum monoamines in freely moving rats. *J. Neural. Transm* **41(Suppl),** 349–355.

31. O'Reilly, R. L. and Davis, B. A. (1994) Phenylethylamine and schizophrenia. *Prog. Neuro-Psychopharmacol. Biol. Psychiatry* **18,** 63–75.

32. Amrein, R., Allen, S. R., Vranesic, D., and Stahl, M. (1988) Antidepressant drug therapy: associated risks. *J. Neural. Transm.* **26(Suppl),** 73–86.

33. Da Prada, M., Zurcher, G., Wuthrich, I., and Haefely, W. E. (1988) On tyramine, food, beverages and the reversible MAO inhibitor moclobemide. *J. Neural. Transm.* **26(Suppl),** 31–56.

34. Korn, A., Wagner, B., Moritz, E., and Dingemanse, J. (1996) Tyramine pressor sensitivity in healthy subjects during combined treatment with moclobemide and selegeline. *Eur. J. Clin. Pharmacol.* **49,** 273–278.

35. Glover, V. (1993) Trials and tribulations with tribulin. *Biogenic Amines* **9,** 443–452.

36. Glover, V., Halket, J. M., Watkins, P. J., Clow, A., Goodwin, B. L., and Sandler, M. (1988) Isatin: identity with the purified monoamine oxidase inhibitor tribulin. *J. Neurochem.* **51,** 656–659.

37. Medvedev, A. E., Goodwin, B. L., Halket, J., Sandler, M., and Glover, V. (1995) Monoamine oxidase A-inhibiting components of urinary tribulin: purification and identification. *J. Neur. Transm.* **9,** 225–237.

38. Bhattacharya, S. K., Banerjee, P. K., Glover, V., and Sandler, M. (1991) Augmentation of rat brain endogenous monoamine oxidase inhibitory activity (tribulin) by electroconvulsive shock. *Neurosci. Lett.* **125,** 65–68.

39. Barden, N., Reul, J. M. H. M., and Holsboer, F. (1995) Do antidepressants stabilize mood through actions on the hypothalamic-pituitary-adrenocrtical system? *TINS* **18,** 6–11.

**5**

# Calcium Channel Antagonists in Mood Disorders

Olgierd Pucilowski

## 1. INTRODUCTION

Mood disorders have been the focus of intense research and drug development efforts. Since the introduction of imipramine, the first tricyclic antidepressant, in 1956, numerous compounds have been developed and others have faded from the market. In fact, until the advent of the selective serotonin reuptake inhibitors (SSRIs) in the late 1980s, there had been no clear advances in the pharmacotherapy of mood disorders. Despite the fact that SSRIs offer convenient dosing and a wider therapeutic index than tricyclic antidepressants (TCAs) or monoamine oxidase inhibitors, the tricyclic antidepressants remain a mainstay of treatment in major depressive disorder and in the depressed phase of bipolar illness. Lithium salts retain their prominent position in the therapy and prophylaxis of bipolar disorder. Two anticonvulsants, sodium valproate and carbamazepine, are acknowledged as alternatives in the almost 50% of bipolar patients who do not respond satisfactorily to lithium *(1)*. Pharmacotherapy remains the most effective mode of treatment in mood disorders, despite the potential toxicity associated with all of these drugs *(2)*. However, out of a concern for serious adverse effects of classic antidepressants, some physicians, especially those involved with primary care in an outpatient setting, may tend to undermedicate patients. It is therefore understandable that efforts continue to develop alternative medications to those already existing. Such drugs would have to be both clinically efficient and safer to use than those currently available. This chapter describes evidence, from both animal experiments and clinical trials, that a group of drugs designated as the calcium channel antagonists has shown promise in this respect. These drugs are similar to lithium in their pharmacodynamic profile of psychotropic activity, but differ in their mechanism of action at the cellular level.

## 2. PHARMACOLOGY OF CALCIUM CHANNEL ANTAGONISTS

Calcium channel antagonists first emerged as a therapeutic group in the 1970s. Compounds clustered here share a common mechanism of action, but are chemically very heterogeneous. In general, three main classes can be distinguished, based on chemical

*From:* Antidepressants: *New Pharmacological Strategies*
*Edited by: P. Skolnick, Humana Press Inc., Totowa, NJ*

structure: 1,4-dihydropyridines, phenylalkylamines, and benzothiazepines. These classes are represented by three drugs that were the prototypes of each respective category: nifedipine, verapamil, and diltiazem. The first two were developed by the German firms Knoll and Bayer, respectively, as coronary vasodilators. Verapamil became available in Europe in 1962 *(3)*. Nifedipine was synthesized in 1967 and marketed in the early 1970s *(4)*. The work of Fleckenstein and his group at the University of Freiburg provided subsequent evidence that blockade of cellular calcium influx is the primary mechanism of action of nifedipine and verapamil *(5)*. The third calcium channel antagonist, diltiazem, was developed by the Japanese firm Tanabe Seiyaku as a potential psychotropic agent, with an antidepressant/anxiolytic profile. Only later studies revealed its marked cardiovascular activity and calcium channel blocking properties *(6)*.

Since the introduction of the first calcium channel antagonists, many more compounds have been developed within each of the three classes. Synthesis of new 1,4-dihydropyridine derivatives has been particularly prolific *(7)*, but phenylalkylamine and benzothiazepine classes have grown as well (Table 1) *(8,9)*. Compounds with agonistic properties, i.e., which increase calcium transmembrane current, have also been synthesized within the 1,4-dihydropyridine class (e.g., Bay K8644) *(10)*. Calcium channel antagonists interfere with transmembrane entry of calcium—a ubiquitous second messenger in excitable cells involved in mediation of a multitude of cellular functions, including stimulus–bioelectrical response, excitation–contraction coupling, excitation–secretion coupling, and gene expression. It affects, via calcium-dependent enzymes, cell integrity, adhesion, and excitability *(11,12)*.

Over the past two decades, calcium channel blockers have become valuable drugs in the treatment of several cardiovascular system disorders, such as tachyarrhythmias, exertional and variant angina, hypertension, and migraine *(13)*. Among dihydropyridine compounds, nimodipine is highly nonpolar and thus easily crosses the blood–brain barrier. This drug may be effective in treatment of stroke, subarachnoid hemorrhage, and prevention of cerebral ischemia *(13)*. Calcium channel antagonists are characterized by high therapeutic indices, and are generally well tolerated *(14)*. The most frequently reported adverse effects follow high-dose therapy with dihydropyridines, and are caused by excessive vasodilatation. These include dizziness, hypotension, nausea, headache, hot flushes, and sedation. Dihydropyridine-induced hypotension causes compensatory tachycardia that may increase oxygen demand. Together with selective coronary vasodilatation of nonischemic vessels (i.e., coronary steal syndrome), this may aggravate myocardial ischemia. Phenylalkylamine and benzothiazepine calcium channel antagonists are less likely to aggravate ischemia, but have significant negative chrono- and dromotropic effects, and are contraindicated in patients with SA or AV nodal conduction disturbances *(13)*. Calcium channel antagonists have a propensity to increase plasma levels of digoxin *(15)*, and verapamil has been reported to potentiate lithium *(16,17)* and carbamazepine toxicity *(18)*. Perhaps because of their marked antidopaminergic properties *(13)*, verapamil and diltiazem may occasionally cause Parkinsonism *(20–22)*. Overall, although calcium channel antagonists are not without adverse effects, these are much less serious and distressing to patients than those of classical antidepressant drugs or lithium.

Numerous studies have been aimed at characterization of the pharmacodynamic profile of calcium channel antagonists with respect to potential psychotropic activity. They

**Table 1
Calcium Channel Antagonists**

| Phenylalkylamines | 1,4-Dihydropyridines | Benzothiazepines | Other |
|---|---|---|---|
| Verapamil | Nifedipine | Diltiazem | Flunarizine[a,b] |
| Gallopamil | Nicardipine | Clentiazem | Cinnarizine[a] |
| Tiapamil | Niludipine | Diclofurime | Lidoflazine[a] |
| Anipamil | Nimodipine | | Bepridil[a] |
| Levemopamil | Nitrendipine | | |
| | | | Caroverine[a] |
| Falipamil | Nisoldipine | | Dotarizine[a] |
| | Azidopine | | Prenylamine[a] |
| | Darodipine | | |
| | Felodipine | | Fluspirilene[c] |
| | Isradipine | | Pimozide[c] |
| | Lacidipine | | |
| | Oxodipine | | |
| | Ryodipine | | |

[a]Nonselective calcium channel modulators; [b]T-type calcium channel antagonist; [c]Neuroleptic drugs of the diphenylbutylpiperidine class, which bind at a specific recognition site on a calcium channel separate from classic calcium antagonists.

exert wide-ranging but relatively modest effects on animal behavior tested under normal or low-stress experimental conditions. When animals are exposed to more severe environmental or drug challenges, their effects become much more pronounced. The most consistent animal data support potential antidepressant and antimanic activity of calcium channel antagonists, although other psychotropic properties have also been demonstrated (*see* refs. *19, 23, and 24* for reviews). On a clinical level, the evidence for a potential use of these drugs in mood disorders is not conclusive. Nonetheless, given the limited number of compounds and patients involved, the data obtained in bipolar disorder may be viewed as encouraging *(22,25)*. The success that calcium channel antagonists enjoy as antiarrhythmic and hypotensive drugs may be one of the reasons for the relative lack of interest on the part of the pharmaceutical industry in their use as psychotropic medications *(26)*. Paradoxically, recent debate over the potentially grave consequences of nifedipine treatment in coronary heart disease *(27; cf. 28)* may help to advance the cause of novel (i.e., psychiatric) indications for these well-known drugs. This chapter is a review of preclinical and clinical data on the antidepressant and antimanic effects of calcium channel antagonists, and an attempt to critically assess highlights and pitfalls of this research. A hypothesis explaining the mode of action of calcium channel antagonists in mood disorders is presented and suggestions pertaining to future research are given.

## 2.1. Voltage-Activated Calcium Channels

Increases in intracellular calcium concentration can result either from enhanced influx (through membrane voltage-activated or receptor-activated channels) or calcium release from intracellular stores in reticulum and mitchondria (as a result of stimulation of membrane metabotropic receptors acting via phospholipase C/inositol phosphate pathway). There are at least four different types of voltage-activated calcium channels, designated as L, N, T, and P, and all these types of channels have been demonstrated in the nervous system (*29*; Table 2). The primary mechanism of action of calcium channel antagonists is the inhibition of transmembrane ion fluxes across the L-type, which stands for "long-lasting," or "slow," calcium channels. There are many other, less selective, calcium antagonists whose action is not restricted to any particular channel subtype. Calcium channel antagonists of the phenylalkylamine, dihydropyridine, and benzothiazepine classes act at recognition sites specific for a particular class. These recognition sites are localized on the ion pore-forming transmembrane protein designated $\alpha_1$, one of four subunits ($\alpha_1$, $\alpha_2/\sigma$, $\beta$, $\gamma$) forming the L-type calcium channel structure. The $\alpha_1$ subunit has four repetitive motifs, each containing six membrane-spanning regions (termed S1–S6). The S6 region of the fourth motif is suspected to contain phenylalkylamine (intracellular) and dihydropyridine (extracellular) binding sites *(30,31)*; the S4 region of each motif is thought to contain the voltage sensor of the channel. All subunit proteins of the L-type calcium channel have been cloned. There are several forms of $\alpha_1$-subunit and their expression is tissue specific. The brain $\alpha_1$-subunit is of either C or D type, of which type C is similar to the cardiac muscle subunit (95% homology), but they both differ from the skeletal muscle subunit (*see* ref. *32* for review). Further diversity among brain C-type subunits is derived from an alternative splicing process *(33)*.

Classification of calcium channels as voltage-activated is based on the observation that their conductance increases in response to membrane depolarization. Voltage-activated calcium channels can be differentiated into low-voltage-activated (T,P) and high-voltage-activated (L,N) subtypes, depending on their respective sensitivity to depolarization of holding potential. Both the L- and N-types give rise to high-voltage-activated currents as they become activated by relatively large changes in membrane potential, positive to −10 and −30 mV, respectively *(11)*. These two channels differ considerably in other aspects of their electrophysiological function, as well as pharmacology, localization, and physiological role. L-Type channels remain open for a long time (inactivation time constant >500 ms), but N-type channels inactivate more rapidly (50–80 ms). Single-channel conductance is greater for L- than N-type channels (18–25 pS [picoSiemens] vs 11–20 pS, respectively). L-Type channels are ubiquitous in excitable cells throughout the organism. In the nervous system, they are located post-synaptically on dendrites and soma *(34,35)*. Autoradiographic studies with radiolabeled calcium channel ligands have revealed that their recognition sites are primarily localized within the olfactory bulbs, hippocampus, amygdala, and frontal cortex *(36–38)*. N-Type channels are primarily confined to presynaptic neuronal terminals, where they are involved in neurotransmitter release. Binding sites for the N-type channel-selective ligand ω-conotoxin are concentrated in areas rich in synaptic connections and show a similar, but not identical, distribution to the dihydropyridine recognition sites *(39,40)*.

Because of their high conductance and slow inactivation, L-type channels substantially contribute to neuronal calcium entry in regions where they are abundant. In hippocampal neurons, for example, 79% of the total calcium influx has been attributed to L-type calcium channels *(41)*. Even in these regions, however, calcium channels do not constitute the exclusive source of intracellular calcium. Increases in intracellular calcium concentration can occur in response to the opening of low-voltage-activated chan-

**Table 2**
**Voltage-Activated Calcium Channel Subtypes**

| | L<br>(Long-lasting) | N<br>(Neuronal) | T<br>(Transient) | P<br>(Purkinje) |
|---|---|---|---|---|
| Activation threshold | High | High | Low | Medium |
| Inactivation rate | Slow | Moderate | Fast | Fast |
| Conductance | 18–25 pS | 11–20 pS | 8 pS | 10–12 pS |
| Ligands | Dihydropyridines<br>Phenylalkylamines<br>Benzothiazepines | $\omega$-conotoxin | Amiloride<br>Octanol<br>Flunarizine | Aga-toxin |
| Function | E-C coupling in muscle; neuroadaptation | Neurotransmitter release | Spike activity in SA node, neurons, and endocrine cells | Neurotransmitter release |

Note: currently available T-type calcium channel ligands are not specific; aga-toxin is a funnel web spider toxin.

nels (T,P), receptor-activated channels of the NMDA-type, or secondary to metabotropic receptor-dependent activation of phospholipase C/inositol phosphate pathway, such as follows agonist interaction with muscarinic $M_1$, adrenergic $\alpha_1$, or serotonergic 5-$HT_{1c}$ receptors *(42)*. Thus calcium channel antagonists, though capable of limiting calcium flux, do not block calcium trafficking altogether. Dependence on the membrane potential is another important characteristic of the interaction between the antagonist and calcium channel receptor. Calcium channels are especially sensitive to inhibition under open and inactivated states that are promoted by depolarization (*see* ref. *11*). Conversely, actions of the calcium channel agonist Bay K8644 are attenuated by depolarization *(43)*. Taken together, these data indicate that the administration of calcium channel antagonists could prevent excessive influx of calcium. This inhibitory effect would be most pronounced in depolarized neuronal cells.

## 3. ANTIDEPRESSANT PROPERTIES OF CALCIUM CHANNEL ANTAGONISTS

### 3.1. Preclinical Studies

There is a range of animal models available for detection of drugs with potential antidepressant properties. The most widely used methods rely on the development of some behavioral deficit in animals subjected to prolonged, unavoidable stress of varying intensity. It can be argued that all these models involve some degree of conditioning (learning), whereby an animal aborts a normal behavioral response when unable to avoid a noxious stimulus (i.e., based on the theory of learned helplessness) *(44)*. The response under study is usually, but not always, directly related to a particular stressor, e.g., an escape reaction elicited by footshock or immersion in water. The most widely used test for detecting antidepressant-like effect of a drug is the forced-swim test in rats. All classical and newer antidepressant drugs significantly decrease rats' immobility in the test, although SSRIs are less efficient than tricyclic and tetracyclic compounds (*see* ref. *45* for review).

All calcium channel antagonists tested, i.e., nifedipine, nitrendipine, nicardipine, Goe 5438, diltiazem, and verapamil, are as effective as imipramine or desipramine in decreasing the immobility time of rats (or mice) in a forced-swim test *(19,46–51)*. The specificity of this effect is confirmed by the observation that a calcium agonist, Bay K8644, increases immobility, an effect blocked by nifedipine pretreatment *(50)*. Decreased immobility was also described following direct microinjections of diltiazem, nicardipine, and verapamil into the anterior hypothalamus in rats, suggesting that this structure may be involved in antidepressant-like properties of calcium channel antagonists *(46)*. As seen with classical antidepressants, prolonged treatment with calcium channel antagonists results in an increased efficacy in the forced-swim test *(47,48,52)*. Furthermore, calcium channel antagonists appear to potentiate the efficacy of tricyclic antidepressants in this test *(50, but cf. 48)*.

The antidepressant-like effects of calcium channel antagonists were also demonstrated in other animal models. Nifedipine was found to decrease foot shock-induced locomotor deficit (immobility) in a shuttle box *(53)* or in an open field *(49)*, and diltiazem showed a similar effect in the latter study. Again, the effect of imipramine in this test was potentiated by nifedipine and attenuated by Bay K8644 pretreatment *(54)*.

There appears to be an underlying genetic predisposition to mood disorders *(55)*. Nevertheless, efforts to develop a valid animal model of depression have, in general, neglected that fact, with one notable exception—the Flinders Sensitive Line (FSL) rat. These rats were originally selectively bred for increased sensitivity to organophosphates, but have been subsequently demonstrated to exhibit several physiological, behavioral, and biochemical characteristics resembling depressed humans *(56)*. For example, FSL rats show markedly more pronounced behavioral deficits in response to acute and chronic stressors than do control Flinders resistant line (FRL) rats and outbred strains *(56–58)*. In a forced-swim test, the immobility time of the FSL rats is typically twice as long as that of FRL rats, even when exposed to a water tank for the first time *(56)*. We have demonstrated that antidepressants, both tricyclics, such as imipramine and desipramine, and the SSRI sertraline, effectively decrease immobility in FSL rats after chronic, but not acute, administration *(57,59)*. A significant decrease in immobility was also observed following 14-d treatment with nicardipine and verapamil, suggesting that calcium channel antagonists are effective in this genetic animal model of increased susceptibility to depressive-like reactions *(57)*. In another experiment, we have found that FSL rats are submissive when competing for water in pairs with FRL rats *(60)*. It had been earlier demonstrated that tricyclic (imipramine, desipramine) and atypical (mianserin, salbutamol) antidepressant drugs can disinhibit behaviorally submissive rats in food and water competition paradigms *(61–63)*; consequently, submissive behavior has been proposed to be yet another manifestation of a depressive-like state *(61)*. Verapamil administration to submissive FSL rats markedly increased their drinking time without influencing water intake (thirst) in individually tested rats *(60)*.

In summary, all calcium channel antagonists tested so far show efficacy comparable to classical antidepressant drugs in different animal models of depression, as well as in screening tests for antidepressant-like properties (e.g., the forced-swim test). Calcium channel antagonists retain or increase their antidepressant-like effect under prolonged treatment conditions and potentiate the behavioral effects of classical antidepressants when given concomitantly. Although calcium channel inhibitors have not yet been tested in more sophisticated animal models of depression, e.g., separation in primates or in chronic mild stress-induced anhedonia *(44,64)*, the data presented above conclusively indicate that they have potential as clinically useful antidepressants.

## 3.2. Clinical Studies

The most significant limitation of currently available animal models of depression is that they do not reflect the periodic character of affective alterations in mood disorders, and thus can only claim to model a single episode of a depressive disorder. Although depressive symptomatology in major depression and bipolar disorder has similar manifestations, there is increasing evidence that the etiologies of the two are different. Genetic factors, for example, are believed to play a much larger, albeit still unspecified, role in bipolar disorder than in major depression *(55)*.

Two early case reports on the use of verapamil (240–320 mg/d) in patients with unipolar depression have been encouraging *(65,66)*. Likewise, Jacques and Cox *(67)* described complete remission of delusional depression in a patient on verapamil (360 mg/d, 5 wk), initially prescribed to combat an arrhythmia that developed after electroconvulsive treatment. In an open study with 23 depressed patients, Hoschl and colleagues *(68)* reported marked improvement in nine and slight improvement in six subjects following 2 wk of verapamil (240–400 mg/d) treatment. In a subsequent double-blind, placebo-controlled study in a sample of 26 patients with recurrent major depression, the effect of a 5-wk regimen of verapamil treatment was not different from that of a placebo *(69)*.

So far, verapamil is the only calcium channel antagonist that has been used in clinical trials to treat depression. There are reports of depressive symptoms associated with nifedipine treatment for cardiovascular indications *(70,71)*, but as a group, calcium channel antagonists do not precipitate depressive symptoms *(72,73)*.

## ANTIMANIC PROPERTIES OF CALCIUM CHANNEL ANTAGONISTS

### 4.1. Preclinical Studies

Face validity, i.e., how well an animal's behavior in a given model resembles the human affective state, has been a starting point for the development of preclinical models of mania. A particular behavioral response in a laboratory animal that serves as a manic-equivalent is usually elicited by drugs that act to enhance dopaminergic transmission. This reflects the currently predominant hypothesis that links manic symptoms to overactivity in central dopaminergic synapses. The symptoms typically modeled include hyperactivity, irritability, elation, or euphoria. These pharmacological models, however, make no claim to provide insight into the etiopathology of mania, unlike the animal modeling efforts in depression that helped to advance a theory of prolonged, unavoidable stress as a likely contributing factor in the etiology of depressive symptomatology (*see* refs. *44* and *64*).

Dopaminomimetic drugs, both direct receptor agonists (e.g., apomorphine) and indirectly acting amine releasers (amphetamines) or reuptake inhibitors (cocaine), produce dose-dependent stages of behavioral stimulation. Hyperactivity seen at lower doses is transformed into various manifestations of stereotyped motor activities (sniffing, licking, gnawing) at increased doses. Even higher doses of agonists are required to elicit bursts of jumping and irritable aggression *(74,75)*.

There is considerable pharmacological data on the interaction of calcium channel antagonists with dopamine-related behavioral stimulation. Cocaine-induced locomotor stimulation is antagonized by dihydropyridine calcium channel antagonists, such as isradipine, nicardipine, and nimodipine *(76–78)*, but not by verapamil *(78,79)*. Experiments with diltiazem yielded conflicting results (cf. *76,79*). Similarly, although some calcium channel antagonists (darodipine, isradipine, nifedipine) suppress amphetamine-induced hyperactivity, others produce either equivocal (nicardipine, nimodipine, nitrendipine) or no effect (nisoldipine, diltiazem, verapamil) *(76–78,80–82)*. All calcium channel antagonists tested, i.e., nicardipine, nifedipine, nimodipine, and verapamil, reportedly cause a dose-dependent suppression of yawning and penile erections elicited by apomorphine, a nonselective (i.e., $D_1$ and $D_2$) dopamine receptor agonist *(83*; but cf. *84)*. There is a significant suppression of apomorphine-induced stereotyped behavior observed in rats pretreated with such calcium channel antagonists as nitrendipine and diltiazem *(76)*, but not with nicardipine, nifedipine, nimodipine, nisoldipine, or verapamil *(76,85,86)*. Stereotyped grooming behavior, elicited in rats by the selective $D_1$ receptor agonist SKF 38393, is markedly inhibited by nicardipine, nifedipine, and verapamil *(87)*. Nicardipine and diltiazem also effectively attenuate irritable fighting behavior elicited by apomorphine injections in pairs of highly aggressive rats *(88,89)*. Overall, these data suggest that, although there may be some antidopaminergic effect exerted by calcium channel antagonists, it is neither robust nor consistent. Biochemical and receptor studies do not resolve these discrepancies, but suggest that a direct dopaminergic blocking property is not likely to explain the antidopaminergic properties of this class of drugs (*see* ref. *19* for review). The relevance of these findings for understanding the pathophysiology of manic/bipolar disorder is limited, since the behavioral arousal in pharmacological models of mania is not cyclic.

The occurrence of manic/depressive episodes shows a pattern of increasing frequency over time. Two phenomena, behavioral sensitization to dopaminergic stimulants and electrophysiological kindling, have been proposed as nonhomologous models of this tendency *(90)*. Calcium channel antagonists have been shown to be active in both these models. They interfere with the induction and expression of behavioral sensitization to amphetamine-induced stereotypy *(91)* and cocaine-induced locomotion *(92,93)*. Calcium channel antagonists also appear to exert at least partial protection against seizure activity elicited by kindling of the hippocampus *(94)* and amygdala *(95)*.

Euphoria is yet another aspect of manic symptomatology that has attracted extensive research efforts. At the experimental level, euphoria has been conceptualized as positive reinforcement, achieved via stimulation of the mesotelencephalic dopaminergic pathway. This is also a mechanism shared by various euphoria-inducing drugs and is thought to be a major contributing factor to their addictive properties *(96,97)*. Recent evidence, obtained with the conditioned place-preference and self-administration paradigms, indicates that calcium channel antagonists attenuate the positive reinforcing (i.e., euphorigenic) effects of amphetamine *(81,98)* and cocaine *(99–101)*. Because calcium channel blockers do not seem to possess reinforcing or aversive properties *(98,102,103)*, their potential clinical use should not be tainted by undue concerns about their abuse potential or patient compliance.

### 4.2. Clinical Studies

Carman and Wyatt (1979) reported a decreased severity and frequency of agitation in an open study of three manic patients treated with calcitonin, a nonspecific calcium antagonist *(104)*. In the early 1980s, the first reports of possible antimanic effects of L-type calcium channel antagonists began to emerge *(105–109)*. Although verapamil was used in most clinical studies in bipolar disorder, diltiazem *(105)* and nimodipine *(110–112)* have also been found to be effective. The antimanic properties of verapamil have been confirmed in double-blind, placebo-controlled studies *(69,113,114)* and its clinical efficacy found equal to that of lithium *(69,115,116)*. Patients who responded to verapamil were either responsive to lithium or had not been previously treated with lithium. Those who did not initially respond to lithium alone benefited from conjoint therapy with lithium and verapamil *(117)*. A combination of verapamil and chlorpromazine has also been found very effective, and without added side effects, in severe manic episodes *(118)*. In bipolar disorder resistant to lithium, verapamil treatment has generally been found less effective (cf. *25*).

Verapamil, either alone *(119)* or in combination with an antidepressant *(120)*, produced remission in patients with rapid-cycling bipolar disorder. Nimodipine has also been found very effective in rapid-cycling patients. Manna *(111)* reported that in nine patients enrolled in an open study, a combination of nimodipine (90 mg/d) and lithium was more effective than either drug alone in reducing the frequency and intensity of manic episodes. Nimodipine was equally effective as a conjoint treatment and more effective than lithium alone in decreasing the number of depressive episodes *(111)*. In a double-blind, placebo-controlled study, Pazzaglia et al. *(112)* have reported that five of nine rapid-cycling, lithium-resistant patients responded favorably to nimodipine.

## 5. TOWARD THE ROLE OF CALCIUM IN MOOD DISORDERS

### 5.1. Calcium Trafficking and Its Role as a Second Messenger

Calcium channel antagonists have been proposed as possible alternatives or supplements to classical antidepressant drugs, primarily on the basis of overwhelmingly positive evidence from animal studies. These drugs are probably better described as mood stabilizers, since they are similar to lithium in their clinical profile. Both preclinical and clinical data strongly support their antimanic effectiveness. Antidepressant effects, strongly suggested by animal experiments, appear equivocal in clinical trials carried out in major depression, but are more evident in depressed bipolar patients *(22,25)*. Calcium channel antagonists also appear to have a protective, mood stabilizing effect when used in maintenance therapy in bipolar disorder. Limited evidence suggests that nimodipine, but not verapamil, may be effective in treatment-resistant cases of bipolar disorder *(112,121)*. However, a combination of verapamil with lithium may be effective in cases not responding to lithium alone *(117)*.

The effectiveness of calcium channel antagonists in animal models of depression and mania has been demonstrated for several compounds from different chemical classes. Although the psychiatric use of these drugs has been much more limited, there is preliminary evidence that compounds from each major class of calcium channel antagonist (i.e., verapamil, diltiazem, and nimodipine) show promising therapeutic effects in mood disorders.

If these drugs are indeed effective in stabilizing altered mood states, can we learn about the mechanisms underlying mania and depression from their mode of action? Unfortunately, the dynamics of transmembrane calcium fluxes at the central neuronal level in humans and behaving animals are currently inaccessible, with available techniques establishing only whether or not there is a change in calcium concentration in a disease state (or a disease model). Indeed, elevated intracellular free calcium concentrations of blood platelets and lymphocytes from patients during both manic and depressed phases of a bipolar disorder have been found *(122–124)*, and stimulation of serotonergic (5-HT$_2$-type) receptors has yielded greater increases in the intraplatelet calcium concentration from depressed and bipolar patients than from healthy controls *(128)*. In unipolar depressed patients and bipolar patients rendered euthymic by various treatments (e.g., TCAs, lithium, carbamazepine), calcium concentrations in platelets have not been different from healthy controls *(122–127)*. In patients in the remission phase of recurrent major depression, serum and intraplatelet calcium were elevated in those treated with lithium, but not with TCAs *(125)*. Because platelets are considered to be a valid peripheral model of central serotonergic neurons, a parallel increase in intraneuronal calcium is suspected in bipolar disorder and perhaps in major depression.

Intracellular calcium plays a crucial role in multiple neuronal processes involved in such essential brain functions as excitability, re- and degeneration, and the synthesis, release, and reuptake of neurotransmitters/neuropeptides. Given its vital functional importance, it is obvious that cellular calcium homeostasis is tightly and multidirectionally controlled. The transmembrane calcium gradient increases in response to postsynaptic depolarization. Both receptor-activated and voltage-activated channels contribute to this increase. Stimulation of membrane metabotropic receptors activates, via G proteins, phospholipase C and breakdown of phosphatidylinositol into inositol triphosphate (IP3) diacylglycerol. IP3 binds to its receptor on endoplasmic reticulum and triggers the release of calcium *(129)*. The IP3 receptor is a major intraneuronal calcium release channel, but there are other IP3-insensitive calcium stores in neurons that respond to caffeine and ryanodine, which are ryanodine receptor ligands *(130)*. Diacylglycerol, together with free cytoplasmic calcium, activates protein kinase C (PKC), which, among its many effects, can increase or decrease the activity of voltage-activated calcium channels *(131)*. From the cytoplasm, calcium is taken up by several calcium-binding proteins; in the nervous system, these include calmodulin, neurocalcin, and S-100 protein *(132)*. After complexing with an appropriate protein, calcium can activate a variety of enzymes, some of which modulate the activity of calcium channels. For example, reversible activation/deactivation of calcium channels in the skeletal muscle has been attributed to the sequential action of calcium/calmodulin-dependent kinase, which phosphorylates, and PKC, which dephosphorylates, dihydropyridine receptors *(133)*. Phosphorylation of $\alpha_1$- and $\beta$-subunits of L-type calcium channel can as well be mediated by cAMP-dependent protein kinase A *(131)*. Neurotransmitters acting through adenylate cyclase-coupled membrane receptors ($\beta$-adrenergic, D$_1$ dopaminergic, 5-HT$_1$ serotonergic) can modify voltage-activated channel function through this mechanism. Nitric oxide synthase is another important intracellular enzyme that is calcium-dependent, at least in neurons *(134)*. Nitric oxide synthase is a putative neurotransmitter that has been implicated in the modulation of several physiological and behavioral functions disturbed in mood disorders, including circadian rhythms

*(135)*, memory *(136)*, consummatory behavior *(137,138)*, anxiety *(139)*, and drug reward *(140)*.

Calcium as a second messenger is unique in that it can both enhance and attenuate its own actions by facilitating the activity of intracellular enzymes, e.g., kinases and phosphatases, in a tightly controlled manner. Excessive concentration of intracellular calcium caused by prolonged depolarization (e.g., persistent NMDA receptor activation) can lead to serious disruption of intracellular metabolism and cell death.

## 5.2. Effects of Drugs on Calcium Handling

In addition to calcium channel antagonists, several other drugs that modify mood (or parallel behavior in animal models), as well as electroconvulsive shock treatment, have been shown to modify cellular calcium trafficking. Tricyclic antidepressants have calmodulin-blocking properties *(141)* and appear to inhibit voltage-activated calcium channels *(142)* and calcium-activated potassium channels *(143)* at clinically relevant concentrations. Electroconvulsive therapy inhibits calcium transport from cerebrospinal fluid to the brain *(144)*, decreases PKC activity *(145)*, and either decreases or increases the density of dihydropyridine recognition sites, depending on the brain area *(146–148)*. Dihydropyridine binding is not changed by acute or chronic imipramine administration, but there is an increase in receptor density ($B_{max}$) following chronic treatment with the atypical antidepressants citalopram (a SSRI) or chlorprotixene *(149)*. Although lithium does not appear to directly affect transmembrane calcium fluxes *(150)*, it is an inhibitor of calmodulin *(107)* and PKC *(151)*. It also acts to normalize intracellular calcium concentrations, supposedly by interfering with the phosphatidylinositol cycle *(42,129,152)*.

Dubovsky *(25)* hypothesized that the alternating mood states in bipolar disorder are a consequence of opposite changes in neuronal excitability secondary to increased intracellular calcium levels. Thus, the treatment of mood disorders may be accomplished by targeting calcium trafficking. According to this hypothesis, a manic state would be induced by the exaggerated neuronal excitability that accompanies transient increase in intracellular calcium concentrations, while long-lasting elevation would lead to disruption of neuronal function, manifested as depression. To prevent neurodegenerative processes associated with persistently high levels of intraneuronal calcium, compensatory mechanisms restore calcium levels to normal. This hypothesis assumes the existence of a dynamic imbalance in intracellular calcium handling that could be a consequence of genetic makeup or environmental insult. While voltage-activated calcium channels of the L-type are among the primary suspects, their dysfunction should not be a prerequisite for proving the calcium–mood disorder hypothesis. Because the L-type channel plays such a major role in calcium entry, and its opening is triggered by membrane depolarization (secondary to either stimulation of ionotropic receptors, of which NMDA receptors cause a direct calcium entry, metabotropic receptors, or G protein/adenylate cyclase-linked membrane receptors), calcium channel antagonists may be clinically effective, even if the calcium-handling defect lies upstream (calcium-dependent transmitter release, receptor-activated channels) or downstream (intracellular calcium-activated mechanisms) of their direct point of action on calcium homeostasis.

Calcium, as a second messenger, plays a pivotal role in a variety of cellular actions of several neurotransmitters and hormones. This increases the likelihood that many mood-altering drugs modulate, at one point or the other, neuronal calcium balance. It may, as well, indicate that interference with calcium-related processes is nonspecific, rather than the primary mechanism of action of any antidepressant or mood-stabilizing drug. For this notion to be true, the etiologic involvement of calcium alterations in mood disorders needs to be proven first. This is for the time being beyond technical reach of current clinical science. Still, indirect molecular evidence could be provided on the possible alterations of particular calcium channel subunit proteins in brain regions thought to be involved in affective mechanisms (e.g., amygdala, septum, hippocampus). The presence of mRNA of particular channel subunits could be visualized by *in situ* hybridization, using specific probes, and the effect of drugs/stressors on gene expression could be compared. It has been shown that calcium channel antagonist recognition sites, albeit widely expressed in the brain, appear to be more concentrated in the limbic system. These would be the likely sites to look for possible alterations in the molecular structure of L-type channels. The utility of increasingly popular studies of the early response genes c-*fos* and c-*jun*, important for the generation of AP-1, one of the transcription factors, are more problematic in this respect. Although calcium appears necessary for transcription of early response genes and AP-1 function, these genes modulate such a plethora of cellular events that it is virtually impossible to relate their altered function to any particular neuronal process.

Another important point relates to the hypothetical specificity of the action of calcium channel antagonists in mood disorders, or, for that matter, in any psychiatric disorder that could involve altered function of L-type calcium channels. Calcium channel antagonists, as mentioned before, do not elicit any significant alterations in the behavior of healthy humans or normal control animals. This is probably because other mechanisms compensate for decreased calcium entry caused by blockade of these channels under normal conditions. Behavioral effects become manifest when neuronal function is subjected to drug or environmental challenges, such as exposure to mood-altering drugs (alcohol, stimulants, opiates) or unavoidable stressors *(19,23)*. We hypothesized that this is because only particular subsets of neurons are subject to prolonged depolarization and secondary adaptive changes in L-type calcium channel function with such treatments. These include increased density/affinity of receptor sites and changes in the mode of channel operation that favors prolonged opening and a short inactivation phase. Prolonged opening of L-type channels should be particularly important for the neuronal action of verapamil and diltiazem because their recognition sites appear to be localized on the intracellular side of the channel pore. Overall, these adaptive (or maladaptive, if they last long enough to cause an excessive increase in intracellular calcium) changes would promote calcium entry, but would also facilitate the action of calcium channel antagonists. Calcium channel antagonists could act to protect central neurons from excessive calcium entry and thus prevent calcium-related hyperexcitability (mania?) or hypoexcitability (depression?). Because of their greater propensity to act at the open/depolarized channels, they could hypothetically be unique, disorder-specific drugs. Thus, although their receptors are spread throughout the brain, only those sites altered by a maladaptive disease process would become effective targets for calcium channel antagonists. On the other hand, drugs that act at recognition sites that are not voltage-dependent, such as NMDA-receptor-linked calcium channels, would not be targeted by their antagonists in a similar site-specific manner, and thus would be expected to produce desirable therapeutic, as well as undesirable, effects.

### 5.3. Calcium Channel Antagonist–Neurotransmitter Interaction

Antidepressant drugs are known to interfere with either amine reuptake (tricyclics, tetracyclics, SSRIs) or metabolic breakdown (MAO inhibitors), both processes leading to increased concentrations of monoamines in the synaptic cleft. Tricyclic and tetracyclic compounds are, in general, nonselective, which means that uptake inhibition applies to all monoamines, i.e., norepinephrine, dopamine, and serotonin, albeit to a different extent. Likewise, some psychotropic effects of calcium channel antagonists may be secondary to their interaction with aminergic synaptic transmission. Indeed, many calcium channel antagonists were found to block aminergic receptors and/or inhibit their reuptake *(19)*. However, the affinity of calcium channel antagonists for aminergic receptors ($M_1/M_2$, $\alpha_1$, 5-HT$_1$, 5-HT$_2$) is in the micromolar range, but they have a 10-fold higher affinity for their calcium channel recognition sites *(153–156)*. Moreover, although all calcium channel antagonists, by definition, block their respective receptors on L-type calcium channels, only some of these drugs also interact with aminergic receptors (cf. *19*). Consequently, behavioral effects that might have been visualized as related to direct interference with aminergic transmission, rather appear to be secondary to calcium-related alterations in postsynaptic responsiveness to aminergic stimulation.

Perhaps the most important synaptic mechanism related to classical tri- and tetracyclic antidepressant activity is the suppression of amine reuptake. Some (i.e., verapamil, bepridil), but not all (e.g., diltiazem, nitrendipine), calcium channel antagonists have indeed been shown to inhibit dopamine, norepinephrine, and serotonin reuptake in rat brain synaptosomes. The micromolar concentrations required to produce these effects *(157,158)* correspond to plasma levels reported clinically (e.g., ref. *159*). Nonetheless, their efficacy in this respect is 10-times lower than that of TCAs *(157)*. We have recently demonstrated that two structurally unrelated calcium channel antagonists, nicardipine and verapamil (but not diltiazem), are competitive inhibitors of choline uptake in human erythrocytes (Bidzinski and Pucilowski, unpublished), a putative peripheral model of CNS cholinergic function *(160)*. Their calculated inhibition constants of 2.1 and 22.7 $\mu M$, respectively, are well within the efficacy range of most potent antidepressants *(161)*. Inhibition of choline reuptake depletes neuronal precursor levels and impairs acetylcholine resynthesis. This effect, together with a direct antimuscarinic action, explains the existence of marked anticholinergic properties that contribute significantly to a wide array of untoward effects produced by tricyclic and tetracyclic antidepressants *(2)*. However, some theorists stress the possibility that anticholinergic effects may also underlie the antidepressant properties of these drugs *(162,163)*. The antimuscarinic effect does not parallel antidepressant effectiveness, but some evidence suggests that the strength of choline reuptake inhibition does *(161,164)*. It is thus possible that calcium channel antagonists that markedly impair choline reuptake, such as, e.g., nicardipine and verapamil, could have more pronounced antidepressant-like effects in the clinic than those that lack this activity.

Unfortunately, no systematic comparative clinical studies of different calcium channel antagonists have as yet been undertaken. Most clinical studies utilized verapamil, the least selective calcium antagonist *(19)*, or diltiazem, which, although producing the fewest adverse effects among the calcium antagonists, possesses relatively poor kinetic properties. Compounds from the dihydropyridine family that easily penetrate cell membranes seem most desirable for clinical trials, since they act at the same receptor sites, but differ in bioavailability and some possibly calcium-unrelated nonspecific effects, which, as we have just argued, may turn out to be beneficial in clinical terms. There are other than antimanic/antidepressant psychotropic properties of calcium channel antagonists that could make them attractive in the therapy of mood disorders. The available experimental evidence, in some cases corroborated by clinical data, suggests that these drugs may have anxiolytic *(165,166)* and memory-enhancing *(167)* effects. They could also prove to be of value in cases of mood disorders complicated by alcohol and drug abuse *(19,23,81,98–101,103,167,168)*, migraine and epilepsy *(13,19,94,95,169)*, or several somatic disorders/manifestations known to respond favorably to calcium channel antagonists (e.g., hypertension, stress peptic ulcer). Thus, there is a need for a concerted effort to test a wider range of calcium channel antagonists in the clinic and compare their efficacy with that of established mood-altering drugs, i.e., lithium, carbamazepine, and valproate.

## REFERENCES


1. Post, R. M., Ketter, T. A., Pazzaglia, P. J., George, M. S., Marangell, L., and Denicoff, K. (1993) New developments in the use of anticonvulsants as mood stabilizers. *Neuropsychobiology* **27,** 132–137.
2. Potter, W. Z., Rudorfer, M. V., and Manji, H. (1991) The pharmacological treatment of depression. *New Engl. J. Med.* **325,** 633–642.
3. Vos, R. (1991) Verapamil: dying drug or sleeping beauty? in *Drugs Looking for Diseases* (Vos, R., ed.), Kluver Academic, Dordrecht, pp. 123–303.
4. Kazda, S. (1991) The story of nifedipine, in *Adalat* (Lichtlen, P. R. and Reale, A., eds.). Springer, Berlin, pp. 9–26.
5. Fleckenstein, A. (1983) History of calcium antagonists. *Cir. Res.* **52,**(**Suppl. 1**), 3–16.
6. Chibata, I., Iwasawa, Y., Kobayashi, H., Nagao, T., Noda, K., Senuma, M., Shinaki, T., Suzuki, H., Takeda, M., Tanaka, T., and Yamada S., eds. (1987) *Diltiazem.* Tanabe Seiyaku, Osaka.
7. Bossert, F. and Vater, W. (1989) 1,4-Dihydropyridines—A basis for developing new drugs. *Med. Res.* Rev. **9,** 291–324.
8. Spedding, M. and Paoletti, R. (1992) Classification of calcium channels and the sites of action of drugs modifying channel function. *Pharmacol. Rev.* **44,** 363–376.
9. Wehringer, E. (1987) $Ca^{2+}$ channel ligands: synthetic approaches, in *Structure and Physiology of the Slow Inward Calcium Channel* (Venter, J. C. and Triggle, D., eds.), Alan R. Liss, New York, pp. 1–27.
10. Bechem, M., Hebisch, S., and Schramm, M. (1988) $Ca^{2+}$ agonists: new, sensitive probes for $Ca^{2+}$ channels. *Trends Pharmacol. Sci.* **9,** 257–261.
11. Scott, R. H., Pearson, H. A., and Dolphin, A. C. (1991) Aspects of vertebrate neuronal voltage-activated calcium currents and their regulation. *Prog. Neurobiol.* **36,** 485–520.
12. Triggle, D. J. (1993) Calcium, calcium channels, and calcium antagonists. *Drugs Develop.* **2,** 3–13.
13. Fisher, M. and Grotta, J. (1993) New uses for calcium channel blockers. Therapeutic implications. *Drugs* **46,** 961–975.

14. Raftery, E. B. (1984) Cardiovascular drugs withdrawal syndromes. A potential problem with calcium antagonists? *Drugs* **28,** 371–374.

15. Hermann, P. and Morselli, P. L. (1985) Pharmacokinetics of diltiazem and other calcium entry blockers. *Acta Pharmacol. Toxicol.* **57(Suppl. 2),** 10–20.

16. Dubovsky, S. L., Franks, R. D., and Allen, S. (1987) Verapamil: a new antimanic drug with potential interactions with lithium. *J. Clin. Psychiatr.* **48,** 371–372.

17. Price, W. A. and Giannini, A. J. (1986) Neurotoxicity caused by lithium-verapamil synergism. *J. Clin. Pharmacol.* **26,** 717–719.

18. MacPhee, G. J. A., Thompson, G. G., McInnes, G. T., and Brodie, M. J. (1986) Verapamil potentiates carbamazepine neurotoxicity: a clinically important inhibitory interaction. *Lancet* **1,** 700–703.

19. Pucilowski, O. (1992) Psychopharmacological properties of calcium channel inhibitors. *Psychopharmacology* **109,** 12–29.

20. Dick, R. S. and Barold, S. S. (1989) Diltiazem induced parkinsonism. *Am. J. Med.* **87,** 95–96.

21. Garcia-Albea, E., Jimenez-Jimenez, F. J., Ayuso-Peralta, L., Cabrera-Valdivia, F., Vaquero, A., and Tejeiro, J. (1993) Parkinsonism unmasked by verapamil. *Clin. Neuropharmacol.* **16,** 263–265.

22. Hoschl, C. (1991) Do calcium antagonists have a place in the treatment of mood disorders? *Drugs* **42,** 721–729.

23. Little, H. J. (1995) The role of calcium channels in drug dependence. *Drug Alcohol Depend.* **38,** 173–194.

24. Silverstone, P. H. and Grahame-Smith, D. G. (1992) A review of the relationship between calcium channels and psychiatric disorders. *J. Psychopharmacol.* **6,** 462–482.

25. Dubovsky, S. L. (1993) Calcium antagonists in manic-depressive illness. *Neuropsychobiology* **27,** 184–192.

26. Dubovsky, S. L. (1994) Why don't we hear more about the calcium antagonists? Editorial. *Biol. Psychiatry* **35,** 149–150.

27. Furberg, C. D., Psaty, B. M., and Meyer, J. V. (1995) Nifedipine. Dose-related increase in mortality in patients with coronary heart disease. *Circulation* **92,** 1326–1331.

28. Opie, L. H. and Messerli, F. H. (1995) Nifedipine and mortality. Grave defects in the dossier (editorial). *Circulation* **92,** 1068–1073.

29. Tsien, R. W., Ellinor, P. T., and Horne, W. A. (1991) Molecular diversity of voltage-dependent $Ca^{2+}$ channels. *Trends Pharmacol. Sci.* **12,** 349–354.

30. Striessnig, J., Glossmann, H., and Catterall, W. A. (1990) Identification of a phenylalkylamine binding region within the alpha1 subunit of skeletal muscle $Ca^{2+}$ channels. *Proc. Natl. Acad. Sci. USA* **87,** 9108–9112.

31. Striessnig, J., Murphy, B. J., and Catterall, W. A. (1991) Dihydropyridine receptor of L-type $Ca^{2+}$ channels: identification of binding domains for [$^3$H](+)-PN200-110 and [$^3$H]azidopine within the $\alpha$1 subunit. *Proc. Natl. Acad. Sci. USA* **88,** 10,769–10,773.

32. Mori, Y., Niidome, T., Fujita, Y., Mynlieff, M., Dirksen, R. T., Beam, K. G., Iwabe, N., Miyata, T., Furutama, D., Furuichi, T., and Mikoshiba, K. (1993) Molecular diversity of voltage-dependent calcium channels. *Ann. NY Acad. Sci.* **707,** 87–108.

33. Snutch, T. P., Tomlinson, W. J., Leonard, J. P., Gilbert, M. M. (1991) Distinct calcium channels are generated by alternative splicing and are differentially expressed in the mammalian CNS. *Neuron* **7,** 45–57.

34. Sanna, E., Head, G. A., and Hanbauer, I. (1986) Evidence for a selective localization of voltage sensitive $Ca^{2+}$ channels in nerve cell bodies of corpus striatum. *J. Neurochem.* **47,** 1552–1557.

35. Westenbroek, R. E., Ahlijanian, M. K., and Catterall, W. A. (1990) Clustering of L-type calcium channels at the base of major dendrites in hippocampal pyramidal neurones. *Nature* **347,** 281–284.

36. Cortes, R., Supavilai, P., Karobath, M., and Palacios J. M. (1984) Calcium antagonist binding sites in the rat brain: quantitative autoradiographic mapping using the 1,4-dihydropyridines [3H] PN 200-110 and [3H] 108-068. *J. Neural Transm.* **60,** 169–197.

37. Ferry, D. R., Goll, A., Gadow, C., and Glossmann, H. (1984) (−)3H-Desmethoxyverapamil labelling of putative calcium channels in brain: autoradiographic distribution and allosteric coupling to 1,4-dihydropyridine and diltiazem sites. *Naunyn-Schmiedeberg's Arch. Pharmacol.* **327,** 183–187.

38. Murphy, K. M. M., Gould, R. J., and Snyder, S. H. (1982) Autoradiographic visualization of [3H]nitrendipine binding sites in rat brain: localization to synaptic zones. *Eur. J. Pharmacol.* **81,** 517–519.

39. Kerr, L. M., Filloux, F., Olivera, B. M., Jackson, H., and Wamsley, J. K. (1988) Autoradiographic localization of calcium channels with [125I]ω-conotoxin in rat brain. *Eur. J. Pharmacol.* **146,** 181–183.

40. Takemura, M., Fukui, H., and Wada, H. (1987) Different localization of receptors for ω-conotoxin and nitrendipine in rat brain. *Biochem. Biophys. Res. Commun.* **149,** 982–988.

41. Thayer, S. A., Murphy, S. N., and Miller, R. J. (1986) Widespread distribution of dihydropyridine-sensitive calcium channels in the central nervous system. *Mol. Pharmacol.* **30,** 505–509.

42. Baraban, J. M., Worley, P. F., and Snyder, S. H. (1989) Second messenger system and psychoactive drug action: focus of the phosphoinositide system and lithium. *Am. J. Psychiatry* **146,** 1251–1260.

43. Kass, R. S. (1987) Voltage-dependent modulation of cardiac calcium channel current by optical isomers of Bay K8644: implications for channel gating. *Circ. Res.* **61,** I1–I5.

44. Willner, P. (1990) Animal models of depression: an overview. *Pharmacol. Ther.* **45,** 425–455.

45. Borsini, F. and Meli, A. (1988) Is the forced swim test a suitable model for revealing antidepressant activity? *Psychopharmacology* **94,** 147–160.

46. Bidzinski, A., Jankowska, E., and Pucilowski, O. (1990) Antidepressant-like action of nicardipine, verapamil and hemicholinium-3 injected into the anterior hypothalamus in the rat forced swim test. *Pharmacol. Biochem. Behav.* **36,** 795–798.

47. Czyrak, A., Mogilnicka, E., and Maj, J. (1989) Dihydropyridine calcium channel antagonists as antidepressant drugs in mice and rats. *Neuropharmacology* **28,** 229–233.

48. Eroglu, L. and Esin, Y. (1990) Effects of long-term nifedipine treatment in rats. *Psychiatry Res* **32,** 203–205.

49. Kostowski, W., Dyr, W., and Pucilowski, O. (1990) Activity of diltiazem and nifedipine in some animal models of depression. *Pol. J. Pharmacol. Pharm.* **42,** 121–128.

50. Mogilnicka, E., Czyrak, A., and Maj, J. (1987) Dihydropyridine calcium antagonists reduce immobility in the mouse behavioral despair test; antidepressants facilitate nifedipine action. *Eur. J. Pharmacol.* **138,** 413–416.

51. Tazi, A., Farh, M., and Hakkou, F. (1991) Psychopharmacological profile of a calcium channel antagonist, nifedipine. *Fundam. Clin. Pharmacol.* **5,** 229–236.

52. Czyrak, A., Mogilnicka, E., Siwanowicz, J., and Maj, J. (1990) Some behavioral effects of repeated administration of calcium channel antagonists. *Pharmacol. Biochem. Behav.* **35,** 557–560.

53. Geoffroy, M., Mogilnicka, E., Nielsen, M., and Rafaelsen, O. J. (1988) Effect of nifedipine on the shuttlebox escape deficit induced by inescapable shock in the rat. *Eur. J. Pharmacol.* **154,** 277–283.

54. Martin, P., Laurent, S., Massol, J., Childs, M., and Puech, A. L. (1989) Effects of dihydropyridine drugs on reversal by imipramine of helpless behavior in rats. *Eur. J. Pharmacol.* **162**, 185–188.
55. Risch, N. and Botstein, D. (1996) A manic depressive history. *Nature Gen.* **12**, 351–353.
56. Overstreet, D. H. (1993) The Flinders sensitive line rat: a genetic animal model of depression. *Neurosci. Biobehav. Rev.* **17**, 51–68.
57. Overstreet, D. H., Pucilowski, O., Rezvani, A. H., and Janowsky, D. S. (1995) Administration of antidepressants, diazepam and psychomotor stimulants further confirms the utility of Flinders Sensitive Line rats as an animal model of depression. *Psychopharmacology* **121**, 27–37.
58. Pucilowski, O., Overstreet, D. H., Rezvani, A. H., and Janowsky, D. S. (1993) Chronic mild stress-induced anhedonia: greater effect in a genetic rat model of depression. *Physiol. Behav.* **54**, 1215–1220.
59. Pucilowski, O. and Overstreet, D. H. (1993) Effect of chronic antidepressant treatment on responses to apomorphine in selectively bred rat strains. *Pharmacol. Biochem. Behav.* **32**, 471–475.
60. Pucilowski, O., Overstreet, D. H., Rezvani, A. H., and Janowsky, D. S. (1990) Effect of verapamil on submissive behavior in genetically bred hypercholinergic rats in a water competition test. *Eur. J. Pharmacol.* **187**, 507–511.
61. Kostowski, W., Plewako, M., and Bidzinski, A. (1984) Brain serotonergic neurons: their role in a form of dominance-subordination behavior in rats. *Physiol. Behav.* **33**, 365–371.
62. Malatynska, E. and Kostowski, W. (1984) The effect of antidepressant drugs on dominance behavior in rats competing for food. *Pol. J. Pharmacol. Pharm.* **36**, 531–540.
63. Plewako, M. and Kostowski, W. (1984) The effect of lesions of the locus coeruleus and treatment with drugs affecting brain noradrenergic neurotransmission on dominant–subordinate behavior in rats competing for water. *Pol. J. Pharmacol. Pharm.* **36**, 555–560.
64. Willner, P., Muscat, R., and Papp, M. (1992) Chronic mild stress-induced anhedonia: a realistic animal model of depression. *Neurosci. Biobehav. Rev.* **16**, 525–534.
65. Hoschl, C. (1983) Verapamil for depression? *Am. J. Psychiatry* **140**, 1100.
66. Pollack, M. H. and Rosenbaum, J. F. (1987) Verapamil in the treatment of recurrent unipolar depression. *Biol. Psychiatry* **22**, 779–782.
67. Jacques, R. M. and Cox S. J. (1991) Verapamil in major (psychotic) depression. *Br. J. Psychiatry* **158**, 124–125.
68. Hoschl, C., Blahos, J., and Kabes, J. (1986) The use of calcium channel blockers in psychiatry, in *Biological Psychiatry 1985* (Shagass, C. E., Josiassen, R. C., and Bridger, W. H., eds.). Elsevier, New York, pp. 329–331.
69. Hoschl, C. and Kozeny, J. (1989) Verapamil in affective disorders: a double-blind, controlled study. *Biol. Psychiatry* **25**, 128–140.
70. Eccleston, D. and Cole, A. J. (1990) Calcium-channel blockade and depressive illness. *Br. J. Psychiatry* **156**, 889–891.
71. Hullett, F. J., Potkin, S. G., Levy, A. B., and Ciasca, R. (1988) Depression associated with nifedipine-induced calcium channel blockade. *Am. J. Psychiatry* **145**, 1277–1279.
72. Long, T. D. and Kathol, R. G. (1993) Critical review of data supporting affective disorder caused by nonpsychotropic medication. *Ann. Clin. Psychiatry* **5**, 259–270.
73. Patten, S. B., Williams, J. V., and Love, E. J. (1995) Self-reported depressive symptoms in association with medication exposures among medical inpatients: a cross-sectional study. *Can. J. Psychiatry* **40**, 264–269.
74. Creese, I. and Iversen, S. D. (1975) The pharmacological and anatomical substrates of the amphetamine response in the rat. *Brain Res.* **83**, 419–436.

75. Pucilowski, O. (1987) Monoaminergic control of affective aggression. *Acta Neurobiol. Exp.* **47,** 213–238.
76. Ansah, T.-A., Wade, L. H., and Shockley, D. C. (1993) Effects of calcium channel entry blockers on cocaine and amphetamine-induced motor activities and toxicities. *Life Sci.* **53,** 1947–1956.
77. Moore, N. A., Rees, G., Sanger, G., and Awere, S. (1993) Effect of L-type calcium channel modulators on stimulant-induced hyperactivity. *Neuropharmacology* **32,** 719–720.
78. Pani, L., Kuzmin, A., Diana, M., De Montis, G., Gessa, G. L., and Rossetti, Z. L. (1990) Calcium receptor antagonists modify cocaine effects in the central nervous system differently. *Eur. J. Pharmacol.* **190,** 217–221.
79. Mecke, E., Kauppila, T., Carlson, S., and Pertovaara, A. (1991) Differential effects of verapamil, a calcium channel antagonist, on morphine and cocaine-induced analgesia and locomotor behavior in rats. *Neurosci. Res. Commun.* **9,** 137–141.
80. Grebb, J. A. (1986) Nifedipine and flunarizine block amphetamine-induced behavioral stimulation in mice. *Life Sci.* **38,** 2375–2381.
81. Pucilowski, O., Plaznik, A., and Overstreet, D. H. (1995) Isradipine suppresses amphetamine-induced conditioned place preference and locomotor stimulation in the rat. *Neuropsychopharmacology* **12,** 239–244.
82. Renwart, N., Frances, H., and Simon, P. (1986) The calcium entry blockers: anti-manic drugs? *Prog. Neuropsychopharmacol. Biol. Psychiatry* **10,** 717–722.
83. Argiolas, A., Melis, M. R., and Gessa, G. L. (1989) Calcium channel inhibitors prevent apomorphine- and oxytocin-induced penile erection and yawning in male rats. *Eur. J. Pharmacol.* **166,** 515–518.
84. Bourson, A. and Moser, P. C. (1990) Yawning induced by apomorphine, physostigmine or pilocarpine is potentiated by dihydropyridine calcium channel blockers. *Psychopharmacology* **100,** 168–172.
85. Kostowski, W. and Krzascik, P. (1992) Effect of certain calcium channel inhibitors on D2 receptor-mediated responses: haloperidol-induced catalepsy and apomorphine-induced locomotor changes. *Biog. Amines* **8,** 277–287.
86. Shah, A. B., Poiletman, R. M., and Shah, N. S. (1983) The influence of nisoldipine—a "calcium entry blocker"—on drug-induced stereotyped behavior in rats. *Prog. Neuropsychopharmacol. Biol. Psychiatry* **7,** 165–173.
87. Kostowski, W., Krzascik, P., and Pucilowski, O. (1990) Effect of calcium channel inhibitors on D-1 receptor-mediated responses: SKF 38393-induced grooming and SCH 23390-induced catalepsy in rats. *Biog. Amines* **7,** 49–56.
88. Pucilowski, O. and Eichelman, B. (1991) Nicardipine protects against chronic ethanol- or haloperidol-induced behavioral supersensitivity to apomorphine-induced aggression. *Neuropsychopharmacology* **5,** 55–60.
89. Pucilowski, O. and Kostowski, W. (1988) Diltiazem suppresses apomorphine-induced fighting and pro-aggressive effect of withdrawal from chronic ethanol or haloperidol in rats. *Neurosci. Lett.* **93,** 96–100.
90. Post, R. M. and Weiss, S. R. B. (1989) Non-homologous animal models of affective illness: Clinical relevance of sensitization and kindling, in *Animal Models of Depression* (Koob, G., Ehlers, C., and Kupfer, D. J., eds.), Birkhauser, Boston, pp. 30–54.
91. Karler, R., Turkanis, S. A., Partlow, L. M., and Calder, L. D. (1991) Calcium channel blockers and behavioral sensitization. *Life Sci.* **49,** 165–170.
92. Martin-Iverson, M. T. and Reimer, A. R. (1994) Effects of nimodipine and/or haloperidol on the expression of conditioned locomotion and sensitization to cocaine in rats. *Psychopharmacology* **114,** 315–320.

93. Reimer, A. R. and Martin-Iverson, M. T. (1994) Nimodipine and haloperidol attenuate behavioral sensitization to cocaine but only nimodipine blocks the establishment of conditioned locomotion induced by cocaine. *Psychopharmacology* **113,** 404–410.

94. Vezzani, A., Wu, H. Q., Stasi, M. A., Angelico, P., and Samanin, W. (1988) Effect of various calcium channel blockers on three different models of limbic seizures in rats. *Neuropharmacology* **27,** 451–458.

95. Yamada, N. and Bilkey, D. K. (1991) Nifedipine has paradoxical effect on the development of kindling but not on kindled seizures in amygdala-kindled rats. *Neuropharmacology* **30,** 501–505.

96. Koob, G. F. (1992) Neural mechanisms for drug reinforcement. *Ann. NY Acad. Sci.* **654,** 171–191.

97. Wise, R. A. (1990) The role of reward pathways in the development of drug dependence, in *International Encyclopedia of Pharmacology and Therapeutics*, sect. 130 (Balfour, D. J. K., ed.), Pergamon, Elmsford, NY, pp. 23–58.

98. Pucilowski, O., Garges, P. L., Rezvani, A. H., Hutheson, S., and Janowsky, D. S. (1993) Verapamil suppresses d-amphetamine-induced place preference conditioning. *Eur. J. Pharmacol.* **240,** 89–92.

99. Kuzmin, A., Zvartau, E., Gessa, G. L., Martellotta, M. C., and Fratta, W. (1992) Calcium antagonists isradipine and nimodipine suppress cocaine and morphine intravenous self-administration in drug-naïve mice. *Pharmacol. Biochem. Behav.* **41,** 497–500.

100. Martellotta, M. C., Kuzmin, A., Muglia, P., Gessa, G. L., and Fratta, W. (1994) Effects of the calcium antagonist isradipine on cocaine intravenous self-stimulation in rats. *Psychopharmacology* **113,** 378–380.

101. Pani, L., Kuzmin, A., Martellotta, M. C., Gessa, G. L., and Fratta, W. (1991) The calcium antagonist PN 200-110 inhibits the reinforcing properties of cocaine. *Brain Res. Bull.* **26,** 445–447.

102. Calcagnetti, D. J. and Schechter, M. D. (1994) Isradipine produces neither a conditioned place preference nor aversion. *Life Sci.* **54,** PL81–PL86.

103. Pucilowski, O., Rezvani, A. H., and Overstreet, D. H. (1996) The role of taste aversion in calcium channel inhibitor-induced suppression of saccharin and alcohol drinking in rats. *Physiol. Behav.* **59,** 319–324.

104. Carman, J. S. and Wyatt, R. J. (1979) Use of calcitonin in psychotic agitation or mania. *Arch. Gen. Psychiatry* **36,** 72–75.

105. Caillard, V. (1985) Treatment of mania using a calcium antagonist. Preliminary trial. *Neurosychobiology* **14,** 23–26.

106. Dubovsky, S. L., Franks, R. D., Lifschitz, M. L., and Coen, P. (1982) Effectiveness of verapamil in the treatment of a manic patient. *Am. J. Psychiatry* **139,** 502–504.

107. Dubovsky, S. L. and Franks, R. D. (1983) Intracellular calcium ions in affective disorders: a review and a hypothesis. *Biol. Psychiatry* **18,** 781–797.

108. Giannini, A. J., Houser, W. L., Loiselle, R. H., and Price, W. A. (1984) Antimanic effects of verapamil. *Am. J. Psychiatry* **139,** 502–504.

109. Gitlin, M. J. and Weiss, J. (1984) Verapamil as maintenance treatment in bipolar illness: a case report. *J. Clin. Psychopharmacol.* **4,** 341–343.

110. Brunet, G., Cerlich, B., Robert, P., Dumas, S., Souetre, E., and Darcourt, G. (1990) Open trial of a calcium antagonist, nimodipine, in acute mania. *Clin. Neuropharmacol.* **13,** 224–228.

111. Manna, V. (1991) Disurbi affectivi bipolari e ruolo del calcio intraneuronale: effetti terapeutici del trattamento con sali di litio e/o calcio antagonista in pazienti con rapida inversione di polarita. *Minerva Medica* **82,** 757–763.

112. Pazzaglia, P. J., Post, R. M., Ketter, T. A., Goerge, M. S., and Marangell, L. B. (1993) Preliminary controlled trial of nimodipine in ultra-rapid cycling affective dysregulation. *Psychiatr. Res.* **49,** 257–272.

113. Dose, M., Emrich, W. L., Cording-Tommel, C., and von Zerssen, D. (1986) Use of calcium antagonists in mania. *Psychoneuroendocrinology* **11,** 241–243.

114. Dubovsky, S. L., Franks, R. D., Allen, S., and Murphy J. (1986) Calcium antagonists in mania: a double-blind study of verapamil. *Psychiatr. Res.* **18,** 309–320.

115. Giannini, A. J., Loiselle, R. H., Price, W. A., and Giannini, M. C. (1985) Comparison of antimanic efficacy of clonidine and verapamil. *J. Clin. Pharmacol.* **25,** 307–308.

116. Garza-Trevino, E. S., Overall, J. E., and Hollister, L. E. (1992) Verapamil versus lithium in acute mania. *Am. J. Psychiatry* **149,** 121–122.

117. Brotman, A. W., Farhadi, A. M., and Gelenberg, A. J. (1986) Varapamil treatment of acute mania. *J. Clin. Psychiatry* **47,** 136–138.

118. Lenzi, A., Marazziti, D., Rafaelli, S., and Cassano, G. B. (1995) Effectiveness of the combination verapamil and chlorpromazine in the treatment of severe manic or mixed patients. *Prog. Neuropsychopharmacol. Biol. Psychiatry* **19,** 519–528.

119. Wehr, T., Sack, D., Rosenthal, N., and Cowdry, R. (1988) Rapid cycling affective disorder: contributing factors and treatment responses in 51 patients. *Am. J. Psychiatry* **145,** 179–184.

120. Jacobsen, F. M., Sack, D. A., and James, S. P. (1987) Delirium induced by verapamil. *Am. J. Psychiatry* **144,** 248.

121. Kennedy, S., Ozersky, S., and Robillard, M. (1986) Refractory bipolar illness may not respond to verapamil. *J. Clin. Psychopharmacol.* **6,** 316–317.

122. Dubovsky, S. L., Christiano, J., Daniell, L. C., Franks, R. D., Murphy, J., Adler, L., Baker, N., and Harris, A. (1989) Increased platelet intracellular calcium concentration in patients with bipolar affective disorders. *Arch. Gen. Psychiatry* **46,** 632–638.

123. Dubovsky, S. L., Lee, C., Christiano, J., and Murphy, J. (1991) Elevated platelet intracellular calcium concentration in bipolar depression. *Biol. Psychiatry* **29,** 441–450.

124. Dubovsky, S. L., Murphy, J., Thomas, M., and Rademacher, J. (1992) Abnormal intracellular calcium ion concentration in platelets and lymphocytes of bipolar patients. *Am. J. Psychiatry* **149,** 118–120.

125. Bothwell, R. A., Eccleston, D., and Marshall, E. (1994) Platelet intracellular calcium in patients with recurrent affective disorders. *Psychopharmacology* **114,** 375–381.

126. Dubovsky, S. L., Lee, C., Christiano, J., and Murphy, J. (1991) Lithium decreases platelet intracellular calcium ion concentrations in bipolar patients. *Lithium* **2,** 167–174.

127. Dubovsky, S. L., Thomas, M., Hijazi, A., and Murphy, J. (1994) Intracellular calcium signalling in peripheral cells of patients with bipolar affective disorder. *Eur. Arch. Psychiatry Clin. Neurosci.* **243,** 229–234.

128. Eckert, A., Gann, H., Riemann, D., Aldenhoff, J., and Muller, W. E. (1994) Platelet and lymphocyte free intracellular calcium in affective disorders. *Eur. Arch. Psychiatr. Clin. Neurosci.* **243,** 235–239.

129. Danoff, S. K. and Ross, C. A. (1994) The inositol triphosphate receptor gene family: implications for normal and abnormal brain function. *Prog. Neuropsychopharmacol. Biol. Psychiatry* **18,** 1–16.

130. Miller, R. J. (1991) The control of neuronal $Ca^{2+}$ homeostasis. *Prog. Neurobiol.* **37,** 255–285.

131. Shearman, M. S., Sekiguchi, K., and Nishizuka, Y. (1989) Modulation of ion channel activity: a key function of the protein kinase C enzyme family. *Pharmacol. Rev.* **41,** 211–237.

132. Hidaka, H. and Okazaki, K. (1993) Neurocalcin family: a novel calcium-binding protein abundant in bovine central nervous system. *Neurosci. Res.* **16,** 73–77.
133. Hosey, M. M., Borsotto, M., and Lazdunsky, M. (1986) Phosphorylation and dephosphorylation of the dihydropyridine-sensitive voltage-dependent $Ca^{2+}$ channel in skeletal muscle membranes by cAMP- and $Ca^{2+}$-dependent processes. *Proc. Natl. Acad. Sci. USA* **83,** 3733–3737.
134. Moncada, S., Palmer, R. M. J., and Higgs, E. A. (1991) Nitric oxide: physiology, pathophysiology, and pharmacology. *Pharmacol. Rev.* **43,** 109–142.
135. Pape, H. C. and Mager, R. (1992) Nitric oxide controls oscillatory activity in thalamocortical neurons. *Neuron* **9,** 441–448.
136. Chapman, P. F., Atkins, C. M., Allen, M. T., Haley, J. E., and Steinmetz, J. E. (1992) Inhibition of nitric oxide synthesis impairs two different forms of learning. *Neuroreport* **3,** 567–570.
137. Calapai, G., Squadrit, F., Altavilla, D., Zingarelli, B., Campo, G. M., Cilia, M., and Caputi, A. P. (1992) Evidence that nitric oxide modulates drinking behavior. *Neuropharmacology* **31,** 761–764.
138. Morley, J. E. and Flood, J. F. (1991) Evidence that nitric oxide modulates food intake in mice. *Life Sci.* **49,** 707–711.
139. Quock, R. M. and Nguyen, E. (1992) Possible involvement of nitric oxide in chlordiazepoxide-induced anxiolysis in mice. *Life Sci.* **51,** PL255–PL260.
140. Rezvani, A. H., Grady, D. R., Peek, A. E., and Pucilowski, O. (1995) Inhibition of nitric oxide synthesis attenuates alcohol consumption in two strains of alcohol-preferring rats. *Pharmacol. Biochem. Behav.* **50,** 265–270.
141. Mannhold, R. (1984) Calmodulin—structure, function and drug action. *Drugs Future* **9,** 677–690.
142. Ogata, N., Yoshii, M., and Narahashi, T. (1989) Psychotropic drugs block voltage-gated ion channels in neuroblastoma cells. *Brain Res.* **476,** 140–144.
143. Kamatachi, G. L. and Ticku, M. K. (1991) Tricyclic antidepressants inhibit $Ca^{2+}$-activated $K^+$ efflux in cultured spinal cord neurons. *Brain Res.* **545,** 59–65.
144. Barkai, A. J. and Nelson, H. D. (1990) Alterations by antidepressants of cerebrospinal fluid formation and calcium distribution dynamics in the intact rat brain. *Biol. Psychiatry* **22,** 892–898.
145. Chen, G., Manji, H., Bitran, J. A., Masana, M. I., and Potter, W. Z. (1991) Down regulation of protein kinase C isoenzyme by chronic ECS. *Biol. Psychiatry* **29(Suppl. 9A),** 80A.
146. Antkiewicz-Michaluk, L., Michaluk, J., Romanska, I., and Vetulani, J. (1990) Effects of repetitive electroconvulsive treatment on reactivity to pain and on [3H]nitrendipine binding sites in cortical and hippocampal mebranes. *Psychopharmacology* **101,** 240–243.
147. Gleiter, C. H., Cain, C. J., Weiss, S. R. B., Post, R. M., and Marangos, P. J. (1989) Differential effects of acute and repeated electrically and chemically induced seizures on [3H]-nimodipine and [125I]-ω-conotoxin GVIA binding in rat brain. *Epilepsia* **30,** 487–492.
148. Bolger, G. T., Weissmen, B. A., Bacher, J., and Isaac, L. (1987) Calcium antagonist binding in cat brain tolerant to electroconvulsive shock. *Pharmacol. Biochem. Behav.* **27,** 217–221.
149. Antkiewicz-Michaluk, L., Romanska, I., Michaluk, J., and Vetulani, J. (1991) Role of calcium channels in effects of antidepressant drugs on responsiveness to pain. *Psychopharmacology* **105,** 269–274.
150. Koenig, M. L. and Jope, R. S. (1988) Effects of lithium on synaptosomal $Ca^{2+}$ fluxes. *Psychopharmacology* **96,** 267–272.

151. Molchan, S. E., Manji, H., Chen, G., Dou, L., Little, J., Potter, W. Z., and Sunderland, T. (1993) Effects of chronic lithium treatment on platelet PKC isoenzymes in Alzheimer's and elderly control subjects. *Neurosci. Lett.* **162,** 187–191.

152. Berridge, M. J. (1989) Inositol triphosphate, calcium, lithium and cell signalling. *JAMA* **262,** 1834–1841.

153. DeFeudis, F. V. (1987) Interactions of $Ca^{2+}$ antagonists at 5-HT$_2$ and H$_2$ receptors and GABA uptake sites. *Trends Pharmacol. Sci.* **8,** 200–201.

154. Fairhurst, A. S., Whittaker, M. L., and Ehlert, F. J. (1980) Interactions of D600 (methoxyverapamil) and local anesthetics with rat brain alpha-adrenergic and muscarinic receptors. *Biochem. Pharmacol.* **29,** 155–162.

155. Green, A. R., DeSouza, R. J., Davies, E. M., and Cross, A. J. (1990) The effect of $Ca^{2+}$ antagonists and hydralazine on central 5-hydroxytryptamine biochemistry and function in rats and mice. *Br. J. Pharmacol.* **99,** 41–46.

156. Morgan, P. F., Tamborska, E., Patel, J., and Marangos, P. J. (1987) Interactions between calcium channel compounds and adenosine systems in brain of rat. *Neuropharmacology* **26,** 1693–1699.

157. Brown, N. L., Sirugue, O., and Worcel, M. (1986) The effects of some slow channel blocking drugs on high affinity serotonin uptake by rat brain synaptosomes. *Eur. J. Pharmacol.* **123,** 161–165.

158. MacGee, R., Jr. and Schneider, J. E. (1979) Inhibition of high affinity synaptosomal uptake systems by verapamil. *Mol. Pharmacol.* **16,** 877–885.

159. Dominic, J. A., Bourne, D. W. A., Tan, T. G., Kirsten, E. B., and McAllister, R. J., Jr. (1981) The pharmacology of verapamil. III. Pharmacokinetics in normal subjects after intravenous administration. *J. Cardiovasc. Pharmacol.* **3,** 25–38.

160. Stoll, A. L., Cohen, B. M., and Hanin, I. (1991) Erythrocyte choline concentrations in psychiatric disorders. *Biol. Psychiatr.* **29,** 309–321.

161. Bidzinski, A. (1988) The effect of some antidepressants and neuroleptics on choline uptake in human erythrocytes. *New Trends Exp. Clin. Psychiatry* **4,** 111–118.

162. Janowsky, D. S., El-Yousef, M. K., Davis, J. M., and Sekerke, H. J. (1972) A cholinergic-adrenergic hypothesis of mania and depression. *Lancet* **2,** 632–635.

163. Janowsky, D. S. and Risch S. C. (1987) Acetylcholine mechanisms in affective disorders, in *Psychopharmacology: The Third Generation of Progress* (Meltzer, H. Y., ed.), Raven, New York, pp. 527–534.

164. Bidzinski, A., Puzynski, S., and Mrozek, S. (1989) Choline transport in erythrocytes of healthy controls and patients with endogenous major depression. *New Trends Exp. Clin. Psychiatry* **5,** 179–185.

165. Jankowska, E., Pucilowski, O., and Kostowski, W. (1991) Chronic oral treatment with diltiazem or verapamil decreases isolation-induced activity impairment in elevated plus maze. *Behav. Brain Res.* **43,** 155–158.

166. Disterhoft, J. F., Moyer, J. R., Jr., Thompson, L. T., and Kowalska, M. (1993) Functional aspects of calcium-channel modulation. *Clin. Neuropharmacol.* **16(Suppl. 1),** S12–S24.

167. Pucilowski, O., Rezvani, A. H., and Janowsky, D. S. (1992) Suppression of alcohol and saccharin preference in rats by a novel $Ca^{2+}$ channel inhibitor, Goe 5438. *Psychopharmacology* **107,** 447–452.

168. Pucilowski, O., Rezvani, A. H., Overstreet, D. H., and Janowsky, D. S. (1994) Calcium channel inhibitors attenuate consumption of ethanol, sucrose and saccharin solutions in rats. *Behav. Pharmacol.* **5,** 494–501.

169. Post, R. M. and Silberstein, S. D. (1994) Shared mechanisms in affective illness, epilepsy, and migraine. *Neurology* **44(Suppl. 7),** S37–S47.

**6**

# Functional NMDA Antagonists

## *A New Class of Antidepressant Agents*

### Ramon Trullas

## 1. INTRODUCTION

One of the most intriguing phenomena in the pharmacological treatment of affective disorders is the availability of a wide variety of antidepressant drugs with no apparent structural relationship, but possessing similar clinical efficacies. Despite the similarity in therapeutic outcome, in vitro studies suggest that each class of antidepressant drugs predominantly modulates different neurotransmitters via either enzyme or reuptake inhibition. These effects can be either selective (i.e., restricted to a particular transmitter or isozyme) or nonselective (exemplified by tricyclic antidepressants [TCAs] and first generation monoamine oxidase inhibitors [MAOIs]). The increased synaptic availability of neurotransmitter evoked by reuptake or enzyme inhibition has generally been considered to be responsible for the therapeutic effects of the various antidepressant drugs. However, it has long been recognized that a serious drawback of this hypothesis is the observation that neurotransmitter receptor agonists, which selectively bind at either postsynaptic catecholamine or indoleamine receptors, are generally devoid of significant antidepressant effects. This implies that the neurochemical processes of antidepressant drug action are more likely to be related to mechanisms that induce permanent molecular and cellular changes in neural function, rather than with specific modifications of synaptic activity *(1)*.

The diversity of immediate neurochemical effects exhibited by the different classes of antidepressants has long been considered as evidence to suggest that affective disorders may be the final outcome of independent alterations in different neurotransmitter systems. However, it has also been argued that depression is not a homogeneous entity and that the concept of a unitary mode of action of antidepressant drugs is questionable. Nonetheless, research performed in the last two decades has not consistently evinced a direct relationship between alterations in specific neurotransmitter systems and subtypes of affective disorders. Quite the contrary, the evidence obtained so far seems to provide equivalent support to each of the different theories on the biochemical basis of affective disorders, which include noradrenergic, serotonergic, dopaminergic, cholinergic, and peptidergic hypotheses.



From: Antidepressants: *New Pharmacological Strategies*
Edited by: *P. Skolnick, Humana Press Inc., Totowa, NJ*

In a feedback-regulated system like the brain, in which there are complex loops of excitatory and inhibitory interactions *(2,3)*, the fact that structurally different drugs acting through diverse neurotransmitter systems have similar antidepressant effects is not surprising. Rather, it substantiates the notion that psychological states and, specifically, mood disorders are not mediated by a single neurotransmitter. Instead, affective disorders may be considered a consequence of a neurotransmitter imbalance in brain areas of the limbic system linked with emotions. In that case, it seems likely that restoration of homeostatic balance among neurochemical systems in certain areas of the brain would probably play a key role in the therapeutic effects of neuropsychiatric drugs. Research in our laboratory is focused on the mechanisms involved in the induction of enduring changes in excitatory–inhibitory neurotransmitter balance by antidepressant drugs, with the hypothesis that agents that selectively modulate these mechanisms may exhibit efficacious antidepressant effects.

The efficacy of currently available antidepressant drugs is relatively low, compared with other treatments for neuropsychiatric disorders. There is no antidepressant that is effective in all depressed patients, and, in general, only ~70% of patients will exhibit some response to any given antidepressant, with merely one-half of these experiencing a full response *(4)*. Combination therapies administering, for instance, TCAs with non-selective inhibitors of monoamine oxidase have been used to improve antidepressant efficacy, but controlled studies indicate that there is no significant therapeutic advantage over treatments with a single antidepressant drug *(5)*. Thus, there is a need for a new generation of antidepressants that have higher efficacy and that retain the gains already achieved in terms of fewer side effects (as with the last generation of antidepressant drugs).

Agents that selectively act on mechanisms implicated in neural plasticity potentially have the ability to modulate different neurotransmitter systems simultaneously. Such agents could represent a new class of antidepressants with higher efficacy. One of the potential targets is the *N*-methyl-D-aspartate (NMDA) receptor, a fast acting, use-dependent, ligand-gated ion channel that mediates glutamatergic neurotransmission under conditions of strong depolarization *(6)* and is involved in the induction of permanent changes in neural function. The majority of excitatory synapses in the central nervous system (CNS) appear to use glutamate as their neurotransmitter, and the influx of calcium through the NMDA subtype of glutamate receptors initiates long-term modifications of synaptic and cellular responses that regulate neural plasticity *(7)*. Recent studies have shown that NMDA receptor antagonists modulate both catecholamine and serotonin neurotransmission *(8–14)* and antagonize stress-induced increases in dopamine metabolism *(15,16)*. These latter findings, coupled with neuroanatomical evidence that glutamatergic afferents innervate both monoaminergic cell bodies *(17)* and terminal fields *(18)*, are consistent with the idea that glutamatergic pathways can simultaneously regulate the activity of several monoaminergic systems.

The ability of glutamatergic pathways to modulate neuronal plasticity may explain another factor involved in the therapeutic effects of antidepressants. Despite the diversity of immediate neurochemical actions across antidepressant drugs, there is one common factor. For all known antidepressants, clinical improvement appears only after chronic drug administration. The lag between the onset of antidepressant drug treatment and the therapeutic response to it has long been accepted as evidence that the immediate neurochemical actions of antidepressants cannot be directly responsible for the therapeutic effects of these drugs *(1)*. The delayed onset of the therapeutic effects of antidepressants observed only after chronic drug administration might thus be mediated by a mechanism based on neuronal plasticity, with the ability to restore neurotransmitter imbalance in areas of the limbic system linked with emotions. The behavioral and neural adaptation seen after repeated exposure to antidepressant drugs, in many respects, is similar to learning. In fact, the activation of the neurochemical processes involved in learning may be one of the mechanisms used by antidepressant drugs to generate their therapeutic effect. Glutamate receptors play a key role in behavioral and neural plasticity phenomena *(19,20)* and, in particular, the NMDA receptor subtype of excitatory amino acid receptors mediates long-term synaptic responses induced by glutamate *(21)*. Research performed in this laboratory is based on the hypothesis that the therapeutic action of the diverse antidepressant drugs is mediated by a mechanism related to other neuronal plasticity processes like learning and memory. The present chapter will attempt to summarize preclinical studies demonstrating that ligands acting to reduce transmission at NMDA receptors, a glutamate-gated calcium channel involved in memory and learning, exhibit antidepressant and anxiolytic properties.

## 2. ANTIDEPRESSANT-LIKE EFFECTS OF NMDA RECEPTOR ANTAGONISTS

There is considerable evidence indicating that stressful experiences may contribute to, provoke, or exacerbate clinical depression *(22,23)*. The effectiveness of aversive experiences to induce pathology is dependent on the organism's ability to cope with these stressors. Consequently, behavioral depression seems to manifest only when there is no control over stressful stimuli, or when the stressful events are perceived as uncontrollable. From this perspective, affective disorders are not a direct outcome of stress, but the result of an inability to activate the neurochemical mechanism(s) necessary to generate adaptive responses in the face of uncontrollable stressors *(22,24)*. Antidepressant drugs would activate such mechanism(s) *(24)*. In support of this concept, recent studies have shown that NMDA receptors are involved in both acute and long-term adaptive neurochemical and behavioral effects of stress *(25–28)*.

In specific regions of the CNS *(21,29)*, activation of the NMDA subtype of glutamate receptor is required for the development of a lasting increase in synaptic efficacy known as long-term potentiation (LTP). Exposure to inescapable, but not escapable, stress impairs the induction of LTP in the $CA_1$ cell body layer of hippocampus *(26,27)*, a region that contains a high density of NMDA receptors *(30)*. Since inescapable stress also induces a syndrome of behavioral depression in animals that is antagonized by clinically effective antidepressants *(31–33)*, it was hypothesized that specific pathways subserved by NMDA receptors could also be involved in modulating the behavioral deficits induced by inescapable stress *(34)*.

This hypothesis was tested by evaluating the effects of substances that reduce transmission at NMDA receptors (reviewed in ref. *7*) in the forced-swim *(35)* and tail suspension *(36)* tests, which are commonly used to detect drugs with antidepressant properties. These tests were designed to detect potential antidepressant agents *(36,37)*, based on the abilities of clinically effective antidepressants to reduce the immobility that animals typically display after active and unsuccessful escape attempts when subjected to inescapable stressors. The predictive validity of these tests is relatively higher than observed in other animal tests that have been used to screen putative antidepressants (e.g., infant–mother separation, reserpine-induced ptosis, and olfactory bulbectomy) *(38,39)*.

Supporting the hypothesis that NMDA receptor activation is involved in the behavioral deficits observed after inescapable stressors, a competitive NMDA antagonist (AP-7) *(40)*, an NMDA receptor-coupled channel blocker (dizocilpine, MK-801) *(41)*, and a partial agonist at the strychnine-insensitive glycine modulatory site of the NMDA receptor (ACPC) *(42)*, all reduced the duration of immobility in the forced-swim test with efficacies comparable to TCAs like imipramine (Table 1). Since the predictive validity of the forced-swim test for detecting antidepressants may be lower in mice than in rats *(43)*, the effects of one of these compounds, ACPC, were also examined in the tail suspension test *(36)*. ACPC also reduced immobility in this test in a dose-dependent fashion *(34)*.

The forced-swim test is generally predictive of clinical antidepressant activity; motor stimulants can produce false positives *(44)*. However, the reductions in immobility observed with NMDA receptor antagonists in both the swim and tail suspension tests appear to be independent of any nonspecific effects on motor activity. Thus, AP-7 produces a profound reduction in swim-induced immobility, without significantly affecting ambulation time in an open field (Table 1). Furthermore, ACPC and MK-801 significantly reduce immobility in the forced-swim and tail suspension tests at doses below those required to significantly stimulate motor activity (Table 1). Reductions in immobility in the forced-swim test reach a plateau at 200 mg/kg of ACPC after systemic administration, but ambulation time continues to increase as a function of dose *(34;* Table 1). Motor stimulants detected as false positives in the forced-swim test can reduce immobilities to values approaching zero, which is not the case for some of the NMDA receptor antagonists, like ACPC and AP-7. Furthermore, studies examining the effects of orally and parenterally administered ACPC indicate that there is a temporal dissociation between reductions in immobility in the forced-swim test induced by this compound and increased motor activity in an open field *(45)*.

Subsequent studies from several laboratories have confirmed that NMDA receptor antagonists, acting at either competitive or modulatory sites, reduce immobility in the forced-swim test independently of their effects on motor activity. Thus, the competitive NMDA antagonists CGP 37849 and CGP 39551 exhibit a positive response in the forced-swim test in both mice and in rats, at doses that either do not have any effect or decrease motor activity *(46)*. Similar results have been reported in rats, with uncompetitive NMDA receptor antagonists like MK-801 *(47,48)*.

More recent studies have demonstrated that both a competitive NMDA antagonist (CGP 37849) and a functional NMDA antagonist at the glycine site (ACPC) reduce immobility in the forced-swim test after both intraperitoneal and intrahippocampal administration *(49)*, indicating that the hippocampus is involved in the anti-immobility effect of these compounds. In this last report, neither CGP 37849 nor ACPC altered exploratory activity in the open field in rats at doses effective in the forced-swim test after either intrahippocampal or systemic administration *(49)*. These observations add further evidence to suggest that the effect of NMDA receptor antagonists in the forced-

**Table 1**
**Effects of NMDA Receptor Antagonists and Imipramine on Swim-Induced Immobility and Ambulatory Activity in an Open Field**

| Drug | Dose | Immobilidy | No. of animals | %δ | Activity | No. of animals | %δ |
|---|---|---|---|---|---|---|---|
| Control | — | 144 ± 6 | (28) | | 145 ± 5 | (9) | |
| AP-7 | 40 | 141 ± 21 | (7) | | 144 ± 5 | (4) | |
| | 80 | 153 ± 9 | (10) | | 167 ± 16 | (3) | |
| | 100 | 93 ± 8[a] | (20) | −35 | 153 ± 5 | (11) | |
| | 200 | 55 ± 13[a,b] | (11) | −62 | 134 ± 10 | (17) | |
| Control | — | 140 ± 12 | (8) | | 150 ± 10 | (8) | |
| MK-801 | 0.1 | 80 ± 10[a] | (8) | −43 | 144 ± 4 | (8) | |
| | 0.5 | 11 ± 4[a,b] | (8) | −92 | 201 ± 3[a,b] | (8) | +34 |
| | 1 | 64 ± 23[a] | (8) | −54 | 170 ± 14 | (8) | |
| Control | — | 136 ± 7 | (40) | | 147 ± 3 | (26) | |
| ACPC | 25 | 119 ± 15 | (11) | −13 | | | |
| | 50 | 102 ± 15 | (10) | −25 | 149 ± 2 | (5) | |
| | 100 | 101 ± 15[a] | (10) | −26 | 156 ± 3 | (10) | |
| | 200 | 47 ± 17[a] | (10) | −65 | 166 ± 8 | (10) | |
| | 400 | 68 ± 21[a] | (10) | −50 | 180 ± 5[a] | (10) | +22 |
| Control | — | 144 ± 9 | (10) | | | | |
| IMI | 10 | 103 ± 15 | (5) | −28 | | | |
| | 20 | 58 ± 17[a] | (5) | −60 | | | |
| | 30 | 45 ± 14[a] | (7) | −69 | | | |

MK-801 and AP-7 were administered systemically to male NIH/HSD mice 15 and 30 min before testing, respectively. Controls received an equivalent volume of the corresponding vehicle. Immobility during forced swim and ambulatory activity in the open field are in seconds. Data are given as ± SEM, with the number of animals in parentheses. Doses are in mg/kg.

[a]Significantly different from control group; [b]Significantly different from all other groups ($p < 0.05$, Student-Newman-Keuls test).

From ref. *34*.

swim test is independent of alterations in spontaneous motor activity. The lack of effect of ACPC on motor activity in rats contrasts to previous results obtained in mice *(34,45,50)*, and suggests that the hyperactivity induced by ACPC may be species-specific. Taken together, these results imply that the reduction of immobility produced by NMDA antagonists in the forced-swim test is not determined by unspecific effects in motor activity. In support of this interpretation, recent studies have demonstrated that eliprodil, an NMDA antagonist acting at polyamine sites, produces a dose-dependent reduction in immobility in the forced-swim test, and that it depresses motor activity in the open field *(51)*.

Pharmacological studies have reported a synergism between the effects of classical antidepressants and NMDA receptor antagonists *(47)*, suggesting an interaction between excitatory amino acid neurotransmission and monoaminergic systems. Thus, the combined treatment with MK-801 + imipramine induces a stronger effect in Porsolt's test than administration of either drug alone *(47)*. Similarly, the combined treatment with imipramine and antagonists at the glycine site of the NMDA receptor potentiates the effects of each one of these drugs in the forced-swim test *(52)*.

Additional support for the potential antidepressant properties of NMDA receptor antagonists emerges from studies investigating the effects of amantadine and memantine. These compounds were originally introduced as putative anticholinergic agents for the treatment of Parkinson's disease *(53)*. However, more recent studies have shown that they are in fact NMDA receptor channel blockers *(54–56)*. Amantadine is commonly used for the treatment of Parkinson's disease and memantine is currently under clinical evaluation for the treatment of the dementia syndrome *(57)*. Although these drugs are not used as antidepressants, clinical observations indicate that they have mood-elevating effects *(58)*. In agreement with this observation, using a reverse validation strategy, recent studies have shown that both amantadine and memantine reduce immobility in the forced-swim test without significant alterations in spontaneous motor activity, a result that confirms the predictive validity of this test, considering the antidepressant-like effects of these compounds in patients *(58)*. Likewise, these results indicate that the putative antidepressant drug action of NMDA receptor antagonists may be reliably predicted with preclinical tests. In summary, all these observations support the hypothesis that substances capable of reducing neurotransmission at the NMDA receptor complex may be effective antidepressant drugs.

Nonetheless, the potential antidepressant-like properties of NMDA receptor antagonists predicted in most of the experiments summarized above have been investigated in normal or naïve animals using preclinical tests to detect antidepressants. The objective of these tests is the screening of drugs and they do not model any aspect of clinical depression. The procedures in drug-screening tests require only one or two doses of drug, and there is evidence indicating that the efficacy of some types of functional NMDA antagonists in the forced-swim test is reduced following chronic administration *(50)*. Furthermore, these preclinical tests are performed on normal animals, and clinical experience with antidepressants indicates that these drugs are devoid of mood-elevating effects in normal subjects *(59)*. Thus, although drug-screening procedures like the forced-swim and tail suspension tests display a high predictive validity for the detection antidepressants *(43)*, the prediction that compounds active on those tests are antidepressants is strengthened if these compounds also show positive effects in studies using animal models of depression *(60,61)*.

Studies performed using animal models of depression like chronic mild stress (CMS) *(39,60–62)* confirm the antidepressant-like properties of NMDA receptor antagonists (*see* Chapter 12). In this model, rats subjected to a variety of mild stressors for a prolonged period of time show a pronounced decrease in consumption of a palatable sucrose solution. This effect is proposed to model anhedonia, and can be effectively reversed by chronic, but not acute, treatment with clinically effective antidepressant drugs *(60,61)*. Competitive (CGP 37849, CGP 40116) and noncompetitive (MK-801) NMDA receptor antagonists are as effective as imipramine in reversing stress-induced anhedonia *(63)*. Similar results have recently been observed with the partial agonist at the glycine site, ACPC *(64)*. Chronic treatment with ACPC reverses, in a dose-dependent manner, the reduction in sucrose consumption induced by CMS, and the magnitude of this effect is comparable to that observed following similar administration of imipramine *(64)*. In these experiments, the onset of the antianhedonic action of ACPC is achieved much earlier than with imipramine (*see* Chapter 12). This finding may have significant clinical implications, and the possibility that the apparent rapid onset of antidepressant-like action produced by ACPC is a common feature of NMDA antagonists certainly merits further investigation *(64)*.

The ability of NMDA receptor antagonists to reverse a behavioral deficit after chronic, but not acute, administration, in an animal model of depression that exhibits face, construct, and predictive validity *(60,62)*, reinforces the predicted antidepressant-like actions of these compounds in drug-screening tests. In these latter tests, NMDA antagonists are active after acute administration, but lose effectiveness after chronic treatment *(50)*. In contrast, ACPC and competitive NMDA receptor antagonists are active in the CMS model after chronic administration and (similar to classical antidepressants *[63,64]*) this activity is sustained for at least 1 wk after cessation of treatment, substantiating the possibility that these compounds may be clinically effective antidepressants.

## 3. THERAPEUTIC INDEX AND PHARMACOLOGICAL PROFILE OF NMDA RECEPTOR ANTAGONISTS: PARTIAL AGONISTS AT THE GLYCINE SITE

The NMDA receptor antagonists exhibiting antidepressant-like properties may be grouped in four classes, depending on the recognition site they bind to on NMDA receptors: (1) glutamate recognition site (competitive antagonists); (2) channel blockers (uncompetitive antagonists); (3) glycine site (functional antagonists); and (4) polyamine recognition site (polyamine antagonists) *(10)*. The pharmacological profile exhibited by each of these four classes of compounds is remarkably different, especially in reference to what are generally recognized as undesirable side effects *(10)*. Both competitive NMDA receptor antagonists and use-dependent channel blockers exhibit muscle relaxant, ataxic, and amnesic, as well as psychomimetic and neurotoxic, side effects in preclinical rodent models *(65–72)*. In particular, NMDA receptor channel blockers elicit a profound behavioral activation and increase heart rate and blood pressure *(73)*. The long list of undesirable side effects and low safety margins of class 1 and 2 of NMDA receptor antagonists raises serious questions about their potential clinical utility *(73,74)*, especially for chronic or subchronic use.

In contrast, there is now a substantial amount of evidence demonstrating that functional NMDA receptor antagonists acting at the modulatory glycine and polyamine sites possess more favorable margins of safety and exhibit a pharmacological profile that does not include the side effects associated with the other two classes of NMDA receptor antagonists *(10,75)*. For instance, functional NMDA antagonists at the glycine site do not increase cerebral glucose metabolism, and are devoid of neurotoxic effects in retrosplenial and cingulate cortices associated with the use of channel blockers and competitive NMDA antagonists *(76,77)*. In addition, recent studies have shown that there is no crossgeneralization between phencyclidine and antagonists or partial agonists at the glycine site as discriminative cues in animal behavior tests *(78,79)*, indicating that glycine site antagonists are not psychomimetic like NMDA receptor channel blockers.

Furthermore, compounds within the same class, e.g., functional NMDA antagonists at the glycine site, may exhibit different behavioral pharmacology and produce ataxia, depending on whether they are antagonists or partial agonists with high or low efficacy *(80,81)*. These differences in pharmacological profile among functional glycine site antagonists may depend on receptor subunit composition. NMDA receptors are neurotransmitter-activated calcium channels composed of two families of NMDA receptor subunits: NR1 (A–G) and NR2 (A–D) *(82–85)*. Although the stoichiometry of native receptors has not yet been determined with certainty, most native NMDA receptors appear to be pentameric complexes of NR1 and NR2 subunits *(6)*. Nonetheless, there is physiological evidence showing that native presynaptic NMDA receptors involved in modulation of neurotransmitter release may also contain only NMDA-R1 homomeric subunits *(86)*. Expression studies in *Xenopus* oocytes have shown that antagonists at the glycine site exhibit a particularly high affinity for homomeric NMDAR1 receptors *(87)*. Moreover, recent experiments have shown that ACPC, being a partial agonist, exhibits both a lower maximal effect (i.e., efficacy) and a higher potency in presynaptic NMDA receptor assays, compared to assay preparations that contain either postsynaptic or mixed populations of NMDA receptor subtypes *(88)*. Consequently, under the high extracellular glycine concentrations found in vivo, partial agonists at the glycine site, like ACPC, may act as glycine antagonists preferentially at presynaptic homomeric NMDA receptors *(88)*. Thus, regional and developmental differences in the subunit composition of native NMDA receptors, as well as differing ligand affinities at distinct combinations of NMDA receptor subunits, may underlie the functional and pharmaco-logical heterogeneity exhibited by NMDA antagonists *(87,89)*, and explain the restricted pharmacological profile of NMDA antagonists at the glycine site. One of this labora-tory's objectives is to examine the pharmacological effects of partial agonists at the glycine modulatory site of the NMDA receptor that may exhibit some of the properties of competitive NMDA receptor antagonists and channel blockers, but with fewer side effects and a more selective neuropharmacological profile.

Among the different chemical classes of compounds binding at the glycine/NMDA site, with functional NMDA antagonist properties now available *(80,90–96)*, partial agonists like ACPC do not exhibit the ataxia and motor disruption produced by full glycine antagonists *(80)*. Preclinical and clinical toxicological studies demonstrate that ACPC does not alter body weight and is devoid of any serious adverse effects *(97)*. Moreover, in contrast to other highly charged NMDA antagonists *(98)*, ACPC is active after oral administration in preclinical tests *(45)*. Serum levels of ACPC follow linear kinetics after systemic, iv, and oral administration, and orally and parenterally admin-istered ACPC are equipotent in reducing immobility in the forced-swim test *(45)*. Time-course studies of the effects of ACPC in the forced-swim test indicate that the time of peak effect ($t_{max}$) is 1.6, 1.8, and 3.2 h after systemic, intravenous, and oral treatments, respectively, and the duration of action of ACPC in this preclinical test is significantly longer after oral administration (7.5 h) compared to systemic or intravenous treatments (3 h) (Fig. 1). These results indicate that ACPC may be administered orally and suggest that this compound is a good candidate for the clinical assessment of the putative anti-depressant-like effects of functional NMDA-receptor antagonists at the glycine site.

## 4. ANXIOLYTIC EFFECTS OF NMDA RECEPTOR ANTAGONISTS

Both clinical and epidemiological studies indicate that major depression and gener-alized anxiety disorder display substantial co-morbidity *(99–102)*. A recent survey in the United States shows that most cases of lifetime major depressive disorder are secondary to anxiety disorders *(99)*. In addition, an epidemiological study conducted in several

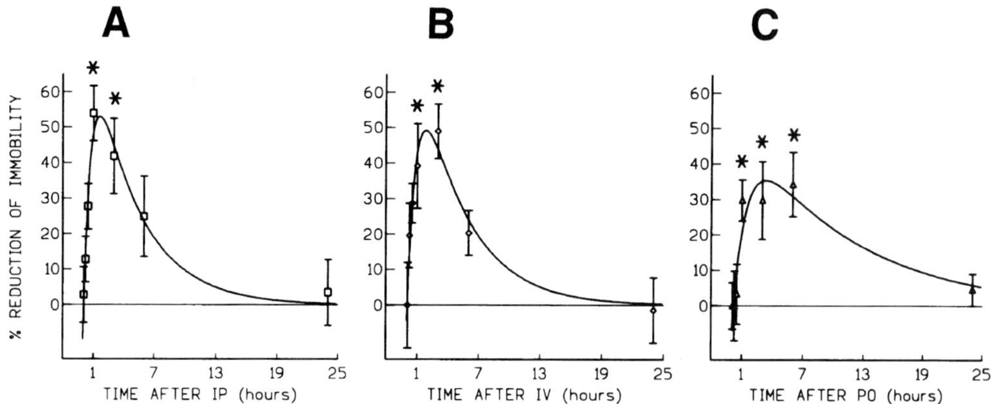

**Fig. 1.** Time-course of antidepressant-like effects of ACPC in the forced-swim test. ACPC (200 mg/kg) was administered ip (**A**), iv (**B**), or po (**C**). The forced-swim test was performed at different intervals. Data are mean ± SEM of % reduction in immobility ($n$ = 8 animals per group). Immobility time over the last 4 min of a 6-min test in control animals was 141 ± 4 s ($n$ = 84). The data were fitted to the combined absorption–disposition first-order exponential function: $y = B.e^{-kcl.x} - A.e^{-kabs.x}$, where A and B are theoretical maximum effects at time 0. *Significantly different from control values, $p < 0.05$, Student-Newman-Keuls test.

European countries and the United States suggests that the onset of anxiety disorders generally precedes that of depression *(101)*. The high levels of comorbidity observed between major depression and generalized anxiety suggests that these disorders are influenced by similar genetic factors *(102)*. Accordingly, it is reasonable to predict that antidepressant drugs acting on neurochemical mechanisms directly involved in major affective disorders should also have anxiolytic effects. Preclinical studies suggest that this may be the case for NMDA receptor antagonists.

There is now compelling evidence that competitive (2-amino-7-phosphonoheptanoic acid [AP 7], 3-[{±}-2-carboxypiperazin-4-yl]-propyl-1-phosphonate [CPP], CGS 19755, NPC 12626, and CGP 37849) and uncompetitive (dizocilpine [MK-801], phencyclidine) NMDA receptor antagonists are active in a number of models used to predict antianxiety activity, including conflict paradigms, elevated plus-maze, separation-induced ultrasonic vocalization, social interaction, and anxiety-related antipredator defensiveness tests *(81,103–116)*. In most of these models, competitive NMDA antagonists and channel blockers appear to be as efficacious as conventional anxiolytics (i.e., 1,4-benzodiazepines). However, these NMDA antagonists produce amnesia, muscle relaxation, and motor incoordination, effects that are also often associated with the benzodiazepines *(107,117)*.

Recent studies using antisense oligodeoxynucleotides (ODN), directed to mRNA encoding the NMDAR1 subunit of the NMDA receptor mRNA, suggest that the broad side effect profile of NMDA receptor antagonists in animal models of anxiety may be related to a lack of receptor subtype selectivity *(118)*. Thus, the administration of NMDAR1 antisense, but not sense, ODN produces anxiolytic-like effects in the elevated plus-maze model of anxiety, similar to anxiolytic benzodiazepines and NMDA receptor antagonists. However, although both benzodiazepines and NMDA receptor antagonists increase spontaneous motor activity in the elevated maze *(119)*, administration of NMDAR1 antisense ODN does not significantly modify this measure. The specific anxiolytic-like activity of NMDAR1 antisense indicates that an action at particular sites of the NMDA receptor may produce selective pharmacological effects. Moreover, these results, showing that inhibiting translation of the NMDAR1 subunit elicits an anxiolytic-like effect in the elevated plus-maze, are consistent with both the reported anxiolytic properties of NMDA receptor antagonists and the hypothesis that NMDA receptors are also involved in the development and/or expression of anxiety-related behaviors *(118)*.

Similar to the effects of other classes of NMDA receptor antagonists, several studies have reported that glycine-site antagonists and partial agonists exhibit anxiolytic effects in animal models of anxiety. In these tests, glycine-site ligands also display higher pharmacological specificity than competitive NMDA antagonists, and appear to be devoid of sedative and muscle relaxant effects *(108)*. Furthermore, recent studies have shown that, unlike both competitive antagonists (e.g., AP-7) and use-dependent channel blockers (e.g., dizocilpine), functional NMDA receptor antagonists at the glycine site do not impair learning and memory in a single-trial, passive-avoidance learning *(81)* at doses that are anxiolytic. In this procedure, animals are trained to avoid a dark compartment by being shocked immediately after they enter. When the animal is subsequently placed in the same situation, the latency to enter the dark compartment increases as a function of the shock intensity. Latencies are tested either within 1 h after training in the presence of drug treatment to assess anxiolytic effects or 24 h later to determine drug effects on memory. Anxiolytic compounds generally decrease the latency to enter the dark compartment when testing is performed within 1 h after training in the presence of the drugs, but not when testing is performed 24 h later, unless they have amnesic effects. For comparison, amnestic agents, like pentylenetetrazole (PTZ), reduce latencies at both 1 h and 24 h after training. Treatment with the glycine-site ligands, 1-aminocyclopropanecarboxylic acid (ACPC), 7-chlorokynurenic acid (7KYN), and D-cycloserine (DCS), as well as with the anxiolytic benzodiazepine diazepam (DZP), immediately after training, reduces step-through latencies when testing is performed within 1 h after administration. This reduction in step-through latency illustrates the anxiolytic-like effects of these compounds. In contrast, posttraining administration of glycine-site ligands or DZP does not modify step-through latencies when tested 24 h later, demonstrating that none of these compounds produce retrograde amnesia (Fig. 2). However, the administration of competitive NMDA antagonists, channel blockers, and diazepam before training reduces step-through latencies 24 h after training. This finding indicates that all these compounds produce anterograde amnesia and disrupt acquisition of avoidance learning. This effect is not observed with anxiolytic doses of glycine-site ligands (Fig. 3). In summary, all NMDA receptor antagonists exhibit anxiolytic properties similar to benzodiazepines in a passive-avoidance test. However, the results observed in these latter experiments suggest that only glycine-site ligands may act as anxiolytics without altering learning and memory *(81)*.

Both the anxiolytic potential of functional NMDA antagonists acting via the glycine site and the absence of apparent side effects of glycinergic ligands, compared

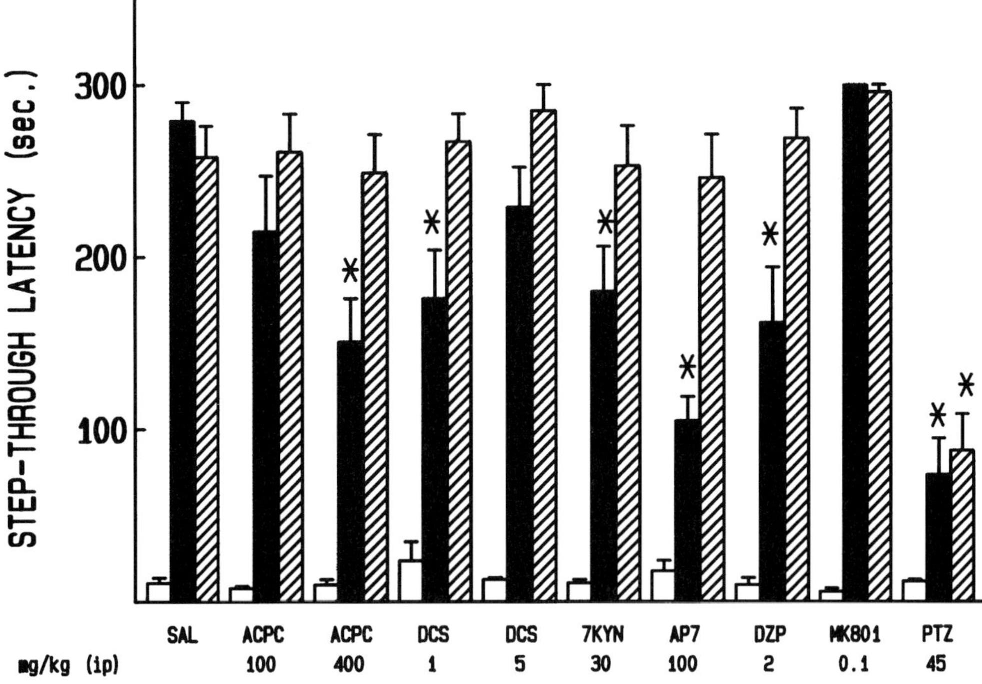



**Fig. 2.** Effects of NMDA receptor complex ligands, diazepam, and pentylenetetrazole on step-through latencies in a passive-avoidance test. All compounds were administered ip immediately after avoidance training with a 0.625-mA shock. Testing took place both 30 min and 24 h after drug administration. Unfilled bars: latency of nonshocked animals. Solid bars: latency 30 min after training. Hatched bars: latency 24 h after training. The number of animals per group was: SAL = 19, ACPC 100 = 10, ACPC 400 = 15, DCS 1 = 8, DCS 5 = 12, 7KYN = 10, AP7 = 8, DZP = 10, MK801 = 8, PTZ = 9. *Significantly different from the respective SAL control ($p < 0.05$ Mann-Whitney U-test). Reprinted with permission from ref. *81*.

to competitive NMDA antagonists, have been confirmed by a number of studies in several animal models of anxiety *(80,81,108,112,120,121)*. Thus, both 5,7-dichlorokynurenic acid (5,7-DCKA) and MDL 102,288, potent antagonists at the glycine site, suppress separation-induced ultrasonic vocalizations with an efficacy comparable to benzodiazepines and lack prominent muscle-relaxant side effects *(80,112)*. Similarly, (+)HA-966, a low-efficacy partial agonist at the glycine site, exhibits anxiolytic-like properties in conflict, social interaction, and elevated plus-maze models *(121,122)*. Likewise, (+)HA-966 appears to be devoid of the undesirable side effects associated with conventional benzodiazepines and shows no adverse interactions with ethanol *(121)*. Finally, the high-efficacy partial agonist ACPC exhibits anxiolytic properties in the elevated maze after both acute and chronic administration *(45,50,119)*. Particularly, ACPC has been reported also to be active after acute administration in passive-avoidance *(81)*, potentiated-startle *(123)*, isolation-induced ultrasonic vocalizations test *(113)*, and Vogel's test *(105)*. In summary, functional NMDA receptor antagonists at the glycine site may represent a novel class

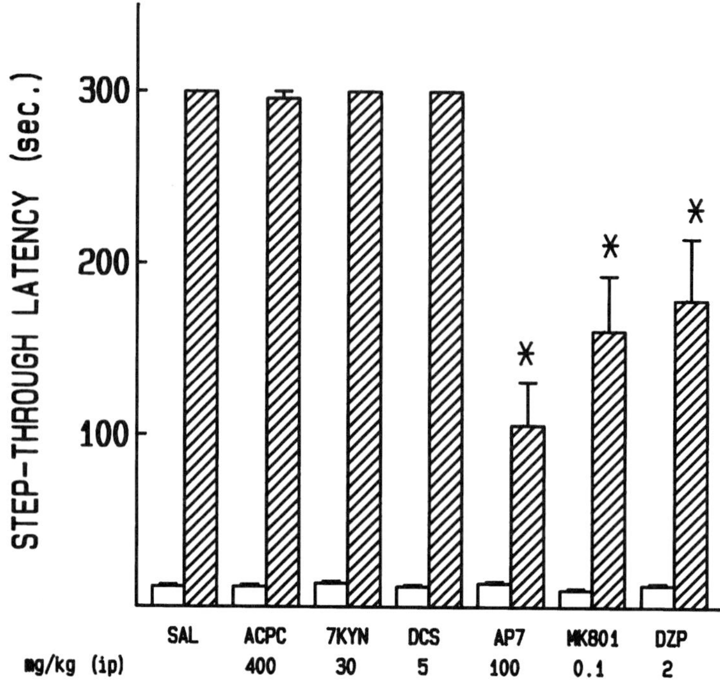



**Fig. 3.** Effects of pretraining administration of NMDA receptor complex ligands and DZP on step-through latencies in a passive-avoidance test. Drugs were administered (ip) 30 min before training. Training was performed with a 0.625-mA shock. Testing was performed 24 h later. Unfilled bars: latency of nonshocked animals. Hatched bars: latency 24 h after training ($n = 10$ for each group). *Significantly different from the control group ($p < 0.05$ Mann-Whitney U-test). Reprinted with permission from ref. *81*.

of anxiolytic and antidepressant agents, with an anxiolytic profile that differs from benzodiazepines.

## 5. NEUROPROTECTIVE ACTIONS OF FUNCTIONAL NMDA RECEPTOR ANTAGONISTS: A POTENTIAL ADVANTAGE IN THE TREATMENT OF POSTSTROKE DEPRESSION

In addition to the antidepressant and anxiolytic effects manifested in preclinical tests, many studies have shown that functional NMDA receptor antagonists acting at the glycine modulatory site are neuroprotective, without neurotoxic side effects (*see* ref. *75* for review). Glycine/NMDA antagonists reduce neuronal injury in both animal models of brain ischemia and in vitro models of excitotoxicity. For instance, ACPC, similar to other partial agonists and antagonists at the glycine site, inhibits glutamate neurotoxicity in vitro *(124,125)* and reduces the neuronal cell death observed in vivo, using two different ischemia models (global cerebral ischemia and dynorphin A-induced spinal ischemia) *(125–127)*. The antiexcitotoxic effects of ACPC are produced by the ability of this compound to reduce NMDA-mediated events in vivo, since recent studies have shown that acute systemic administration of ACPC inhibits the death of $CA_1$ pyramidal cells and protects against seizures and death induced by intrahippocampal infusion of NMDA *(128)* (Fig. 4).

The neuroprotective profile of glycine/NMDA receptor antagonists may be particularly relevant for the treatment of poststroke depression. For many years, clinicians have recognized that affective disorders frequently accompany stroke *(129)*, and there is evidence indicating that poststroke depressions are not fully explained by the severity of the associated impairment *(130)*. In a prospective study of 103 stroke patients capable of undergoing a psychiatric interview, almost 50% of patients showed a clinically significant mood disorder during the acute poststroke period *(131)*. Longitudinal studies indicate that poststroke depressive disorders can be long-lasting without treatment *(16,132)*. Tricyclic antidepressants are effective in the treatment of poststroke depression *(133)*, but the availability of compounds like glycine/NMDA antagonists that (in addition to anxiolytic and antidepressant properties) prevent excitotoxic neurodegeneration may represent a relevant therapeutic advantage.

## 6. SUMMARY AND CONCLUDING REMARKS

In summary, evidence from antidepressant screening tests, animal models of depression and anxiety, and models of neurodegeneration indicates that functional NMDA antagonists exhibit antidepressant, anxiolytic, and neuroprotective properties. Although the pharmacological side-effect profile of competitive NMDA antagonists restricts their potential clinical use, partial agonists and antagonists at the glycine site of the NMDA receptor appear to be devoid of neurotoxic, ataxic, amnesic, and psychomimetic side effects *(75)*. Although the potential absence of undesirable side effects by glycine/NMDA antagonists predicted in animal models needs to be substantiated by clinical experience, there is already evidence from Phase 1 clinical trials indicating that ACPC lacks psychomimetic-like actions *(97)*. Thus, functional NMDA antagonists at the glycine site may constitute a novel class of neuropharmacological agents with simultaneous antidepressant, anxiolytic, and neuroprotective effects. Some of these agents, like ACPC, exhibit oral efficacy and a good therapeutic index.

The implications of the studies reviewed in this chapter are that modulation of NMDA receptor function at the glycine site may provide a neuropharmacological target for the development of novel and more efficacious antidepressant drugs. In addition, the antidepressant and anxiolytic-like properties of NMDA receptor antagonists provide support for the hypothesis that the NMDA receptor complex may be involved in the (patho)physiology of anxiety and affective disorders. Since glutamatergic pathways innervate both catecholamine and serotonin systems, the heuristic proposal that glutamatergic neurotransmission is involved in the etiopathology of affective disorders *(34,134)* provides a comprehensive formulation that unifies different monoaminergic hypotheses of the neurochemical basis of affective disorders. Further evidence in support of this proposal includes recent studies showing that major depression is accompanied by increases in plasma concentrations of glutamate and serine *(135,136)* and the finding that radioligand binding to NMDA receptors is altered in frontal cortex of suicide victims *(see* Chapter 8) *(137)*. Nonetheless, only controlled clinical assays will ultimately determine if functional NMDA receptor antagonists are a new and more efficacious class of antidepressant-anxiolytic agents, as predicted by preclinical findings.

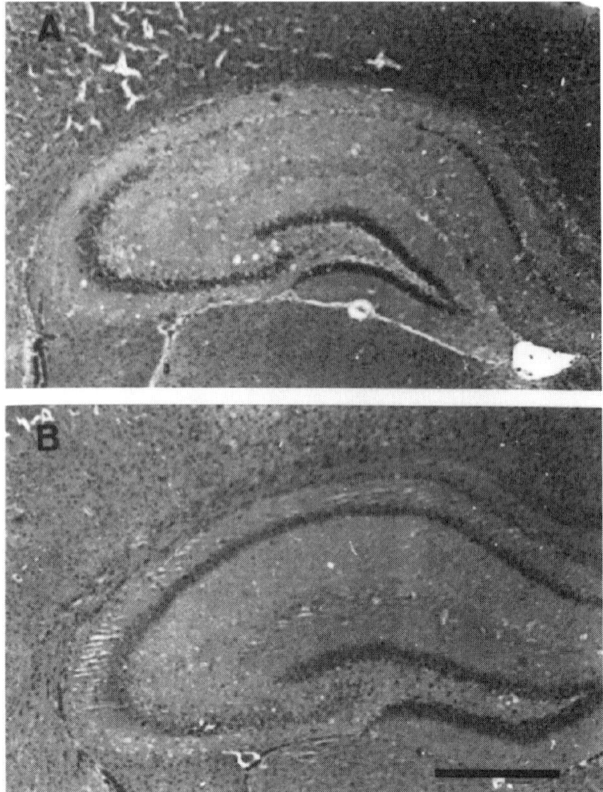

**Fig. 4.** H&E stained 5-μm sections showing the hippocampal neurodegeneration induced by local 6 nmols infusion of NMDA in **(A)** PBS- and **(B)** ACPC-treated mice. ACPC prevents the nearly complete destruction of the $CA_1$ pyramidal cell layer evoked by intrahippocampal administration of NMDA. Sections were 400 μm anterior to the infusion site. Bar = 0.5 mm. Reprinted with permission from ref. *128.*

## ACKNOWLEDGMENTS

Supported in part by a grant from Dirección General de Investigación Científica y Técnica, DGICYT, PB94-00017. I thank Symphony Pharmaceuticals (Malvern, PA) for the gift of ACPC used in some experiments.

## REFERENCES

1. Hyman, S. E, Nestler, E. J. (1996) Initiation and adaptation: a paradigm for understanding psychotropic drug action. *Am. J. Psychiatry* **153,** 151–162.
2. Carlsson, M. and Carlsson, A. (1990) Interactions between glutamatergic and monoaminergic systems within the basal ganglia—implications for schizophrenia and Parkinson's disease. *Trends. Neurosci.* **13,** 272–276.
3. Carlsson, M. and Carlsson, A. (1990) Schizophrenia: a subcortical neurotransmitter imbalance syndrome? *Schizophr. Bull.* **16,** 425–432.
4. Broekkamp, C. L. E., Leysen, D., Peeters, B. W. M. M, and Pinder, R. M. (1995) Prospects for improved antidepressants. *J. Med. Chem.* **38,** 4615–4633.

5. O'Brien, S., McKeon, O., and O'Regan, M. (1993) The efficacy and tolerability of combined antidepressant treatment in different depressive subgroups. *Br. J. Psychiatry* **162,** 362–368.

6. McBain, C. J. and Mayer, M. L. (1994) N-methyl-D-aspartic acid receptor structure and function. *Physiol. Rev.* **74,** 723–760.

7. Cotman, C. W., Kahle, J. S., Miller, S. E., Ulas, J., and Bridges, R. J. (1995) Excitatory amino acid neurotransmission, in: *Psychopharmacology: The Fourth Generation of Progress* (Bloom, F. E. and Kupfer, D. J., eds.). Raven, New York, pp. 75–85.

8. Kalivas, P. W., Duffy, P., and Barrow, J. (1989) Regulation of the mesocorticolimbic dopamine system by glutamic acid receptor subtypes. *J. Pharmacol. Exp. Ther.* **251,** 378–387.

9. Kalivas, P. W. (1995) Interactions between dopamine and excitatory amino acids in behavioral sensitization to psychostimulants. *Drug Alcohol Depend.* **37,** 95–100.

10. Iversen, S. D. (1995) Interactions between excitatory amino acids and dopamine systems in the forebrain: implications for schizophrenia and Parkinson's disease. *Behav. Pharmacol.* **6,** 478–491.

11. Kalivas, P. W. (1993) Neurotransmitter regulation of dopamine neurons in the ventral tegmental area. *Brain Res. Rev.* **18,** 75–113.

12. Whitton, P. S., Biggs, C. S., Pearce, B. R., and Fowler, L. J. (1992) MK-801 Increases extracellular 5-hydroxytryptamine in rat hippocampus and striatum in vivo. *J. Neurochem.* **58,** 1573–1575.

13. Löscher, W., Annies, R., and Hönack, D. (1991) The N-methyl-D-aspartate receptor antagonist MK-801 induces increases in dopamine and serotonin metabolism in several brain regions of rats. *Neurosci. Lett.* **128,** 191–194.

14. Löscher, W. and Hönack, D. (1992) The behavioural effects of MK-801 in rats: involvement of dopaminergic, serotonergic and noradrenergic systems. *Eur. J. Pharmacol.* **215,** 199–208.

15. Serrano, A., D'Angio, M., and Scatton, B. (1989) NMDA antagonists block restraint-induced increase in extracellular DOPAC in rat nucleus accumbens. *Eur. J. Pharmacol.* **162,** 157–166.

16. Morrow, B., Clark, W. A., and Roth, R. H. (1993) Stress activation of mesocorticolimbic dopamine neurons: effects of a glycine/NMDA receptor antagonist. *Eur. J. Pharmacol.* **238,** 255–262.

17. Ennis, M. and Aston Jones, G. (1988) Activation of locus coeruleus from nucleus paragigantocelularis: a new excitatory amino acid pathway in brain. *J. Neurosci.* **8,** 3644–3657.

18. Fagg, G. E. and Foster, A. C. (1983) Amino acid neurotransmitters and their pathways in the mammalian central nervous system. *Neuroscience* **9,** 701–719.

19. Danysz, W., Zajaczkowski, W., and Parsons, C. G. (1995) Modulation of learning processes by ionotropic glutamate receptor ligands. *Behav. Pharmacol.* **6,** 455–474.

20. Rison, R. A. and Stanton, P. K. (1995) Long-term potentiation and N-methyl-d-aspartate receptors: foundations of memory and neurologic disease? *Neurosci. Biobehav. Rev.* **19,** 533–552.

21. Collingridge, G. L. and Bliss, T. V. P. (1987) NMDA receptors: their role in long-term potentiation. *Trends. Neurosci.* **10,** 288–293.

22. Anisman, H. and Zacharko, R. M. (1992) Depression as a consequence of inadequate neurochemical adaptation in response to stressors. *Br. J. Psychiatry* **160,** 36–43.

23. Anisman, H. and Zacharko, R. M. (1982) Depression: the predisposing influence of stress. *Behav. Brain Sci.* **5,** 89–137.

24. Stone, E. A. (1983) Problems with current catecholamine hypotheses of antidepressant agents. *Behav. Brain Sci.* **6,** 535–577.
25. Nowak, G., Redmond, A., Mcnamara, M., and Paul, I. A. (1995) Swim stress increases the potency of glycine at the N-methyl-D-aspartate receptor complex. *J. Neurochem.* **64,** 925–927.
26. Kim, J. J., Foy, M. R., and Thompson, R. F. (1996) Behavioral stress modifies hippocampal plasticity through N-methyl-D-aspartate receptor activation. *Proc. Natl. Acad. Sci. USA* **93,** 4750–4753.
27. Shors, T. J., Seib, T. B., Levine, S., and Thompson, R. F. (1989) Inescapable versus escapable shock modulates long-term potentiation in the rat hippocampus. *Science* **244,** 224–226.
28. Shors, T. J. and Servatius, R. J. (1995) Stress-induced sensitization and facilitated learning require NMDA receptor activation. *Neuroreport.* **6,** 677–680.
29. Bliss, T. V. P. and Collingridge, G. I. (1993) A synaptic model of memory: long-term potentiation in the hippocampus. *Nature* **361,** 31–39.
30. Cotman, C. W., Monaghan, D. T., Ottersen, O. P., and Storm-Mathisen, J. (1987) Anatomical organization of excitatory amino acid receptors and their pathways. *Trends. Neurosci.* **10,** 273–280.
31. Leshner, A. I., Remler, H., Biegon, A., and Samuel, D. (1979) Desmethylimipramine counteracts learned helplessness in rats. *Psychopharmacology* **66,** 207–208.
32. Petty, E. and Sherman, D. (1979) Reversal of learned helplessness by imipramine. *Commun. Psychopharmacol.* **3,** 371–373.
33. Shanks, N. and Anisman, H. (1989) Strain-specific effects of antidepressants on escape deficits induced by inescapable shock. *Psychopathology* **99,** 122–128.
34. Trullas, R. and Skolnick, P. (1990) Functional antagonists at the NMDA receptor complex exhibit antidepressant actions. *Eur. J. Pharmacol.* **185,** 1–10.
35. Porsolt, R. D., Bertin, A., and Jalfre, M. (1977) Behavioral despair in mice: a primary screening test for antidepressants. *Arch. Int. Pharmacodyn.* **229,** 327–336.
36. Steru, L., Chermat, R., Thierry, B., and Simon, P. (1985) The tail suspension test: a new method for screening antidepressants in mice. *Psychopharmacology* **85,** 367–370.
37. Porsolt, R. D. (1981) Behavioral despair, in: *Antidepressants: Neurochemical, Behavioral and Clinical Perspectives* (Enna, S. J., Malick, J. B., and Richelson, E., eds.). Raven, New York, pp. 121–139.
38. Willner, P. (1984) The validity of animal models of depression. *Psychopharmacology* **83,** 1–16.
39. Willner, P. (1991) Animal models as simulations of depression. *Trends Pharmacol. Sci.* **12,** 131–136.
40. Evans, R. H., Francis, A. A., Jones, A. W., Smith, D. A. S., and Watkins, J. C. (1982) The effects of a series of $\Omega$-phosphonic-$\alpha$-carboxylic amino acids on electrically evoked and amino acid-induced responses in isolated spinal cord preparations. *Br. J. Pharmacol.* **75,** 65–75.
41. Wong, E., Kemp, J., Priestley, T., Knight, A., Woodruff, G., and Iversen, L. (1986) The anticonvulsant MK-801 is a potent N-methyl-D-aspartate antagonist. *Proc. Natl. Acad. Sci. USA* **83,** 7104–7108.
42. Marvizon, J. C. G., Lewin, A., and Skolnick, P. (1989) 1-Aminocyclopropanecarboxylic acid: a potent and selective ligand for the glycine modulatory site of the N-methyl-D-aspartate receptor complex. *J. Neurochem.* **52,** 992–994.
43. Borsini, F. and Meli, A. (1988) Is the forced swimming test a suitable model for revealing antidepressant activity? *Psychopharmacology* **94,** 147–160.

44. Panconi, E., Roux, J., Altenbaumer, M., Hampe, S., and Porsolt, R. D. (1993) MK-801 and enantiomers: potential antidepressants or false positives in classical screening models? *Pharmacol. Biochem. Behav.* **46,** 15–20.

45. Trullas, R., Folio, T., Young, A., Miller, R., Boje, K., and Skolnick, P. (1991) 1-aminocyclopropanecarboxylates exhibit antidepressant and anxiolytic actions in animal models. *Eur. J. Pharmacol.* **203,** 379–385.

46. Maj, J., Rogoz, Z., Skuza, G., and Sowinska, H. (1992) The effect of CGP 37849 and CGP 39551, competitive NMDA receptor antagonists, in the forced swimming test. *Pol. J. Pharmacol. Pharm.* **44,** 337–346.

47. Maj, J., Rogoz, Z., Skuza, G., and Sowinska, H. (1992) Effects of MK-801 and antidepressant drugs in the forced swimming test in rats. *Eur. Neuropsychopharmacol.* **2,** 37–41.

48. Maj, J., Rogoz, Z., and Skuza, G. (1992) The effects of combined treatment with MK-801 and antidepressant drugs in the forced swimming test in rats. *Pol. J. Pharmacol. Pharm.* **44,** 217–226.

49. Przegalinski, E., Tatarczynska, E., Deren-Wesolek, A., and Chojnacka-Wojcik, E. (1997) Antidepressant-like effects of a partial agonist at strychnine-insensitive glycine receptors and a competitive NMDA receptor antagonist. *Neuropharmacology,* in press.

50. Skolnick, P., Miller, R., Young, A., Boje, K., and Trullas, R. (1992) Chronic treatment with 1-aminocyclopropanecarboxylic acid desensitizes behavioral responses to compounds acting at the N-methyl-D-aspartate receptor complex. *Psychopharmacology Berl.* **107,** 489–496.

51. Layer, R. T., Popik, P., Olds, T., and Skolnick, P. (1995) Antidepressant-like actions of the polyamine site NMDA antagonist, eliprodil (SL-82.0715). *Pharmacol. Biochem. Behav.* **52,** 621–627.

52. Maj, J., Rogoz, Z., Skuza, G., and Kolodziejczyk, K. (1994) Some central effects of kynurenic acid, 7-chlorokynurenic acid and 5,7-dichloro-kynurenic acid, glycine site antagonists. *Pol. J. Pharmacol.* **46,** 115–124.

53. Doble, A. (1995) Excitatory amino acid receptors and neurodegeneration. *Therapie* **50,** 319–337.

54. Berger, W., Deckert, J., Hartmann, J., Krotzer, C., Kornhuber, J., Ransmayr, G., Heinsen, H., Beckmann, H., and Riederer, P. (1994) Memantine inhibits [3H]MK-801 binding to human hippocampal NMDA receptors. *Neuroreport.* **5,** 1237–1240.

55. Kornhuber, J., Weller, M., Schoppmeyer, K., and Riederer, P. (1994) Amantadine and memantine are NMDA receptor antagonists with neuroprotective properties. *J. Neural Transm.* **43(Suppl.),** 91–104.

56. Olney, J. W., Price, M. T., Labruyere, J., Salles, K. S., Frierdich, G., Mueller, M., and Silverman, E. (1987) Anti-parkinsonian agents are phencyclidine agonists and N-methyl-aspartate antagonists. *Eur. J. Pharmacol.* **142,** 319–320.

57. Gortelmeyer, R. and Erbler, H. (1992) Memantine in the treatment of mild to moderate dementia syndrome. *Arzneimittelforschung.* **42,** 904–913.

58. Moryl, E., Danysz, W., and Quack, G. (1993) Potential antidepressive properties of amantadine, memantine and bifemelane. *Pharmacol. Toxicol.* **72,** 394–397.

59. Pillard, S. C. and Fisher, S. (1978) Normal humans as models for psychopharmacologic therapy, in: *Psychopharmacology: A Generation of Progress* (Lipton, M. A., DiMascio, A., and Kiliam, K. F., eds.). Raven, New York, 783–787.

60. Willner, P., Muscat, R., and Papp, M. (1992) Chronic mild stress-induced anhedonia: a realistic animal model of depression. *Neurosci. Biobehav. Rev.* **16,** 525–534.

61. Papp, M., Willner, P., and Muscat, R. (1991) An animal model of anhedonia: attenuation of sucrose consumption and place preference conditioning by chronic unpredictable mild stress. *Psychopharmacology* (Berlin) **104,** 255–259.

62. Papp, M., Willner, P., and Muscat, R. (1991) An animal model of anhedonia: attenuation of sucrose consumption and place preference conditioning by chronic unpredictable mild stress. *Psychopharmacology* **104,** 255–259.
63. Papp, M. and Moryl, E. (1994) Antidepressant activity of non-competitive and competitive NMDA receptor antagonists in a chronic mild stress model of depression. *Eur. J. Pharmacol.* **263,** 1–7.
64. Papp, M. and Moryl, E. (1997) The effect of glycine partial agonists, ACPC and d-cycloserine, in a chronic mild stress model of depression. *Eur. J. Pharmacol.,* in press.
65. Parada-Turska, J. and Turski, W. A. (1990) Excitatory amino acid antagonists and memory: effect of drugs acting at N-methyl-D-aspartate receptors in learning and memory tasks. *Neuropharmacology* **29,** 1111–1116.
66. Turski, L., Klockgether, T., Sontag, K. H., Herrling, P. L., and Watkins, J. C. (1987) Muscle relaxant and anticonvulsant activity of 3-((±)-2-carboxypiperazin-4-yl)-propyl-1-phosphonic acid, a novel N-methyl-D-aspartate antagonist in rodents. *Neurosci. Lett.* **73,** 143–148.
67. Bennett, D. A., Bernard, P. S., Amrick, C. L., Wilson, D. E., Liebman, J. M., and Hutchison, A. J. (1989) Behavioral pharmacological profile of CGS 19755, a competitive antagonist at N-methyl-D-aspartate receptors. *J. Pharmacol. Exp. Ther.* **250,** 454–460.
68. Morris, R. G. M., Anderson, E., Lynch, G. S., and Baudry, M. (1986) Selective impairment of learning and blockade of long-term potentiation by an N-methyl-D-aspartate receptor antagonist. *Nature* **319,** 774–776.
69. Hargreaves, R. J., Rigby, M., Smith, D., Hill, R. G., and Iversen, L. L. (1993) Competitive as well as uncompetitive N-methyl-D-aspartate receptor antagonists affect cortical neuronal morphology and cerebral glucose metabolism. *Neurochem. Res.* **18,** 1263–1269.
70. Tricklebank, M. D., Singh, L., Oles, R. J., Preston, C., and Iversen, S. D. (1989) The behavioural effects of MK-801: a comparison with antagonists acting non-competitively and competitively at the NMDA receptor. *Eur. J. Pharmacol.* **167,** 127–135.
71. Olney, J. W., Labruyere, J., and Price, M. T. (1989) Pathological changes induced in cerebrocortical neurons by phencyclidine and related drugs. *Science* **244,** 1360–1362.
72. Olney, J. W., Labruyere, J., Wang, G., Wozniak, D. F., Price, M. T., and Sesma, M. A. (1991) NMDA antagonist neurotoxicity—mechanism and prevention. *Science* **254,** 1515–1518.
73. Hargreaves, R. J., Hill, R. G., and Iversen, L. L. (1994) Neuroprotective NMDA antagonists—the controversy over their potential for adverse effects on cortical neuronal morphology. *Acta Neurochir.* (Wien) **60(Suppl),** 15–19.
74. Allen, H. L. and Iversen, L. L. (1990) Phencyclidine, dizocilpine, and cerebrocortical neurons [comment]. *Science* **247,** 221.
75. Leeson, P. D. and Iversen, L. L. (1994) The glycine site on the NMDA receptor: structure-activity relationships and therapeutic potential. *J. Med. Chem.* **37,** 4053–4067.
76. Berger, P., Farrel, D., Sharp, F., and Skolnick, P. (1994) Drugs acting at the strychnine-insensitive glycine receptor do not induce HSP-70 protein in the cingulate cortex. *Neurosci. Lett.* **168,** 147–150.
77. Hargreaves, R. J., Rigby, M., Smith, D., and Hill, R. G. (1993) Lack of effect of L-687,414 ((+)-*cis*-4-methyl-HA-966), an NMDA receptor antagonist acting at glycine site, on cerebral glucose metabolism and cortical neuronal morphology. *Br. J. Pharmacol.* **110,** 36–42.
78. Witkin, J. M., Brave, S., French, D., and Geter-Douglass, B. (1995) Discriminative stimulus effects of R-(+)-3-amino-1-hydroxypyrrolid-2-one, [(+)-HA-966], a partial agonist of the strychnine-insensitive modulatory site of the N-methyl-D-aspartate receptor. *J. Pharmacol. Exp. Ther.* **275,** 1267–1273.

79. Singh, L., Menzies, R., and Tricklebank, M. D. (1990) The discriminative stimulus properties of (+)-HA-966, an antagonist at the glycine/N-methyl-D-aspartate receptor. *Eur. J. Pharmacol.* **186**, 129–132.

80. Kehne, J. H., Baron, B. M., Harrison, B. L. McCloskey, T. C., Palfreyman, M. G., Poirot, M., Salituro, F. G., Siegel, B. W., Slone, A. L., Van Giersbergen, P. L., et al. (1995) MDL 100,458 and MDL 102,288: two potent and selective glycine receptor antagonists with different functional profiles. *Eur. J. Pharmacol.* **284**, 109–118.

81. Faiman, C. P., Viu, E., Skolnick, P. and Trullas, R. (1994) Differential effects of compounds that act at strychnine-insensitive glycine receptors in a punishment procedure. *J. Pharmacol. Exp. Ther.* **270**, 528–533.

82. Moriyoshi, K., Masu, M., Ishii, T., Shigemoto, R., Mizuno, N., and Nakanishi, S. (1991) Molecular cloning and characterization of the rat NMDA receptor. *Nature* **354**, 31–37.

83. Monyer, H., Sprengel, R., Schoepfer, R., Herb, A., Higuchi, M., Lomeli, H., Burnashev, N., Sakmann, B., and Seeburg, P. H. (1992) Heteromeric NMDA receptors: molecular and functional distinction of subtypes. *Science* **256**, 1217–1221.

84. Kutsuwada, T., Kashiwabuchi, N., Mori, H., Sakimura, K., Kushiya, E., Araki, K., Meguro, H., Masaki, H., Kumanishi, T., Arakawa, M., and Mishina, M. (1992) Molecular diversity of the NMDA receptor channel. *Nature* **358**, 36–41.

85. Ishii, T., Moriyoshi, K., Sugihara, H., Sakurada, K., Kadotani, H., Yokoi, M., Akazawa, C., Shigemoto, R., Mizuno, N., Masu, M., and Nakanishi, S. (1993) Molecular characterization of the family of the N-methyl-D-aspartate receptor subunits. *J. Biol. Chem.* **268**, 2836–2843.

86. Wang, J. K. T. and Thukral, V. (1996) Presynaptic NMDA receptors display physiological characteristics of homomeric complexes of NR1 subunits that contain the exon 5 insert in the N-terminal domain. *J. Neurochem.* **66**, 865–868.

87. Laurie, D. J. and Seeburg, P. H. (1994) Ligand affinities at recombinant N-methyl-D-aspartate receptors depend on subunit composition. *Eur. J. Pharmacol-Mol. Pharm.* **268**, 335–345.

88. Clos, M. V., Sanz, A. G., Trullas, R., and Badia, A. (1996) Effect of 1-aminocyclopropanecarboxylic acid on N-methyl-D-aspartate-stimulated [$^3$H]-noradrenaline release in rat hippocampal synaptosomes. *Br. J. Pharmacol.* **118**, 901–904.

89. Seeburg, P. H. (1993) The molecular biology of mammalian glutamate receptor channels. *Trends Pharmacol. Sci.* **14**, 297–302.

90. Boireau, A., Malgouris, C., Burgevin, M. C., Peny, C., Durand, G., Bordier, F., Meunier, M., Miquet, J. M., Daniel, M., Chevet, T., Jimonet, P., Mignani, S., Blanchard, J. C., and Doble, A. (1996) Neuroprotective effects of RPR 104632, a novel antagonist at the glycine site of the NMDA receptor, in vitro. *Eur. J. Pharmacol.* **300**, 237–246.

91. Laird, J. M. A., Mason, G. S., Webb, J., Hill, R. G., and Hargreaves, R. J. (1996) Effects of a partial agonist and a full antagonist acting at the glycine site of the NMDA receptor on inflammation-induced mechanical hyperalgesia in rats. *Br. J. Pharmacol.* **117**, 1487–1492.

92. Jimonet, P., Audiau, F., Aloup, J. C., Barreau, M., Blanchard, J. C., Bohme, A., Boireau, A., Cheve, M., Damour, D., Doble, A., Lavayre, J., Dutrucrosset, G., Randle, J. C. R., Rataud, J., Stutzmann, J. M., and Mignani, S. (1994) Synthesis and SAR of 2H-1,2,4-benzothiadiazine-1,1-dioxide-3-carboxylic acid derivatives as novel potent glycine antagonists of the NMDA receptor-channel complex. *Bioorg. Med. Chem. Lett.* **4**, 2735–2740.

93. Kulagowski, J. J., Baker, R., Curtis, N. R., Leeson, P. D., Mawer, I. M., Moseley, A. M., Ridgill, M. P., Rowley, M., Stansfield, I., Foster, A. C., Grimwood, S., Hill, R. G., Kemp, J. A., Marshall, G. R., Saywell, K. L., and Tricklebank, M. D. (1994) 3'-(Arylmethyl)- and 3'-(aryloxy)-3-phenyl-4-hydroxyquinolin-2(1H)-ones: orally active antagonists of the glycine site on the NMDA receptor. *J. Med. Chem.* **37**, 1402–1405.

94. Rao, T. S., Gray, N. M., Dappen, M. S., Cler, J. A., Mick, S. J., Emmett, M. R., Iyengar, S., Monahan, J. B., Cordi, A. A., and Wood, P. L. (1993) Indole-2-carboxylates, novel antagonists of the N-methyl-D-aspartate (NMDA)-associated glycine recognition sites: in vivo characterization. *Neuropharmacology* **32,** 139–147.

95. Baron, B. M., Harrison, B. L., McDonald, I. A., Meldrum, B. S., Palfreyman, M. G., Salituro, F. G., Siegel, B. W., Slone, A. L., Turner, J. P., and White, H. S. (1992) Potent indole- and quinoline-containing N-methyl-D-aspartate antagonists acting at the strychnine-insensitive glycine binding site. *J. Pharmacol. Exp. Ther.* **262,** 947–956.

96. Carter, A. J. (1992) Glycine antagonists: regulation of the NMDA receptor-channel complex by the strychnine-insensitive glycine site. *Drugs of the Future* **17,** 595–613.

97. Cherkofsky, S. C. (1995) 1-aminocyclopropanecarboxylic acid: mouse to man interspecies pharmacokinetic comparisons and allometric relationships. *J. Pharmaceut. Sci.* **84,** 1231–1237.

98. Chapman, A. (1991) Excitatory amino acid antagonists and therapy of epilepsy, in: *Excitatory Amino Acid Antagonists* (Meldrum, B. S., ed.). Oxford, Blackwell, pp. 265–274.

99. Kessler, R. C., Nelson, C. B., Mcgonagle, K. A., Liu, J., Swartz, M., and Blazer, D. G. (1996) Comorbidity of DSM-III-R major depressive disorder in the general population: results from the US National Comorbidity Survey, 17–30.

100. Sartorius, N., Ustun, T. B., Lecrubier, Y., and Wittchen, H. U. (1996) Depression comorbid with anxiety: results from the WHO study on Psychological Disorders in Primary Health Care. *Br. J. Psychiatry* **168,** 38–43.

101. Merikangas, K. R., Angst, J., Eaton, W., Canino, G., Rubiostipec, M., Wacker, H., Wittchen, H. U., Andrade, L., Essau, C., Whitaker, A., Kraemer, H., Robins, L. N., and Kupfer, D. J. (1996) Comorbidity and boundaries of affective disorders with anxiety disorders and substance misuse: results of an international task force. *Br. J. Psychiatry* **168,** 58–67.

102. Kendler, K. S. (1996) Major depression and generalised anxiety disorder—same genes, (partly) different environments—revisited. *Br. J. Psychiatry* **168,** 68–75.

103. Willets, J., Balster, R. L., and Leander, J. D. (1990) The behavioural pharmacology of NMDA receptor antagonists. *Trends Pharmacol. Sci.* **11,** 423–428.

104. Bennett, D. A. and Amrick, C. L. (1986) 2-Amino-7-phosphonoheptanoic acid (AP7) produces discriminative stimuli and anticonflict effects similar to diazepam. *Life Sci.* **39,** 2455–2461.

105. Przegalinski, E., Tatarczynska, E., Derenwesolek, A., and Chojnacka-Wójcik, E. (1996) Anticonflict effects of a competitive NMDA receptor antagonist and a partial agonist at strychnine-insensitive glycine receptors. *Pharmacol. Biochem. Behav.* **54,** 73–77.

106. Wiley, J. L., Cristello, A. F., and Balster, R. L. (1995) Effects of site-selective NMDA receptor antagonists in an elevated plus-maze model of anxiety in mice. *Eur. J. Pharmacol.* **294,** 101–107.

107. Plaznik, A., Palejko, W., Nazar, M., and Jessa, M. (1994) Effects of antagonists at the NMDA receptor complex in two models of anxiety. *Eur. Neuropsychopharmacol.* **4,** 503–512.

108. Corbett, R. and Dunn, R. W. (1993) Effects of 5,7 dichlorokynurenic acid on conflict, social interaction and plus maze behaviors. *Neuropharmacology* **32,** 461–466.

109. Blanchard, D. C., Blanchard, R. J., Carobrez, A., Veniegas, R., Rodgers, R. J., and Shepherd, J. K. (1992) MK-801 produces a reduction in anxiety-related antipredator defensiveness in male and female rats and a gender-dependent increase in locomotor behavior. *Psychopharmacology* (Berlin) **108,** 352–362.

110. Guimaraes, F. S., Carobrez, A. P., De Aguiar, J. C., and Graeff, F. G. (1991) Anxiolytic effect in the elevated plus-maze of the NMDA receptor antagonist AP7 microinjected into the dorsal periaqueductal grey. *Psychopharmacology* (Berlin) **103,** 91–94.

111. Winslow, J. T., and Insel, T. R. (1991) Infant rat separation is a sensitive test for anxiolytics. *Prog. Neuropsychopharmacol. Biol. Psychiatry* **15,** 745–757.

112. Kehne, J. H., McCloskey, T. M., Baron, B. M., Chi, E. M., Harrison, B. L., Whitten, J. P., and Palfreyman, M. G. (1991) NMDA receptor complex antagonists have potential anxiolytic effects as measured with separation-induced ultrasonic vocalizations. *Eur. J. Pharmacol.* **193,** 283–292.

113. Winslow, J. T., Insel, T. R., Trullas, R., and Skolnick, P. (1990) Rat pup isolation calls are reduced by functional antagonists of the NMDA receptor complex. *Eur. J. Pharmacol.* **190,** 11–21.

114. Conti, L. H., Maciver, C. R., Ferkany, J. W., and Abreu, M. E. (1990) Footshock-induced freezing behavior in rats as a model for assessing anxiolytics. *Psychopharmacology* (Berlin) **102,** 492–497.

115. Moghaddam, B. (1993) Stress preferentially increases extraneuronal levels excitatory amino acids in the prefrontal cortex: comparison to hippocampus and basal ganglia. *J. Neurochem.* **60,** 1650–1657.

116. Armanini, M. P., Hutchins, C., Stein, B. A., and Sapolsky, R. M. (1990) Glucocorticoid endangerment of hippocampal neurons is NMDA-receptor dependent. *Brain Res.* **532,** 7–12.

117. Rupniak, N. M., Boyce, S., Tye, S., Cook, G., and Iversen, S. D. (1993) Anxiolytic-like and antinociceptive effects of MK-801 accompanied by sedation and ataxia in primates. *Pharmacol. Biochem. Behav.* **44,** 153–156.

118. Zapata, A., Capdevila, J. L., Tarrason, G., Adan, J., Martinez, J. M., Piulats, J., and Trullas, R. (1997) Effects of NMDA-R1 antisense oligodeoxynucleotide administration: behavioral and radioligand binding studies. *Brain Res.* **745,** 114–120.

119. Trullas, R., Jackson, B., and Skolnick, P. (1989) Anxiolytic properties of 1-aminocyclopropanecarboxylic acid, a ligand at strychnine-insensitive glycine receptors. *Pharmacol. Biochem. Behav.* **34,** 313–316.

120. Matheus, M. G., Nogueira, R. L., Carobrez, A. P., Graeff, F. G., and Guimaraes, F. S. (1994) Anxiolytic effect of glycine antagonists microinjected into the dorsal periaqueductal grey. *Psychopharmacology* (Berlin) **113,** 565–569.

121. Dunn, R. W., Flanagan, D. M., Martin, L. L., Kerman, L. L., Woods, A. T., Camacho, F., Wilmot, C. A., Cornfeldt, M. L., Effland, R. C., Wood, P. L., and Corbett, R. (1992) Stereoselective R-(plus) enantiomer of HA-966 displays anxiolytic effects in rodents. *Eur. J. Pharmacol.* **214,** 207–214.

122. Corbett, R. and Dunn, R. W. (1991) Effects of HA-966 on conflict, social interaction and plus maze behaviors. *Drug Dev. Res.* **24,** 201–205.

123. Anthony, E. W. and Nevins, M. E. (1993) Anxiolytic-like effects of N-methyl-D-aspartate-associated glycine receptor ligands in the rat potentiated startle test. *Eur. J. Pharmacol.* **250,** 317–324.

124. Boje, K. M., Wong, G., and Skolnick, P. (1993) Desensitization of the NMDA receptor complex by glycinergic ligands in cerebellar granule cell cultures. *Brain Res.* **603,** 207–214.

125. Fossom, L. H., Von Lubitz, D. K. J. E., Lin, R. C. S., and Skolnick, P. (1995) Neuroprotective actions of 1-aminocyclopropanecarboxylic acid (ACPC): a partial agonist at strychnine-insensitive glycine sites. *Neurol. Res.* **17,** 265–269.

126. Von Lubitz, D. K. J. E., Lin, R. C., McKenzie, R. J., Devlin, T. M., McCabe, R. T., and Skolnick, P. (1992) A novel treatment of global cerebral ischemia with a glycine partial agonist. *Eur. J. Pharmacol.* **219,** 153–158.

127. Long, J. B. and Skolnick, P. (1994) 1-Aminocyclopropanecarboxylic acid protects against dynorphin A-induced spinal injury. *Eur. J. Pharmacol.* **261,** 295–301.

128. Zapata, A., Capdevila, J. L., Viu, E., and Trullas, R. (1996) 1-Aminocyclopropane-carboxylic acid reduces NMDA-induced hippocampal neurodegeneration in vivo. *Neuroreport.* **7,** 397–400.
129. Robinson, R. G. (1989) The use of an animal model to study post-stroke depression, in: *Animal Models of Depression* (Koob, G. F., Ehlers, C. L., and Kupfer, D. J., eds.). Birkhauser, Boston, pp. 74–98.
130. Robinson, R. G. and Starkstein, S. E. (1989) Mood disorders following stroke: new findings and future directions. *J. Geriatr. Psychiatry* **22,** 1–15.
131. Robinson, R. G., Starr, L. B., Kubos, K. L., and Price, T. R. (1983) A two-year longitudinal study of post-stroke mood disorders: findings during the initial evaluation. *Stroke* **14,** 736–741.
132. Robinson, R. G., Bolduc, P. L., and Price, T. R. (1987) Two-year longitudinal study of poststroke mood disorders: diagnosis and outcome at one and two years. *Stroke* **18,** 837–843.
133. Lipsey, J. R., Robinson, R. G., Pearlson, G. D., Rao, K., and Price, T. R. (1984) Nortriptyline treatment of post-stroke depression: a double-blind study. *Lancet* **1,** 297–300.
134. Skolnick, P., Layer, R. T., Popik, P., Nowak, G., Paul, I. A., and Trullas, R. (1996) Adaptation of N-methyl-D-aspartate (NMDA) receptors following antidepressant treatment: implications for the pharmacotherapy of depression. *Pharmacopsychiatry* **29,** 23–26.
135. Altamura, C., Maes, M., Dai, J., and Meltzer, H. Y. (1995) Plasma concentrations of excitatory amino acids, serine, glycine, taurine and histidine in major depression. *Eur. Neuropsychopharmacol.* **5,** 71–75.
136. Maes, M., Debacker, G., Suy, E., and Minner, B. (1995) Increased plasma serine concentrations in depression. *Neuropsychobiology* **31,** 10–15.
137. Nowak, G., Ordway, G. A., and Paul, I. A. (1995) Alterations in the N-methyl-D-aspartate (NMDA) receptor complex in the frontal cortex of suicide victims. *Brain Res.* **675,** 157–164.

**7**

# Is an Adaptation of NMDA Receptors an Obligatory Step in Antidepressant Action?

## Nuo-Yu Huang, Richard T. Layer, and Phil Skolnick

## 1. INTRODUCTION

Clinically effective antidepressants are structurally diverse, ranging from phenylcycloalkylamines (e.g., tranylcypromine), and tri- and tetracyclics (e.g., imipramine and mianserin, respectively) to nonfused polyaromatic molecules (e.g., zimelidine) (Fig. 1). Most antidepressants have well-documented effects on either biogenic amine reuptake or metabolism, and it is these neurochemical properties (e.g., blockade of serotonin reuptake, inhibition of monoamine oxidase), rather than structural considerations, that most often form the basis of classification schemes for these agents. Despite the unequivocal evidence that antidepressants can perturb biogenic amine metabolism and disposition, establishing a causal link between these actions and therapeutic response has been problematic *(1–4)*. Thus, although effects on biogenic amine metabolism and reuptake are generally readily demonstrable both in vitro and in vivo, there is an apparent requirement for sustained treatment to achieve an antidepressant effect. This therapeutic lag cannot be defined with certainty, but most double-blind, well-controlled clinical trials indicate a latent period of $\geq 2$ wk, with a general (albeit conservative) estimate of 3–6 wk of drug therapy required to obtain significant therapeutic effects. The symptomatology associated with major depression is complex, and the rating scales used to assess clinical improvement may not be sufficiently sensitive to detect a more rapid improvement within the constraints imposed by a well-controlled trial. Both open trials and clinical impressions suggest that certain individuals may respond more rapidly than the norm (for example, clinical improvement following electroconvulsive therapy (ECT) has been observed following 3–4 seizures—that is, in $\leq 2$ wk *[5]*), but, based on the large number of well-controlled studies demonstrating a therapeutic lag, the hypothesis that one or more adaptive changes must precede a therapeutic response has become a tenet of biological psychiatry.

Since the pioneering investigations of Vetulani and Sulser *(6)*, numerous preclinical studies have documented the ability of chronic antidepressant treatments to produce adaptive changes in a variety of transmitter systems (reviewed in refs. *7–10*). Adaptive

*From:* Antidepressants: *New Pharmacological Strategies*
*Edited by: P. Skolnick, Humana Press Inc., Totowa, NJ*

125

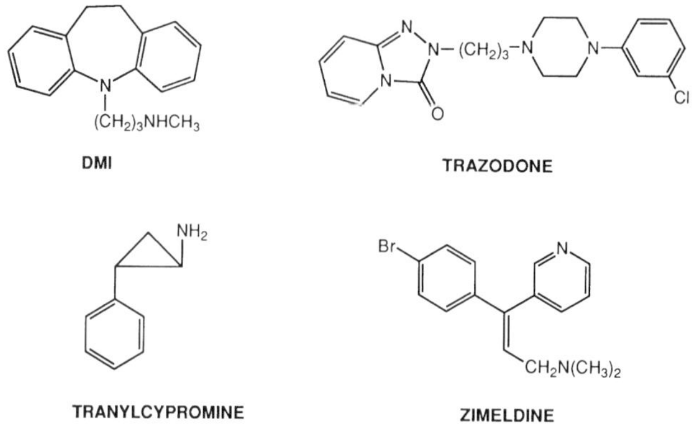

DMI

TRAZODONE

TRANYLCYPROMINE

ZIMELDINE

**Fig. 1.** Structural diversity of antidepressants as illustrated by four representative agents.

changes in monoaminergic receptors (exemplified by β-adrenergic, 5-HT$_2$, and dopamine D$_1$ receptor downregulation) in the rodent central nervous system (CNS) have been observed following chronic administration of a wide range of drugs, ECT, and rapid eye movement sleep deprivation *(8)*. While current theories underlying the therapeutic response to antidepressant treatments are generally based on adaptation of one or more monoaminergic pathways (reviewed in refs. *3* and *11*), none of these adaptive phenomena are uniform across therapies (reviewed in refs. *7* and *12*). Moreover, the adaptation of monoaminergic systems disappears rapidly, in some cases lasting only 24 h after withdrawal of treatment, but remission of depressive symptomatologies can persist for several months after treatment is discontinued *(12)*.

An exact correspondence between chronic drug effects in well animals and depressed individuals might be an unreasonable expectation, but the failure to evince consistent changes in any single transmitter system (e.g., noradrenergic, dopaminergic, serotonergic, GABA-ergic, and peptidergic pathways) following chronic antidepressant treatments indicates that, when present, such perturbations cannot adequately account for either antidepressant activity across therapies or provide a coherent biochemical basis of depressive symptomatology *(4,10,11)*. Based on these findings, it may be hypothesized that antidepressant-induced adaptive changes in any one of several transmitter-related pathways would be sufficient to evoke a therapeutic response, perhaps through a common signal transduction (e.g., a second messenger) mechanism (*see* Chapters 10 and 11). Alternatively, chronic antidepressant treatment may produce an obligatory adaptation in an as-yet undescribed pathway.

The seminal finding that *N*-methyl-D-aspartate (NMDA) antagonists mimic the effects of antidepressants *(13)* in preclinical tests with good predictive validity *(14)* (*see* Chapter 6) led to a series of experiments demonstrating that chronic, but not acute, antidepressant treatment can affect adaptive changes in radioligand binding to NMDA receptors. The consistency of the neurochemical changes observed following antidepressants drawn from every principal therapeutic class, when taken together with the slowly developing nature of these phenomena, strongly suggests that adaptation of NMDA receptors in circumscribed areas of the CNS is an obligatory step in antidepressant action. This chapter will critically review these data and describe new research directions catalyzed by these findings.

## 2. DOWNREGULATION OF CORTICAL β-ADRENOCEPTORS PRODUCED BY CHRONIC TREATMENT WITH FUNCTIONAL NMDA ANTAGONISTS

Chronic (but not acute) treatment of rodents with many antidepressants will reduce the density of β-adrenergic receptors in the CNS *(12,15)*. The relationship between this phenomenon and antidepressant action remains obscure, but β-adrenoceptor downregulation has been among the most consistent preclinical findings across antidepressant therapies *(12)*. Based on the activity of functional NMDA antagonists in preclinical tests predictive of antidepressant action *(13,16)* (*see* Chapter 6), the effects of chronic treatment with 1-aminocyclopropanecarboyxlic acid (ACPC) and dizocilipine were examined on radioligand binding to cortical β-adrenoceptors. In these studies *(17)*, mice were treated for 7 d with either ACPC (a glycine partial agonist) or dizocilpine (MK-801; a voltage-dependent channel blocker), with doses capable of reducing immobility in the forced-swim test *(13)*. Both ACPC and a high dose of dizocilpine (1 mg/kg) significantly reduced [$^3$H]dihydroalprenolol (DHA) to cortical β-adrenoceptors (19 and 21%, respectively [each $p < 0.05$]) compared to vehicle-treated mice. By comparison, the prototypic antidepressant imipramine (15 mg/kg for 7 d) reduced [$^3$H]DHA binding by 23% ($p < 0.05$). The low dose of dizocilpine (0.1 mg/kg) did not significantly affect [$^3$H]DHA binding. Since the high dose of dizocilpine produces a variety of bizarre behavioral effects *(18)* and is toxic following sustained (14 d) administration *(19)*, the specificity of this effect on β-adrenoceptors could be questioned. Nonetheless, a subsequent report *(20)* demonstrated that an intermediate dose of dizocilpine (0.3 mg/kg), administered to rats for 5 wk, reduced [$^3$H]DHA binding by 14%, compared to 25% following a regimen of imipramine (10 mg/kg) ($p < 0.05$ for both drug treatments, compared with saline). Morever, both dizocilpine and imipramine prevented the increase in β-adrenoceptors produced by 8 wk of chronic mild stress. Layer et al. *(21)* recently reported that chronic regimens of eliprodil (e.g., either 20 mg/kg twice daily or 40 mg/kg/d for 14 d), a functional NMDA antagonist acting through polyamine-associated sites *(22,23)*, produced a robust (~30%) reduction in cortical β-adrenoceptors, compared to saline-treated mice. Although the effects of functional NMDA antagonists on β-adrenoceptors have not been examined in great detail, these findings are consistent with behavioral studies indicating that this class of compounds possesses antidepressant properties (*see* Chapters 6 and 12). Moreover, these findings led us to hypothesize that chronic antidepressant treatment might affect NMDA receptors, which resulted in the experiments described below.

## 3. CHRONIC ANTIDEPRESSANT TREATMENTS PRODUCE ADAPTIVE CHANGES IN RADIOLIGAND BINDING TO NMDA RECEPTORS

Like other ligand-gated ion channels, wild-type NMDA receptors are constituted as heterooligomers *(24,25)* that possess multiple, allosterically regulatory sites capable of regulating channel kinetics *(26)*. Several of these sites are potential loci for drug action, and can be readily studied using radioligand-binding techniques. The effects of chronic imipramine, ACPC, and dizocilpine on NMDA receptors were initially examined using three radioligands: [$^3$H]CGP 39653, a competitive NMDA antagonist *(27)*; [$^3$H]5,7-dichlorokynurenic acid (DCKA), a competitive antagonist at strychnine-insensitive glycine sites *(28)*; and the voltage-dependent channel blocker, [³H]MK-801 *(29)*, which binds to the channel in its open state *(30)*. Four groups of mice were injected daily for 14 d with imipramine (15 mg/kg/d), ACPC (200 mg/kg/d), dizocilpine (0.1 mg/kg), or vehicle. Parallel groups of mice received one injection of drug or vehicle. The doses selected for these studies were based on activity (of a single dose) in the forced-swim test *(13)*. Twenty-four hours after the last injection, brains were removed, dissected, and fast-frozen over solid $CO_2$. Tissues were extensively washed (subjected to centrifugation/resuspension) prior to assay to remove endogenous materials (e.g., glycine, glutamate) known to affect radioligand binding to NMDA receptors.

Chronic, but not acute, treatment with imipramine significantly altered radioligand binding to cortical NMDA receptors *(19)*. The salient features of this effect can be summarized as follows: a decrease (~36%) in basal [³H]MK-801 binding that was reversed by the addition of glutamate (100 n*M*); a ~2.5-fold reduction (compared to vehicle-treated mice) in the potency of glycine to inhibit [³H]5,7-DCKA binding to strychnine-insensitive glycine receptors (Fig. 2) (no significant difference in basal [³H]5,7-DCKA binding was observed); a 28% reduction in the proportion of high-affinity glycine sites inhibiting [³H]CGP 39653 binding to NMDA receptors (Table 1) (no significant difference in basal [³H]CGP 39653 binding was observed). These changes were observed in cerebral cortex, but not in hippocampus, striatum, or basal forebrain *(19)*. Based on these observations, it could be concluded that imipramine-induced changes in radioligand binding to NMDA receptors are confined to cerebral cortex. However, this conclusion is based on administering a fixed dose of imipramine for 14 d. Since subsequent studies have shown that these effects of imipramine are both dose- and time-dependent *(see below)*, it is possible that adaptive changes in other brain regions may develop over time. Based on recent *in situ* hybridization studies from our laboratory *(see* Section 4.), this hypothesis merits further investigation.

Basal [³H]MK-801 binding measured under nonequilibrium conditions was reduced by chronic imipramine treatment. The observation that addition of glutamate reversed this effect indicates chronic imipramine impaired the rate at which [³H]MK-801 enters NMDA receptor-coupled cation channels. Interposing a 24-h period between the last dose of imipramine and tissue collection, extensive washing of the tissues, and the regional specificity of this effect indicate that residual imipramine cannot account for the observed reduction in [³H]MK-801 binding. No difference in the potency or efficacy of glycine to enhance [³H]MK-801 binding was observed. This finding is consonant with a lack of effect of imipramine on glycine sites, but other neurochemical indices do not support this interpretation. For example, imipramine treatment reduced the potency of glycine to inhibit [³H]5,7-DCKA binding by ~2.5-fold (from 0.55 ± 0.04 to 1.43 ± 0.2 μ*M*; *p* < 0.05), without affecting basal binding of this radioligand. If DCKA and glycine interact with an identical population of sites, the basal binding of [³H]5,7-DCKA would also be predicted to be affected by imipramine treatment (in fact, other antidepressant treatments do have a significant effect on [³H]5,7-DCKA binding; *see below)*. Despite the evidence suggestive of an apparent competitive interaction between these two glycinergic ligands, saturation studies indicate a higher density of [³H]5,7-DCKA, compared to [³H]glycine binding sites *(28)*. Moreover, point mutation studies in recombinant NMDA receptors indicate glycine agonists and antagonists interact with different domains of the NR1 (ξ)-subunit *(31,32)*. Glycine inhibition

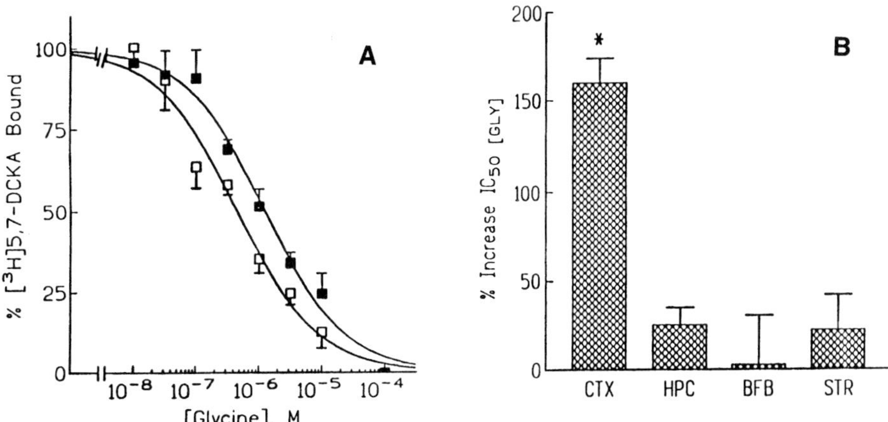

**Fig. 2.** Effect of chronic imipramine treatment (filled symbols, shaded bars) on [³H]5,7-DCKA binding to cortical NMDA receptors. Data are presented as the mean ± SEM of 5–12 animals/group. **(A)** Glycine inhibition of [³H]5,7-DCKA binding to cortical membranes. Data are expressed as percent specific binding using 20.2 ± 0.6 n*M* [³H]5,7-DCKA. The IC$_{50}$ values of glycine were 0.55 ± 0.04 and 1.43 ± 0.2 µ*M* in saline- and imipramine-treated animals, respectively (*p* < 0.05). **(B)** Effect of chronic imipramine treatment on glycine inhibition of [³H]5,7-DCKA binding in cortex (CTX), hippocampus (HPC), basal forebrain (BFB), and striatum (STR). Data are expressed as the percent increase in IC$_{50}$ vs saline and represent the mean ± SEM of 4–6 animals/group. **p* < 0.05 vs saline (Dunnett's *post hoc* comparison). Reprinted from ref. *19.*

**Table 1**
**Effect of Chronic Treatment with Selected Antidepressants on Glycine-Inhibition of [³H]CGP-39653 Binding to Mouse Cortex**

| Treatment | IC$_{50}$ site 1, µ*M* | Site 1, % | IC$_{50}$ site 2, µ*M* |
| --- | --- | --- | --- |
| Saline | 0.95 ± 0.24 | 71 ± 5 | 4113 ± 1129 |
| Imipramine (15) | 0.64 ± 0.10 | 54 ± 3[a] | 3200 ± 900 |
| Amitriptyline (15) | 1.33 ± 0.35 | 44 ± 4[a] | 668 ± 197 |
| Citalopram (15) | Undetectable | 0[a] | 2098 ± 1866 |
| Pargyline (20) | 1.68 ± 0.84 | 33 ± 8[a] | 3412 ± 2271 |

Data are given as mean ± SEM of 4–5 animals per group. Acute treatment was without effect on either basal [³H]CGP-39653 binding or the displacement of this radioligand by glycine (data not shown). [a]*p* < 0.05 vs saline-treated animals. Numbers in parentheses represent mg/kg. Portions of these data were published in refs. *19, 34,* and *67.*

of [³H]CGP-39653 binding is allosteric *(27),* and may reflect the interaction of glycine and glutamate sites evinced in both neurochemical and electrophysiological studies. A reduction in the proportion of high-affinity, glycine-displaceable [³H]CGP 39653 binding sites (from 75 ± 2% to 54 ± 3%; *p* < 0.05) (Table 1), when taken together with a reduction in the potency of glycine to inhibit [³H]5,7-DCKA binding, indicates that chronic imipramine treatment also affects the allosteric coupling between glycine and glutamate sites. Although chronic imipramine did not produce significant changes in either basal [³H]glycine or [³H]CGP 39653 binding, it must be noted that both a fixed dose and dosing interval were employed. However, in other studies using either a larger number of antidepressants or a different species, such changes were evinced *(33,34)*. For example, rats subjected to repeated ECT resulted in both a 38% decrease in the proportion of high-affinity glycine sites displacing [$^3$H]CGP 39653 binding and a 35% decrease in the $B_{max}$ of this radioligand *(33)*. Thus, the preliminary experiments of Nowak et al. *(19)* may be viewed as the first indication that chronic antidepressant treatment can affect NMDA receptors.

Chronic (but not acute) treatment with ACPC produced changes in radioligand binding to NMDA receptors that paralleled those obtained following chronic imipramine treatment. Thus, statistically significant reductions were observed in both the potency of glycine to inhibit [$^3$H]5,7-DCKA binding (~twofold reduction) and in the proportion of high-affinity, glycine-displaceable [$^3$H]CGP 39653 binding sites (~22% reduction). The reduction in basal [$^3$H]dizocilpine binding produced by ACPC (~25%) did not achieve statistical significance. In contrast, no changes in radioligand to NMDA receptors were observed after a low-dose regimen (0.1 mg/kg) dizocilpine. This dose of dizocilpine was active in the forced-swim test *(13)*, but a 7-d regimen was not sufficient to alter β-adrenoceptor density. A higher dose (1 mg/kg) of MK-801, which produced a downregulation of β-adrenoceptors following a 7-d regimen, proved toxic when administered for 14 d, precluding a detailed examination of its effects of radioligand binding to NMDA receptors.

If the effects of chronic treatment of imipramine and ACPC on radioligand binding to NMDA receptors are adaptive and relevant to an antidepressant action, then similar effects should be manifested across antidepressant therapies, independent of chemical structure or in vitro effects, and these changes in radioligand binding to NMDA should develop slowly and persist for some time following cessation of treatment. These hypotheses were examined by treating groups of mice for 14 d with a wide range of antidepressants. The effects of antidepressant treatments on the potency of glycine (IC$_{50}$) to inhibit [$^3$H]5,7-DCKA binding and the proportion of high- and low-affinity glycine sites inhibiting [$^3$H]CGP 39653 binding were examined in cerebral cortical membranes. The doses of antidepressants used in these experiments were based primarily on reports demonstrating either a reduction in cortical β-adrenoceptor density following chronic treatment *(8,9)* or efficacy in the forced-swim test *(14,35)*. The duration of treatment was based on the ability of 14-d regimens of both imipramine and ACPC to alter radioligand binding to NMDA receptors *(19)*. Subsequently, the time required for these effects to develop and dissipate, as well as their dose dependence, was examined using three representative antidepressants: imipramine, citalopram, and ECT.

## 4. PHARMACOLOGICAL SPECIFICITY OF ADAPTIVE CHANGES IN RADIOLIGAND BINDING TO NMDA RECEPTORS

The principal neurochemical measure used in these experiments was the potency of glycine to inhibit [$^3$H]5,7-DCKA binding. For a more limited series of antidepressants, the proportion of high- and low-affinity glycine sites inhibiting [$^3$H]CGP 39653 binding) was also estimated. This latter neurochemical measure was not as extensively used as the former because of both the instability of [$^3$H]CGP 39653 and the larger quantities of tissue required to accurately analyze these data. The criterion for inclusion of an

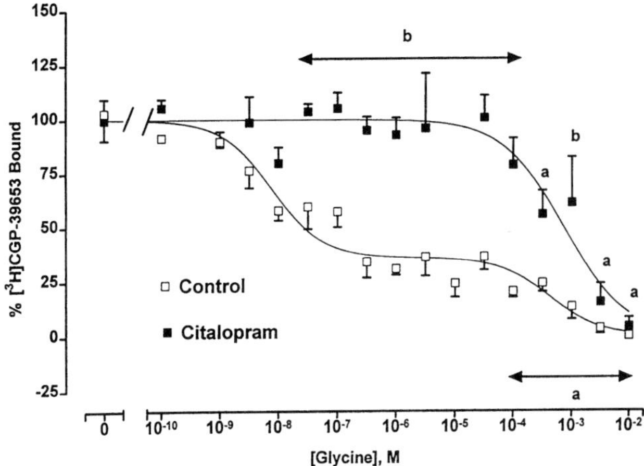



**Fig. 3.** Effect of chronic citalopram treatment on glycine inhibition of [³H]CGP-39653 binding to cortical membranes. Mice were administered either saline or citalopram (20 mg/kg ip) for 14 d. Animals were sacrificed by decapitation 24 h after the last injection. Data are expressed as percent specific binding using 5.3 ± 0.3 n$M$ [³H]CGP-39653. Data were best fit to a two-site model (F[1,76] = 5.1, $p < 0.027$) for saline-treated mice and to a one-site model for citalopram-treated mice (F [1,76] = 3.06, $p = 0.084$), respectively. Symbols: a, $p < 0.05$ vs 10⁻¹⁰$M$ glycine; b, $p < 0.05$ vs saline-treated mice (Bonferoni-corrected univariate F-test). Reprinted from ref. *38*.

antidepressant in this study was efficacy in at least one well-controlled, double-blind study (Table 2) *(36,37)*. The antidepressants evaluated ranged from tricyclics, monoamine oxidase inhibitors (MAOI), and the so-called atypical agents, to ECT. Antidepressants increased the IC$_{50}$ (i.e., reduced the potency) of glycine to inhibit [³H]5,7-DCKA binding, ranging from ~1.8-fold (citalopram) to ~4.3-fold (ECS), as summarized in Table 2. Among the 17 treatments evaluated, only chronic treatment with the MAOI pargyline failed to significantly reduce the potency of glycine in this measure. Since antidepressant-induced reductions in the potency of glycine to inhibit [³H]5,7-DCKA binding appear to be dose- and time-dependent (*see below*), it is possible that either a higher dose or longer duration of treatment with pargyline would result in a statistically significant change. However, since chronic treatment with imipramine and ACPC also reduced the proportion of high-affinity, glycine-displaceable [³H]CGP-39653 binding sites to cerebral cortex *(19,33)* (Table 2), the effect of chronic treatment with pargyline was examined in this measure. In contrast to its modest effects on the IC$_{50}$ of glycine, pargyline produced a more robust reduction (54%) in the proportion of high-affinity, glycine-displaceable [³H]CGP-39653 binding sites than imipramine (Table 2). All four of the antidepressants examined (imipramine, amitryptyline, citalopram, and pargyline) reduced the proportion of high-affinity glycine sites, but it is noteworthy that chronic treatment with citalopram, which produced only a modest increase in the IC$_{50}$ of glycine to inhibit [³H]5,7-DCKA binding, abolished high-affinity, glycine-displaceable binding of [³H]CGP 39653 (Table 2). A recent report *(38)* confirmed this rather remarkable finding (Fig. 3), and further demonstrated that a similar effect was not manifested

**Table 2**
**Effect of Chronic Treatment with Antidepressants and Nonantidepressants on the Potency of Glycine to Inhibit [$^3$H]5,7-DCKA Binding to Mouse Cortex: Comparison with Biochemical and Behavioral Actions**

| Treatment (mg/kg) | ↑ IC$_{50}$ ($p < 0.05$) | FST[a] | ↓ β-AR[b] | Clinical Activity[c] |
|---|---|---|---|---|
| Tricyclics | | | | |
| Imipramine (15) | Yes | Yes | Yes | Yes |
| Desipramine (10) | Yes | Yes | Yes | Yes |
| Amitriptyline (15) | Yes | Yes | Yes | Yes |
| Doxepin (15) | Yes | Yes | Yes | Yes |
| MAOIs | | | | |
| Pargyline (20) | No | Yes | Yes | Yes |
| Tranylcypromine (2) | Yes | Yes | Yes | Yes |
| Clorgyline (5) | Yes | No | Yes | Yes |
| L-deprenyl (2) | Yes | Yes | Yes | Yes |
| D-deprenyl (2) | No | ? | ? | No |
| Atypicals | | | | |
| Alaproclate (20) | Yes | No | No | Yes |
| Mianserin (15) | Yes | No | No | Yes |
| Fluoxetine (20) | Yes | No | No | Yes |
| Nisoxetine (30) | Yes | Yes | Yes | Yes[d] |
| Citalopram (20) | Yes | No | No | Yes |
| Bupropion (40) | Yes | Yes | Yes | Yes |
| Sertraline (25) | Yes | Yes | Yes | Yes |
| Other | | | | |
| Chlorprothixene (2) | Yes | No | No | Yes |
| Chlorpromazine (2) | No | No | No | No |
| Salbutamol (2) | No | No | Yes[e] | ? |
| Chlordiazepoxide (10) | No | No | No | No |
| Chlorpheniramine (30) | No | Yes? | ? | No |
| Scopolamine (0.5) | No | Yes | ? | No |
| Electroconvulsive therapy | Yes | Yes | Yes | Yes |

The drug doses and electroshock regimen employed were based on either reported changes in cortical β-adrenoceptor density ($B_{max}$) following chronic treatment *(8,9)* or efficacy in the forced-swim test *(14,35)*. For those drugs that have no effects on either β-adrenoceptor density or the forced-swim test (e.g., alaproclate, mianserin, fluoxetine, citalopram, chlorprothixene, and chlorpromazine), doses were chosen on the basis of characteristic effects on other neurotransmitter systems reported in the literature. The data in column two (↑ IC$_{50}$) represent the presence of a statistically significant ($p < 0.05$; Dunnett's *t*-test) increase in the IC$_{50}$ of glycine to inhibit [$^3$H]5,7-DCKA binding to mouse cortical membranes. Data are summarized from refs. *19* and *34*. Values in parentheses are the doses in mg/kg/d; all regimens were administered for 14 d. Data represent the mean ± SEM of 5–11 animals per group, except alaproclate (*n* = 3). In saline and sham ECT mouse cortex, basal [$^3$H]5,7-DCKA binding was 1295 ± 48 fmol/mg protein (*n* = 71 from 10 experiments); the IC$_{50}$ and slope factor for glycine in these mice was 0.32 ± 0.02 μ*M* (range of the mean values from 10 experiments: 0.25–0.38 μ*M*) and 0.74 ± 0.03 (range: 0.57–1.05), respectively. In an independent series of experiments, we determined the ability of salbutamol (2 mg/kg for 14 d) to downregulate cortical β-adrenoceptors. This regimen reduced the $B_{max}$ of [$^3$H]dihydroalprenolol by 20% (saline: 168 ± 10 fmol/mg protein; salbutamol: 135 ± 5 fmol/mg protein [*n* = 9/group] *p* < 0.05, compared to in hippocampal membranes following a 2-wk regimen of citalopram. In this study, 68 ± 4% of [3H]CGP 39653 binding was displaced by glycine with high affinity in membranes from saline-treated animals; citalopram treatment reduced this value to 11 ± 11%. Consistent with previous observations *(19,27)*, inhibition of [3H]CGP 39653 data was best fit to a two-site model in 6/6 control subjects; citalopram treatment resulted in data that were best modeled to a one-site (low-affinity) fit in 5/6 subjects. Moreover, as previously reported *(34)*, citalopram treatment resulted in only a modest (50%), albeit statistically significant ($p < 0.05$), reduction in the affinity of glycine to inhibit [3H]CGP 39653 binding. The ability of this diverse group of antidepressants to alter radioligand binding to NMDA receptors is consistent with the hypothesis that chronic treatment with these compounds induces an adaptation of NMDA receptors. Based on the differential effects of several antidepressants (but most evident following citalopram treatment) on glycine inhibition of [3H]5,7-DCKA and [3H]CGP-39653 binding, it may be hypothesized that the lack of a clearcut stoichiometry between these two neurochemical measures may result from multiple, discrete effects of antidepressants on NMDA receptors. The recent observation *(38a)* that chronic ACPC treatment can bidirectionally modulate the levels of mRNA-encoding NMDAR-2 ($\varepsilon$)-subunits in cerebral cortex (reducing the quantity of NMDAR-2C and increasing NMDAR-2A, respectively) is consistent with the hypothesis that compounds possessing antidepressant actions can produce multiple effects at NMDA receptors. The relationship of these neurochemical changes to the therapeutic characteristics of a particular antidepressant (e.g., speed of onset, efficacy) remains unresolved. However, the extent to which clinically effective antidepressants mimic these effects of ACPC on NMDA mRNA levels may provide avenues to explore these issues.

Antidepressant-induced changes in radioligand binding to NMDA receptors appear to reflect an alteration in the strychnine-insensitive glycine receptor and its ability to allosterically modulate other sites on this family of ligand-gated ion channels. The small (~30%), but statistically significant, reductions in basal [3H]5,7-DCKA binding observed following chronic treatment with desipramine, amitriptyline, and chlorprothixene *(34)* are consistent with this hypothesis. In the two instances in which complete dose-response studies were performed (imipramine and citalopram), no alterations in basal [3H]5,7-DCKA binding were observed. Nonetheless, since only one dose of the other antidepressants was employed, it remains to be determined if the inability of the other compounds to affect basal [3H]5,7-DCKA binding is the result of pharmacokinetic or pharmacodynamic factors.

---

**Table 2** *(continued)* saline-treated animals), which is comparable to the reduction in $B_{max}$ observed in this mouse strain following chronic treatment with imipramine *(17)*. [a]FST: Treatment significantly reduces immobility in the mouse forced-swim test. Data reviewed in ref. *35*. [b]↓β-AR: Treatment significantly reduces the density ($B_{max}$) of forebrain β-adrenoceptors. Data reviewed in refs. *8* and *9*. [c]Clinical activity: Treatment possesses antidepressant properties in clinical trials. Data reviewed in refs. *36* and *37*. Only data from double-blind, placebo-controlled studies was accepted as evidence of clinical activity and is presented here for comparison. [d]Ray Fuller, personal communication. [e]Huang, unpublished observations. Although clinically active as an antidepressant in double-blind trials, nisoxetine was withdrawn because of a relatively high incidence of adverse side effects. This summary is from ref. *67*.

Although the breadth of antidepressants that affect radioligand binding to NMDA receptors provides an indication of the robust nature of this phenomenon, the results obtained with several of the nonantidepressant drugs examined begin to address the issue of specificity. For example, the ability of L- but not D-deprenyl to increase the $IC_{50}$ of glycine to inhibit [$^3$H]5,7-DCKA binding (Table 2) provides some evidence that this effect is stereoselective, albeit with the reservation that only one dose of each isomer was examined. The lack of effect of chlorpromazine, a congener of chlorprothixene, is also noteworthy. Chronic treatment with salbutamol, which to date has not been demonstrated to produce an antidepressant action in double-blind clinical trials (but, like other β-adrenoceptor agonists, reduces the density of cortical β-adrenoceptors (*[39]* [*see* Table 2]), did not affect the potency of glycine to inhibit [$^3$H]5,7-DCKA binding. Although the potential for obtaining false positives can never be completely excluded, these data provide an indication that these changes in radioligand binding to NMDA receptors are associated with clinically active antidepressants. Moreover, as a predictor of antidepressant action, these neurochemical measures are significantly better than either changes in β-adrenoceptor density (12/17) or efficacy in the forced-swim test (12/17, 71%; $p < 0.02$, Fisher's exact test), perhaps the most widely used biochemical and behavioral predictors of antidepressant activity, respectively *(9,14,35)*. The predictive validity of antidepressant-induced alterations in radioligand binding to NMDA receptors is even more robust if compounds generally considered false positives or negatives in either the forced-swim test or β-adrenoceptor downregulation are also considered. Under these circumstances, alterations in radioligand binding to NMDA receptors correctly predict the outcome for all treatments examined in this study; the forced-swim test and β-adrenoceptor downregulation are 57% (13/23; $p < 0.002$, Fisher's exact test) and 65% (15/23; 65%; $p < 0.01$, Fisher's exact test) accurate, respectively.

## 5. TIME-COURSE AND DOSE-DEPENDENCE STUDIES

If antidepressant-induced changes in radioligand binding to NMDA receptors are truly adaptive, then they should develop slowly and persist for some time after cessation of treatment. Since 14-d (but not 1-d) regimens of antidepressants increased the $IC_{50}$ of glycine to inhibit [$^3$H]5,7-DCKA binding to cortical membranes *(19,34)*, we examined both the time-course required for this effect to appear and the interval required for its dissipation following cessation of treatment. Three representative antidepressants (imipramine, citalopram, and ECT) were used in this study *(34)*. Imipramine treatment first produced a statistically significant increase in the $IC_{50}$ of glycine after 14 d (15 mg/kg/d) of treatment (264% of control), with a further increase (to 330% of control) noted after 21 d (Fig. 4). The increased $IC_{50}$ of glycine produced by 14 d of treatment with imipramine was present at both 1 and 5 d following cessation of treatment, but was no longer apparent 10 d later (Fig. 4). The dose–response (2.5–30 mg/kg/d) relationship for the effect of a 14-d regimen demonstrated significant increases in the $IC_{50}$ for glycine at 15 and 30 mg/kg/d, with an estimated $ED_{50}$ of ~5 mg/kg *(34)*. Parallel studies with citalopram (20 mg/kg) demonstrated that at least a 10-d regimen was required to significantly increase the $IC_{50}$ of glycine to ~200% of control. Further treatment for 14 or 21 d did not result in a further increase in the $IC_{50}$ of glycine. When administered for 14 d, citalopram

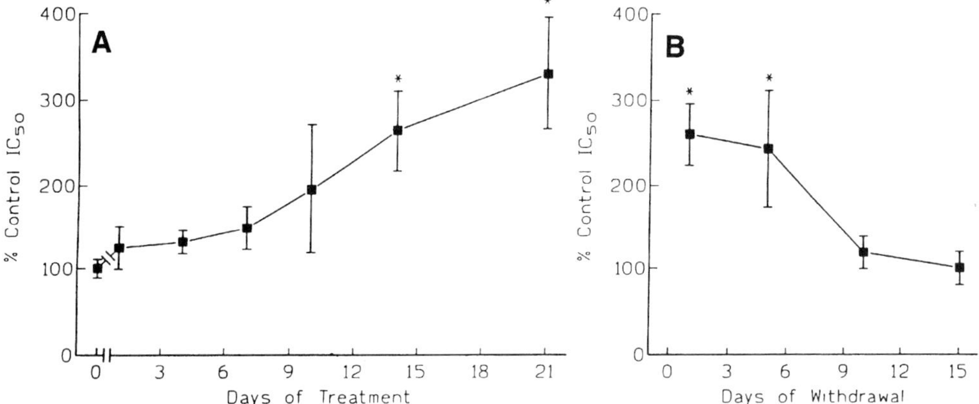

**Fig. 4.** Time-course of imipramine-induced increases in the $IC_{50}$ of glycine to inhibit $[^3H]5,7$-DCKA binding. Mice were treated with imipramine (15 mg/kg ip) for 1–21 d and killed 24 h after the last treatment (left panel), or treated for 14 d and killed 1–15 d after cessation of treatment (right panel). Data are expressed as the percent of basal (control) $IC_{50}$ values and represent the mean ± SEM of 4–15 animals/group. In cortices from saline-treated mice, basal $[^3H]5,7$-DCKA binding was 1229 ± 85 fmol/mg protein and the $IC_{50}$ value of glycine was 0.39 ± 0.04 $\mu M$ ($n$ = 15 animals). Symbol: *, $p < 0.05$ vs saline-treated animals. Reprinted from ref. *34*.

(5–40 mg/kg/d) produced significant increases in the $IC_{50}$ of glycine at doses ≥10 mg/kg, with an $ED_{50}$ of ~5 mg/kg. ECT resulted in a 370% increase in the $IC_{50}$ of glycine by d 7 of treatment. This effect persisted through d 10 following cessation of treatment *(34)*.

The prolonged treatment required to effect significant reductions in the potency of glycine in this measure is indicative of an adaptive response, with the effect of ECT developing more rapidly (7 d vs 10–14 d) than those following drug treatments. Moreover, these effects were remarkably persistent, generally requiring at least 7 d following cessation of treatment to return to control values (Fig. 4). The temporal pattern associated with antidepressant-induced changes in radioligand binding to NMDA receptors are, in general, consonant with the clinical observation that chronic antidepressant administration is required to achieve therapeutic results *(1)*, and that the beneficial effects can persist following discontinuation of treatment *(11,12)*.

## 6. FUNCTIONAL CONSEQUENCES OF ANTIDEPRESSANT-INDUCED ALTERATIONS IN NMDA RECEPTORS

In absolute terms, antidepressant-induced reductions in the potency of glycine to inhibit $[^3H]5,7$-DCKA binding may be considered relatively modest *(19,34,38)*. If strychnine-insensitive glycine receptors were saturated under physiological conditions, the impact of these antidepressant-induced changes on the operation of NMDA receptor-coupled cation channels would be marginal. Several studies have failed to demonstrate a modulation of NMDA receptors by application of glycine *(40)*, but others have shown that both glycine and glycinemimetics augment the effects of NMDA *in situ* *(41)*. For example, glycine can potentiate the actions of NMDA in both neocortical slices *(42)* and thalamic neurons *(43)*. Furthermore, administration of glycine and D-serine (a glycine agonist) elevates cerebellar cyclic GMP levels, an effect that can be blocked by NMDA antagonists *(44–46)*. Based on these relatively modest effects of glycine, compared to the augmentation of glutamate (or NMDA) currents observed in culture *(47)* (in which glycine concentrations can be more precisely regulated), it appears that strychnine-insensitive glycine receptors may approach, but not achieve, full saturation *in situ*. However, under these conditions, even a modest (e.g., 2–4-fold) reduction in agonist potency could have a significant effect on the operation of NMDA receptor-coupled cation channels. Thus, in view of the essential role of glycine to the operation of NMDA receptor-coupled cation channels *(48)*, a reduced affinity of glycine at strychnine-insensitive glycine receptor could result in impaired channel operation. This hypothesis is also consistent with the ~50% reduction in basal (nonequilibrium) [$^3$H]dizocilpine binding to the NMDA receptor-coupled ionophore following chronic imipramine, with no change in the maximum response to either glycine or glutamate *(19)*. Although antidepressant-induced reductions in the proportion of high-affinity, glycine-displaceable [$^3$H]CGP-39653 binding can be dramatic (as in the case of chronic citalopram treatment), it is more difficult to predict the functional consequences of this effect. However, if this neurochemical measure reflects an allosteric coupling of glycine and glutamate recognition sites, then a reduction in the proportion of high-affinity glycine sites coupled to glutamate sites might also result in a diminished frequency of channel opening. If the neuroanatomical localization of the neurochemical effects of imipramine and citalopram can be generalized to other antidepressants, then it may be possible to more directly test this hypothesis using electrophysiological techniques.

## 7. FUTURE DIRECTIONS

Perhaps the most intriguing issue to arise from these studies is the molecular mechanism (or mechanisms) responsible for the antidepressant-induced changes in radioligand binding to NMDA receptors. Studies using both recombinant receptors *(25,49)* and subunit-specific antibodies *(24)* indicate NMDA receptors are heterooligomers composed of NMDAR1 ($\xi$)- and NMDAR2A-D ($\epsilon$1–4)-subunits (reviewed in ref. *50*). Furthermore, in some wild-type receptors, the presence of more than one species of NMDAR2 subunit (e.g., NMDAR2A and NMDAR2B) has been documented *(24)*. Like other ligand-gated ion channels, the affinities of both drugs and transmitters acting at the multiple, allosteric regulatory sites on NMDA receptors is highly dependent upon subunit composition *(50–52)*. Based on this information, alterations in subunit composition could be invoked to explain the ability of chronic antidepressant treatment to alter radioligand binding to NMDA receptors. Such changes are likely to be subtle, but we have embarked on studies on explore this possibility. Our initial experimental strategy has been to examine NMDA receptor mRNA content in cerebral cortex following chronic antidepressant treatments. Groups of mice were administered four drugs (citalopram, imipramine, tranylcypramine, and ACPC) for 2 wk, followed by extraction of RNA from cerebral cortex and hybridization to cDNAs *(53)* for NMDAR-1 and NMDA R2A-C on slot blots (Table 3). Although antidepressants produced modest changes in NMDA subunit mRNA levels relative to saline-treated groups (data were normalized to cyclophilin levels *[54]*), no consistent effects were manifested across treatments. While

**Table 3**
**Effect of Chronic Antidepressant Treatments on NMDA Receptor Subunit RNA Levels**

| | Saline-treated mice, % | | | |
|---|---|---|---|---|
| | $\zeta$ | $\varepsilon1$ | $\varepsilon2$ | $\varepsilon3$ |
| Citalopram | 85 ± 14 | 91 ± 21 | 75 ± 16 | 73 ± 11 |
| Imipramine | 143 ± 27 | 83 ± 15 | 122 ± 24 | 149 ± 36 |
| ACPC | 99 ± 19 | 78 ± 27 | 81 ± 19 | 81 ± 9 |
| Tranylcypromine | 104 ± 15 | 115 ± 25 | 111 ± 8 | 134 ± 35 |

NIH Swiss mice received 14 daily ip injections of saline, citalopram (20 mg/kg), imipramine (15 mg/kg), ACPC (400 mg/kg), or tranylcypromine (2 mg/kg). Doses were selected based on the ability of these agents to effect changes in radioligand binding to NMDA receptors *(19,34,38)*. Total RNA (5 μg) from cerebral cortex was analyzed by slot blot hybridization with cDNA probes for the $\zeta$-, $\varepsilon$-1, $\varepsilon$-2, and $\varepsilon$–3-subunits of the NMDA receptor family *(53)*, and for cyclophilin *(54)* that had been random prime-labeled with $\alpha$-[$^{32}$P]dCTP. The amount of radioactivity that hybridized to the slot-blotted RNA was quantified using densitometric analysis of autoradiograms, with conversion to calibrated optical density units (Image-1 software, Universal Imaging). All NMDA subunit values were corrected to the corresponding cyclophilin value for internal consistency. Data are represented as mean ± SEM of percent respective saline control (*n* = 6/group). A one-way analysis of variance revealed no significant treatment effects.

these studies were in progress, Oretti et al. *(55)* reported that chronic treatment with citalopram failed to alter NMDA receptor subunit mRNA levels in (whole) rat brain. The failure to detect consistent changes in mRNA levels across treatments may indicate that the effects on radioligand binding produced by antidepressants reflect posttranslational modifications of NMDA receptors. Alternatively, antidepressant-induced changes in NMDA receptor composition may be confined to specific areas and layers of cerebral cortex, requiring a higher level of resolution than analysis of slot blots from pooled cortical tissues. Consistent with this latter hypothesis, we (S. Bovetto, L. H. Fossom, and P. Skolnick) have recently shown that chronic ACPC treatment differentially increases the levels of mRNA-encoding NMDAR-2A in cortical subfields, ranging from >30% in occipital cortex to ~17% in frontal cortex. Moreover, within cortical subfields, these increases in NMDAR2A mRNA exhibit a distinct laminar distribution. For example, in parietal cortex, increases of ~40 and ~18% were found in layers I/II and layer VI, respectively. Chronic ACPC also decreased NMDAR-2C mRNA in cortical subfields, and these decreases were unevenly distributed among laminae. Thus, NMDAR-2C mRNA was decreased by ~30% in occipital cortex, ranging from a >40% decrease in layers I/II to ~19% in layer VI, respectively. NMDAR-2C mRNA is not an abundant species in mouse cortex, but studies in recombinant receptors have shown the affinity of glycine is more than one order of magnitude higher in receptors composed of NMDAR1/NMDAR2C subunits than in receptors containing NMDAR2A subunits *(51,52)*. If these changes in mRNA content also reflect differences in protein product, then alterations in subunit composition may well explain the changes in radioligand binding to NMDA receptors produced by chronic treatment with ACPC *(19)*, and, by inference, chronic treatment with antidepressants.

Antidepressant-induced changes in radioligand binding to NMDA receptors could also reflect changes in posttranslational modification, since the subunits that constitute this family of ligand-gated ion channels are well-endowed with consensus sequences that can serve as substrates for PKC (protein kinase C), PKA (protein kinase A), calcium-calmodulin-dependent kinases, and protein tyrosine kinases *(56,57)*. Moreover, there is an emerging body of evidence that suggests that NMDA receptor function is modulated by phosphorylation *(58,59)*. Based on these findings, it may be hypothesized that changes in the activity of one or more of these enzymes may be directly responsible for antidepressant-induced changes in radioligand binding to NMDA receptors. Since there are multiple isoforms of many of these enzymes (and it is likely that not all isoforms of each enzyme have been identified), we are attempting to identify potential changes in gene expression produced by chronic antidepressants, using differential display polymerase chain reaction (PCR) (reviewed in ref. *60*). We have now identified several potential gene candidates whose expression can be either increased or suppressed by chronic treatment with two antidepressants (imipramine and ECT) (Fig. 5). By using multiple antidepressants, rather than a more conventional strategy of comparing saline treatment to a single antidepressant, it may be possible to more accurately identify relevant gene candidates. Although differential display PCR has a relatively high rate of detecting false positives, the antidepressant-induced expression of several candidates has already been confirmed by Northern blot analysis (unpublished data). Thus, by using this technique, it may be possible to identify either those factors controlling expression of NMDA receptor subunits or potential candidates responsible for an altered posttranslational modification of these proteins.

## 8. CONCLUDING REMARKS

Antidepressant-induced adaptation of NMDA receptors may be a means of dampening elevations in intracellular calcium concentration in specific neuronal populations following glutamate release. A review of the literature suggests a role for calcium homeostasis in the pathophysiology of human depression *(61)*. For example, alterations in cerebrospinal fluid (CSF) or serum calcium levels that could not be directly associated with endocrine dysfunction have been reported in depressed patients *(62,63)*. Conversely, reduced plasma, serum, or CSF calcium levels have been reported with improvement of depressive symptomatology after both ECT and pharmacotherapy *(62–64)*. These clinical observations appear to bridge, albeit in a tenuous fashion, the body of findings presented in this review, with preclinical data indicating that antagonists acting at voltage-operated calcium channels (e.g., dihydropyridines, such as nimodipine) possess antidepressant-like actions (*see* Chapter 5).

Further neurochemical, electrophysiological, and behavioral studies will be required to fully appreciate the functional significance of antidepressant-induced changes in radioligand binding to NMDA receptors, but our studies represent the first description of a consistent, selective adaptation of a neurotransmitter-associated receptor produced by all major classes of antidepressants. By comparison, 20 yr after the demonstration that chronic antidepressant treatment results in downregulation of forebrain β-adrenoceptors *(15)*, the physiological and behavioral consequences of this phenomenon remain

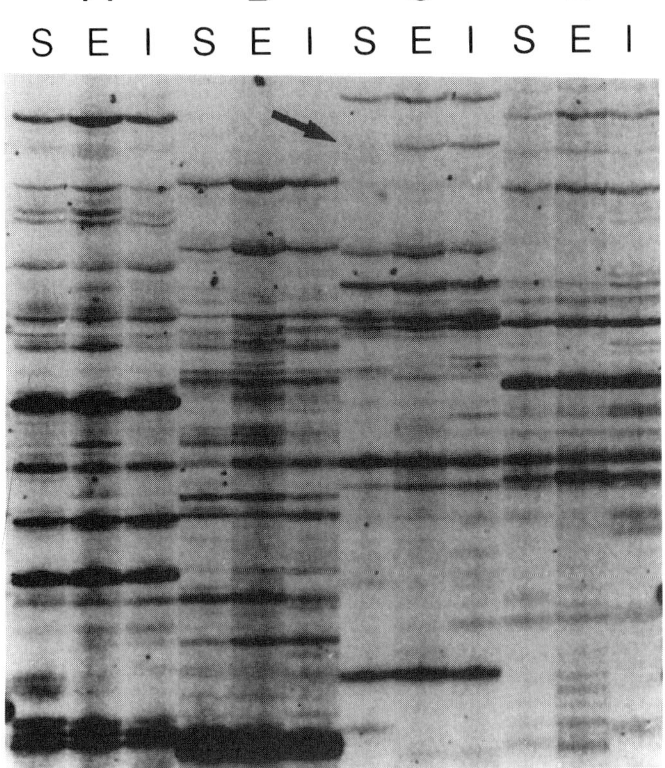


**Fig. 5.** Representative autoradiogram of radiolabeled PCR differential display products from mouse cortical RNA after chronic administration of sham (S), ECT (E), or imipramine (I). A typical differentially expressed candidate gene is marked by the arrow. Note the presence of bands in lanes E and I; a corresponding band in lane S is at the limit of detection. This is a portion of an autoradiogram that was enlarged for clarity. Mice received daily imipramine injections (15 mg/kg ip), ECT (administered via ear clip using an Ugo Basile ECT unit; 100 pulses/s frequency, 0.5-ms pulse width, 0.2-s shock duration, 90-mA current), or sham treatment (ear clips applied without current) for 2 wk. PCR differential display products were obtained using one-base-anchored oligo-dT primers and arbitrary 13-mers (RNAimage™, GenHunter Corporation, Nashville, TN). Lane groups (A–D) represent each treatment condition amplified with a different primer set.

unknown *(10,11)*. The role of monoaminergic pathways in antidepressant-induced adaptation of NMDA receptors is unknown, but clearly merits further investigation, in view of the vast literature linking monoamines to antidepressant action and depression. Nonetheless, it must be noted that Rossby et al. *(65)* (*see* Chapter 11) have recently demonstrated that desmethylimipramine can increase hippocampal mRNA levels of glucocorticoid receptor type II in DSP-4-treated rats. Thus, this effect was evident in the absence of presynaptic noradrenergic terminals, the putative locus of action for desipramine. Although the relationship between glucocorticoid type II receptors and antidepressant action is unclear, this study provides proof of concept that a specific blocker of norepinephrine reuptake can produce effects on gene regulation by a mechanism(s) independent of its effects on synaptic monoamine levels.

In summary, the neurochemical data presented in this review provide compelling evidence that adaptation of NMDA receptors is a common feature of chronic antidepressant treatment, and as such may represent an obligatory step in antidepressant action. Taken together with the ability of functional NMDA antagonists to mimic the actions of antidepressant in preclinical behavioral paradigms (*see* Chapters 6 and 12), these findings suggest that NMDA receptors may also play a role in depressive disorders. Recent studies using autopsy materials from suicide victims *(66)* and animal models (reviewed in Chapter 8) are consistent with this hypothesis.

## REFERENCES


1. Oswald, I., Brezinova, V., and Dunleavy, D. L. F (1972) On the slowness of action of tricyclic antidepressant drugs. *Br. J. Psychiatry* **120,** 673–677.
2. Segal, D.S., Kuczenski, R., and Mandell, A. J. (1974) Theoretical implications of drug-induced adaptive regulation for a biogenic amine hypothesis of effective disorder. *Biol. Psychiatry* **9,** 19–24.
3. Maj, J., Przegalinski, E., and Moglnicka, E. (1984) Hypothesis concerning the mechanism of action of antidepressant drugs. *Rev. Physiol. Biochem. Pharmacol.* **100,** 1–74.
4. Heninger, G. R., Delgado, P. L., and Charney, D. S. (1996) The revised monoamine theory of depression: a modulatory role for monoamines, based on new findings from monoamine depletion experiments in humans. *Pharmacopsychiatry* **29,** 2–11.
5. Fink, M. (1979) *Convulsive Therapy: Theory and Practice.* Raven, New York.
6. Vetulani, J. and Sulser, F. (1975) Action of various antidepressant treatments reduces reactivity of noradrenergic cyclic AMP generating system in limbic forebrain. *Nature* **257,** 495–496.
7. Sugrue, M. F. (1985) Delayed biochemical changes following antidepressant treatment. *Psychopharmacology* **21,** 619–622.
8. Hollister, L. E. (1986) Current antidepressants. *Ann. Rev. Pharmacol. Toxicol.* **26,** 23–37.
9. Caldecott-Hazard, S., Morgan, D. G., DeLeon-Jones, F., Overstreet, D. H., and Janowsky, D. (1991) Clinical and biochemical aspects of depressive disorders: II. Transmitter receptor theories. *Synapse* **9,** 251–301.
10. Vetulani, J. (1991) The development of our understanding of the mechanism of action of antidepressant drugs. *Pol. J. Pharmacol.* **43,** 323–338.
11. Stone, E. A. (1983) Problems with the current catechholamine hypothesis of antidepressant agents: speculations leading to a new hypothesis. *Behav. Brain Sci.* **6,** 555–577.
12. Heninger, G. and Charney, D. (1987) Mechanism of action of antidepressant treatments: implications for the etiology and treatment of depressived disorders. In: *Psycopharmacology: The Third Generation of Progress* (Meltzer, H., ed.). Raven, New York, pp. 535–544.
13. Trullas, R. and Skolnick, P. (1990) Functional antagonists at the NMDA receptor complex exhibit antidepressant actions. *Eur. J. Pharmacol.* **185,** 1–10.
14. Porsolt, R. D., Lenègre, A., and McArthur, R. A. (1991) Pharmacological Models of Depression, in: *Animal Models in Psychopharmacology* (Olivier, B., Mos, J., and Slangen, J. L., eds.). Birkhäuser Verlag, Basel, pp. 137–159.
15. Banerjee, S. P., Kung, L. S., Riggi, S. J., and Chanda, S. K. (1977) Development of beta adrenergic receptor subsensitivity by antidepressants. *Nature* **268,** 455–456.
16. Trullas, R., Folio, T., Young, A., Miller, R., Boje, K., and Skolnick, P. (1991) 1-aminocyclopropanecarboxylates exhibit antidepressant and anxiolytic actions in animal models. *Eur. J. Pharmacol.* **203,** 379–385.

17. Paul, I. A., Trullas, R., Skolnick, P., and Nowak, G. (1992) Down-regulation of cortical β-adrenoceptors by chronic treatment with functional NMDA antagonists. *Psychopharmacology* **106,** 285–287.

18. Evoniuk, G. E., Hertzman, R. P., and Skolnick, P. (1991) A rapid method for evaluating the behavioral effects of dissociative anesthetics in mice. *Psychopharmacology* **105,** 125–128.

19. Nowak, G., Trullas, R., Layer, R. T., Skolnick, P. and Paul, I. A. (1993) Adaptive changes in the N-methyl-D-aspartate receptor complex after chronic treatment with imipramine and 1-aminocyclopropanecarboxylic acid. *J. Pharmacol. Exp. Ther.* **265,** 1380–1386.

20. Klimek, V. and Papp, M. (1994) The effect of MK-801 and imipramine on β-adrenergic and 5-HT2 receptors in the chronic mild stress model of depression in rats. *Pol. J. Pharmacol.* **46,** 67–69.

21. Layer, R. T., Popik, P., Olds, T., and Skolnick, P. (1995) Antidepressant-like actions of the polyamine site NMDA antagonist, eliprodil (SL 82.0715). *Pharmacol. Biochem. Behav.* **52,** 621–627.

22. Carter, C., Benavides, J., Legendre, P., Vincent, J., Noel, F., Thuret, F., Lloyd, K., Arbilla, S., Zivkovic, B., MacKenzie, E., Scatton, B., and Langer, S. (1988) Ifenprodil and SL 82.0715 as cerebral antischemic agents. II. Evidence for N-methyl-D-aspartate receptor antagonist properties. *J. Pharmacol. Exp. Ther.* **247,** 1222–1232.

23. Carter, C., Rivy, J.-P., and Scatton, B. (1989) Ifenprodil and SL 82.0715 are antagonists at the polyamine site of the N-methyl-D-aspartate (NMDA) receptor. *Eur. J. Pharmacol.* **164,** 611–612.

24. Sheng, M., Cummings, J., Roldan, L. A., Jan, Y. N., and Jan, L. Y. (1994) Changing subunit composition of heteromeric NMDA receptors during development of rat cortex. *Nature* **368,** 144–147.

25. Kutsuwada, T., Kashiwabuchi, N., Mori, H., Sakimura, K., Kushiya, E., Araki, K., Meguro, H., Masaki, H., Kumanishi, T., Arakawa, M., and Mishina, M. (1992) Molecular diversity of the NMDA receptor channel. *Nature* **358,** 36–41.

26. Palfreyman, M. G., Reynolds, I. J., and Skolnick, P., eds. (1994) Direct and Allosteric Control of Glutamate Receptors, in *Pharmacology and Toxicology: Basic and Clinical Aspects,* CRC, Boca Raton.

27. Sills, M. G., Fagg, G., Pozza, M., Angst, C., Brundish, D., Hurt, S. D., Wilusz, E. J., and Williams, M. (1991) [3H]CGP 39653: A new N-methyl-D-aspartate antagonist radioligand with low nanomolar affinity in rat brain. *Eur. J. Pharmacol.* **192,** 19–24.

28. Baron, B. M., Siegel, B. W., Sloan, A. L., Harrison, B. L., Palfreyman, M. G., and Hurt, S. D. (1991) [3H]5,7-Dichlorokynurenic acid, a novel radioligand labels NMDA receptor associated glycine binding sites. *Eur. J. Pharmacol.* **206,** 149–154.

29. Wong, E., Knight, A., and Woodruff, G. (1988) [3H]MK-801 labels a site on the N-methyl-D-aspartate receptor channel complex in rat brain membranes. *J. Neurochem.* **50,** 274–281.

30. Huettner, J. and Bean, B. (1988) Block of N-methyl-D-aspartate-activated current by the anticonvulsant MK-801: selective binding to open channels. *Proc. Natl. Acad. Sci. USA* **85,** 1307–1311.

31. Kuryatov, A., Laube, B., Betz, H., and Kuhse, J. (1994) Mutational analysis of the glycine-binding site of the NMDA receptor: structural similarity with bacterial amino acid binding proteins. *Neuron* **12,** 1291–1300.

32. Wafford, K. A., Kathoria, M., Bain, C. J., Le Bourdelles, B., Kemp, J. A., and Whiting, P. J. (1995) Identification of amino acids in the N-methyl-D-aspartate receptor NR1 subunit that contribute to the glycine binding site. *Mol. Pharmacol.* **47,** 374–380.

33. Paul, I. A., Layer, R. T., Skolnick, P., and Nowak, G. (1993) Adaption of the N-Methyl-D-Aspartate receptor complex in rat front cortex following chronic treatment with electroconvulsive shock or imipramine. *Eur. J. Pharmacol.* **247,** 305–312.

34. Paul, I. A., Nowak, G., Layer, R. T., Popik, P., and Skolnick, P. (1994) Adaptation of the N-Methyl-D-Aspartate receptor complex following chronic antidepressant treatments. *J. Pharmacol. Exp. Ther.* **269,** 95–102.

35. Borsini, F. and Meli, A. (1988) Is the forced swim test a suitable model for revealing antidepressant activity? *Psychopharmacology* **94,** 147–160.

36. Hollister, L. E. and Csernansky, J. G., eds. (1990) *Clinical Pharmacology of Psychotherapeutic Drugs.* 3rd ed. Churchill Livingstone, New York.

37. Baldessarini, R. J., ed. (1983) *Biomedical Aspects of Depression and Its Treatment.* American Psychiatric, New York.

38. Nowak, G., Li, Y., and Paul, I. A. (1996) Adaptation of cortical but not hippocampal NMDA receptors after chronic citalopram treatment. *Eur. J. Pharmacol.* **295,** 75–85.

38a. Bovetto, S., Boyer, P. A., Skolnick, P., and Fossom, L. (1997) Chronic administration of a glycine partial agonist alters the expression of NMDA receptor subunit mRNAs. *J. Pharmacol. Exp. Ther.* (submitted).

39. Beer, M., Hacker, S., Poat, J., and Stahl, S. M. (1987) Independent regulation of Beta1 and Beta2 adrenoceptors. *Br. J. Pharmacol.* **92,** 827–824.

40. Kemp, J. A. and Leeson, P. D. (1993) The glycine site of the NMDA receptor—five years on. *Trends Pharmacol. Sci.* **14,** 20–25.

41. Thomson, A. M. (1990) Glycine is a coagonist at the NMDA receptor/channel complex. *Prog. Neurobiol.* **35,** 53–74.

42. Thomson, A. M., Walker, V. E., and Flynn, D. M. (1989) Glycine enhances NMDA-receptor mediated synaptic potentials in neurocortical slices. *Nature* **338,** 422–424.

43. Salt, T. E. (1989) Modulation of the NMDA receptor mediated responses by glycine and d-serine in the rat thalamus in vivo. *Brain Res.* **481,** 403–406.

44. Wood, P. L., Emmett, M. R., Rao, T. S., Mick, S., Cler, J., and Iyengar, S. (1989) In vivo modulation of the N-methyl-D-aspartate receptor complex by D-serine: potentiation of ongoing neuronal. *Neurochemistry* **53,** 979–981.

45. Dansyz, W., Wroblewski, J. T., Brooker, G., and Costa, E. (1989) Modulation of glutamate receptors by phencyclidine and glycine in rat cerebellum: cGMP increases in vivo. *Brain Res.* **479,** 270–276.

46. Rao, T., Cler, J., Emmet, M., Mick, S., Iyengar, S., and Wood, P. (1990) Glycine, glycinamide, and D-serine act as positive modulators of signal tranduction at the N-methyl-D-aspartate (NMDA) receptor in vivo: differential effects on mouse cerebellar cyclic guanosine monophosphate levels. *Neuropharmacology* **29,** 1075–1080.

47. Johnson, J. W. and Ascher, P. (1987) Glycine potentiates the NMDA response in cultured mouse brain neurons. *Nature* **325,** 529–531.

48. Kleckner, N. W. and Dingledine, R. (1988) Requirement for glycine in activation of NMDA-receptors expressed in *Xenopus* oocytes. *Science* **241,** 835–837.

49. Monyer, H., Sprengel, R., Schoepfer, R., Herb, A., Higuchi, M., Lomeli, H., Burnashev, N., Sakmann, B., and Seeburg, P. (1992) Heteromeric NMDA receptors: molecular and functional distinction of subtypes. *Science* **256,** 1217–1221.

50. Seeburg, P. H., Burnashev, N., Kohr, G., Kuner, T., Sprengel, R., and Monyer, H. (1995) The NMDA receptor channel: molecular design of a coincidence detector. *Rec. Prog. Horm. Res.* **50,** 19–34.

51. Meguro, H., Mori, H., Araki, K., Kushiya, E., Kutsuwada, T., Yamazaki, M., Kumanishi, T., Arakawa, M., Sakimura, K., and Mishina, M. (1992) Functional characterization of a heteromeric NMDA receptor channel expressed from cloned cDNAs. *Nature* **357,** 70–74.

52. Wafford, K. A., Bain, C. J., Le Bourdelles, B., Whiting, P. J., and Kemp, J. A. (1993) Preferential co-assembly of recombinant NMDA receptors composed of three different subunits. *NeuroReport* **4,** 1347–1349.

53. Fossom, L. H., Basile, A. S., and Skolnick, P. (1995) Sustained exposure to 1-aminocy-clopropanecarboxylic acid, a glycine partial agonist, alters N-methyl-D-aspartate receptor function and subunit composition. *Mol. Pharmacol.* **48,** 981–987.

54. Milner, R. J. and Sutcliffe, J. G. (1983) Gene expression in rat brain. *Nucleic Acids Res.* **11,** 5497–5520.

55. Oretti, R. G., Spurlock, G., Buckland, P. R., and McGuffin, P. (1994) Lack of effect of antipsychotic and antidepressant drugs on glutamate receptor mRNA levels in rat brains. *Neurosci. Lett.* **177,** 39–43.

56. Planells-Cases, R., Sun, W., Ferrer-Montiel, A. V., and Montal, M. (1993) Molecular cloning, functional expression, and pharmacological characterization of an N-methyl-D-aspartate receptor subunit from human brain. *Proc. Natl. Acad. Sci. USA* **90,** 5057–5061.

57. Lin, Y. J., Bovetto, S., Carver, J. M., and Giordano, T. (1996) The cloning of the human NR2-C NMDA glutamate receptor and its expression in the central nervous system and periphery. *Mol. Brain Res.,* in press.

58. Raymond, L. A., Blackstone, C. D., and Huganir, R. L. (1993) Phosphorylation of amino acid neurotransmitter receptors in synaptic plasticity. *TINS* **16,** 147–153.

59. Ehlers, M. D., Tingley, W. G., and Huganir, R. L. (1995) Regulated subunit distribution of the NR1 subunit of the NMDA receptor. *Science* **269,** 1734–1737.

60. Livesey, F. J. and Hunt, S. P. (1996) Identifying changes in gene expression in the nervous system: mRNA differential display. *TINS* **19,** 84–88.

61. Ortolano, G. A., Swonger, A. K., Kaiser, E. A., and Hammond, R. P. (1983) A calcium hypothesis of antidepressant action. *Med. Hypoth.* **10,** 207–221.

62. Linder, J., Brismar, K., Beck-Friis, J., Saaf, J., and Wetterberg, L. (1989) Calcium and magnesium concentrations in affective disorder: difference between plasma and serum in relation to symptoms. *Acta. Psychiatry Scand.* **80,** 527–537.

63. Carman, J. S., Post, R. M., Goodwin, F. K., Bunney, W. E. (1977) Calcium and electro-convulsive therapy in severe depressive illness. *Biol. Psychiatry* **12,** 5–17.

64. Mellerup, E. T., Bech, P., Sorenson, T., Fuglsang-Frederiksen, Rafaelsen, O. J. (1979) Calcium and electroconvulsive therapy in depressed patients. *Biol. Psychiatry* **14,** 711–714.

65. Rossby, S. P., Nalepa, I., Huang, M., Perrin, C., Burt, A. M., Schmidt, D. E., Gillespie, D. D., and Sulser, F. (1995) Norepinephrine-independent regulation of GRII mRNA in vivo by a tricyclic antidepressant. *Brain Res.* **687,** 79–82.

66. Nowak, G., Ordway, G. A., and Paul, I. A. (1995) Alterations in the N-methyl-D-aspartate (NMDA) receptor complex in the frontal cortex of suicide victims. *Brain Res.* **675,** 157–164.

**8**

# NMDA Receptors and Affective Disorders

## Ian A. Paul

## 1. INTRODUCTION

Recent in vivo and ex vivo findings indicate that the $N$-methyl-D-aspartate receptor complex may be a locus of antidepressant action (*see* Chapters 6 and 7). Functional antagonists of the NMDA receptor complex, including a competitive NMDA antagonist (2-amino-7-phosphonoheptanoic acid [AP-7] *(1)*, a glycine partial agonist (1-aminocyclopropanecarboxylic acid [ACPC] *(2,3)*, and a use-dependent channel antagonist (dizocilpine *(4)*, are as efficacious as tricyclic antidepressants in preclinical tests predictive of antidepressant activity *(5–8)*. Similarly, dizocilpine and the NMDA receptor antagonist CGP-37849 block the behavioral effects of two putative animal models of depression, learned helplessness and chronic mild stress-induced deficits in sucrose consumption *(9–11)*. Moreover, a chronic regimen of either ACPC or dizocilpine produces a reduction in the density (downregulation) of cortical β-adrenoceptors in mice comparable to that produced by the prototypic tricyclic antidepressant imipramine *(12)*. Likewise, Klimek and Papp *(13)* have reported that chronic dizocilpine treatment downregulates cortical β-adrenoceptors, as well as 5-HT$_2$ receptors, in rats. Thus, in both behavioral and biochemical screening procedures, antagonists at the NMDA receptor complex behave in a manner comparable to clinically active antidepressants.

Conversely, considerable evidence has accumulated indicating that chronic treatment with antidepressants results in pharmacodynamic changes in the NMDA receptor complex. Thus, chronic, but not acute, administration of imipramine and electroconvulsive shock alters the ligand-binding properties of both glycine- and glutamate-recognition sites on the NMDA receptor complex *(14–16)*. These effects are manifest as a reduction in both the potency of glycine to inhibit the binding of [$^3$H]5,7-dichlorokynurenic acid (DCKA) *(17)* to strychnine-insensitive glycine receptors and the percentage of high-affinity glycine sites inhibiting the binding of the NMDA antagonist [$^3$H]CGP-39653 *(18)*. Subsequent studies have revealed that chronic, but not acute, treatment of mice with clinically effective antidepressants (including representatives from every principal therapeutic class) produces ~1.8–4.3-fold reductions in the potency of glycine to inhibit [$^3$H]5,7-DCKA binding to strychnine-insensitive glycine receptors. These effects were


*From:* Antidepressants: *New Pharmacological Strategies*
*Edited by: P. Skolnick, Humana Press Inc., Totowa, NJ*

observed after chronic treatment with 16 of 17 (94%) antidepressant agents (*see* Chapter 7). Similar effects were not manifested by chronic treatment with compounds that typically produce false positives in either the forced-swim test (FST) or cortical β-adrenoceptor downregulation. Moreover, antidepressant-induced adaptive changes in the ligand-binding properties of the NMDA receptor complex are dose-dependent, requiring 10–21 d of treatment, and persist for 5–10 d following the cessation of treatment. These findings support the hypothesis *(14,16)* that adaptation of the NMDA receptor complex may be a prerequisite for antidepressant action, and suggest that this ligand-gated ion channel may be involved in the pathophysiology of depression.

Despite this growing body of evidence, the issue remains whether alterations in the functional characteristics of the NMDA receptor complex are related to alterations in affective behaviors. This can be resolved into two major questions. First, is dysfunction of the NMDA receptor complex related to depressive symptomatology? Second, if depressive behavior is accompanied by alterations in the characteristics of the NMDA receptor, is normalization of this behavior by antidepressant treatment accompanied by normalization of the characteristics of the NMDA receptor complex? Unfortunately, to date no peripheral marker of NMDA receptor function has been identified. As a consequence, studies of these two questions must be limited to postmortem examinations of central nervous system (CNS) NMDA receptors. Thus, the second question can only be addressed in studies of animal analogs of depression. If the NMDA receptor complex plays a critical role in the therapeutic action of antidepressants, then it would be predicted that the NMDA receptor in human major depressives will differ from that of nondepressed subjects, and that animal analogs of depression will produce antidepressant-sensitive adaptation of the NMDA receptor.

## 2. HUMAN POSTMORTEM STUDIES

We have hypothesized that glutamatergic pathways are involved in the mechanism of action of antidepressants and, by implication, in the pathophysiology of major depressive disorder. A critical test of this hypothesis is to demonstrate that this pathway and/or the NMDA receptor complex is significantly different in major depressives compared to normal controls. One method for addressing this issue is to examine the NMDA receptor complex in CNS tissue taken from these groups postmortem. Since a diagnosis of major depressive disorder can be rendered in approx 50% of suicide victims *(19,20)*, we hypothesized that dysfunction of NMDA receptors might occur in frontal cortices from human suicide victims.

We have recently published these studies, using tissue from 22 pairs of suicide victims and age- and postmortem interval-matched controls *(21)*. As shown in Table 1, in normal human frontal cortical (Brodmann Area 10) homogenates, glycine displaces the binding of the glutamate receptor antagonist [$^3$H]CGP-39653 in a biphasic manner, similar to its effects in rodent tissue *(14,18)*. In contrast, the high-affinity component of glycine displacement of [$^3$H]CGP-39653 is significantly reduced in frontal cortical homogenates from the majority of suicide victims (Table 1). Moreover, specific binding of 10 n*M* [$^3$H]CGP-39653 was significantly lower (40 ± 14%) in cortices from suicide victims relative to that from age- and postmortem interval-matched control subjects. No differences between control and suicide tissue were observed in the IC$_{50}$

**Table 1**
**Parameters of [³H]CGP-39653 Binding in Frontal Cortical Homogenates from Suicide Victims and Controls**

| Group | Specific Binding, fmol/mg protein | High affinity, % | $IC_{50 \text{ (High)}}$, n$M$ | $IC_{50 \text{ (Low)}}$, m$M$ |
|---|---|---|---|---|
| Control | 262 ± 12 | 45 ± 5 | 118 ± 53 | 0.97 ± 0.34 |
| Suicide | 219 ± 10[a] | 27 ± 6[a] | 307 ± 211 | 2.90 ± 0.94 |

Data represent the mean ± SEM of 22 age- and postmortem-interval matched subjects per group, and were obtained as described. A one-site model was assumed (% high affinity = 0), unless the sum of squares of the model was significantly reduced by employing a two-site model (F-test, InPlot 4.03, San Diego, CA).

[a]$p < 0.05$ vs control (paired, two-tailed Student's $t$-test).

for either component of glycine displacement of [³H]CGP-39653 binding. In addition, no differences between groups were observed in either specific or glycine-displaceable [³H]5,7-DCKA binding or in specific, glycine- or glutamate-stimulated [³H]dizocilpine binding.

The results from this study demonstrate that the glutamate recognition site and the allosteric regulation of this site by glycine *(22)* differ in the frontal cortex of suicide victims, compared with controls dying of natural causes. Thus, both the specific binding of the glutamate site antagonist [³H]CGP-39653 and the proportion of high-affinity, glycine-displaceable [³H]CGP-39653 binding sites were reduced in the frontal cortex of suicide victims, compared to controls. In contrast, neither the binding of the glycine recognition site antagonist [³H]5,7-DCKA nor the binding of [³H]dizocilpine was altered in frontal cortical homogenates from suicide victims, compared to controls. Moreover, the radioligand-binding data obtained in sudden-death control subjects in the present study is in good agreement with that obtained in our earlier studies with rodent tissue *(14–16)*. Thus, these data indicate that suicide is accompanied by selective alterations in the glutamatergic recognition site and its allosteric coupling to the glycine site of the NMDA receptor complex.

In the two other studies that examined the NMDA receptor complex in postmortem tissues, no significant differences in either the apparent affinity ($K_d$) of [³H]dizocilpine or receptor density ($B_{max}$) were observed between suicide victims and control subjects *(23)* or depressed suicide victims and controls *(24)*. The study of Nowak et al. *(21)*, indicating no differences in basal or stimulated [³H]dizocilpine binding between controls and suicide victims, is consistent with these reports. Holemans et al. *(24)* also reported a significant negative correlation between age and receptor density in suicide victims, but not controls. However, in the study of Nowak et al. *(21)*, we found no correlations between age, postmortem interval, or storage time, and either specific [³H]dizocilpine binding or binding in the presence of 30 µ$M$ glycine and/or glutamate analyzed, either as a pool of samples from both control and suicide victims, or analyzed separately. Since the suicide victims in the study of Nowark et al. *(21)* were undifferentiated by diagnosis, it is possible that the negative correlation with age in depressed suicide victims reported by Holemans et al. is specific to major depressives and does not obtain in other diagnostic groups committing suicide. These data lend significant support to the hypothesis that dysfunction of the NMDA receptor complex is involved in the psychopathology underlying human suicide. However, it must be recognized that major depressives comprise only about 50% of suicide victims *(25)*. Further studies using tissue from suicide victims, accompanied by psychiatric autopsy, to determine the nature of premortem psychopathology will be needed to unequivocally link NMDA receptor dysfunction to major depressive disorder. Nonetheless, these data represent the first direct demonstration that changes in the NMDA receptor complex in frontal cortex attend human behaviors linked to major depression.

## 3. ANIMAL ANALOGS OF MAJOR DEPRESSION

### 3.1. Forced-Swim Stress

Exposure to stress has been reported to increase release of the excitatory amino acid neurotransmitters, glycine and glutamate *(26,27)*. Exposure to uncontrollable stress has also been reported to disrupt the development of long-term potentiation, a neurophysiological phenomenon that is critically dependent on the activation of the NMDA subtype of excitatory amino acid receptor *(28)*. Previous studies have demonstrated that NMDA receptor antagonists are active in the forced-swim test, a preclinical behavioral paradigm based on the presentation of uncontrollable stress, which is sensitive to antidepressant treatments *(5–7)*.

The forced-swim test results in an increase in the specific binding of [$^3$H]dihydroalprenolol to cortical β-adrenoceptors that is evident within 24 h of the last swim *(29a)*. Moreover, this effect is evident regardless of whether behavioral adaptation is observed in the test. Similarly, Trullas et al. have demonstrated that swim-stress results in rapid alterations in [$^{35}$S]t-butylbicyclophosphorothionate binding to the cortical γ-aminobutyric acid (GABA)-gated chloride ionophore. Thus, forced-swim stress can result in rapid adaptation of cortical β-adrenergic and GABA$_A$ receptors. Based on these observations, we hypothesized that exposure to the forced-swim test would also result in rapid adaptation of the NMDA receptor complex. We have recently demonstrated that this paradigm can alter the radioligand-binding properties of the NMDA receptor complex *(29)*. In this study, we exposed male Sprague-Dawley rats to a 15-min, ambient temperature conditioning swim followed on the next day by a 5 min test swim, utilizing methods previously described *(29a,30)*. Twenty-four hours after the test-swim exposure, all rats were decapitated and the cortices dissected over ice. The tissue was then processed as previously described to assay glycine-displacement of [$^3$H]5,7-DCKA binding *(14,16)*.

In these experiments, we observed that exposure to two successive daily forced swims resulted in a 56% reduction in the IC$_{50}$ for glycine displacement of [$^3$H]5,7-DCKA (Fig. 1). These data demonstrated that exposure of rats to the forced-swim test paradigm results in a robust increase in the potency of glycine to displace [$^3$H]5,7-DCKA from the strychnine-insensitive glycine recognition site of the NMDA receptor complex in cortical homogenates *(29)*. In contrast, no effects of swim stress were observed on the specific binding of [$^3$H]5,7-DCKA in cortical homogenates. This increase in the potency of glycine following exposure to forced swim is consistent with the hypothesis that this stressor can produce rapid adaptation of the NMDA receptor complex.

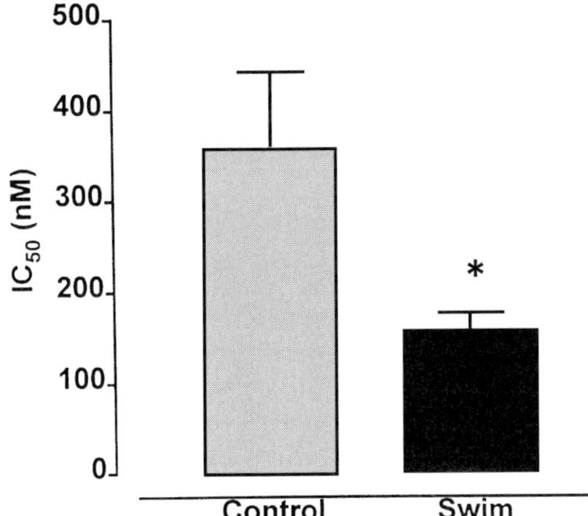

**Fig. 1.** Acute exposure to forced swimming increases the potency of glycine to displace [3H]5,7-DCKA. Data represent the mean ± SEM of 9 rats/group. *$p < 0.05$ vs control, one-tailed *t*-test.

## 3.2. Chronic Mild Stress-Induced Deficits in Sucrose Drinking

We have previously demonstrated that chronic administration of antidepressants results in reductions in the potency of glycine to displace [3H]5,7-DCKA binding to frontal cortical homogenates, as well as reductions in the proportion of high-affinity, glycine-displaceable [3H]CGP-39653 binding to the glutamate recognition site of the NMDA receptor complex *(14–16; see* Chapter 7). Since these treatments were administered to normal rodents, it is not known whether similar effects will obtain in animal analogs of depression, such as a chronic mild stress (CMS) paradigm (*see* Chapter 12) or olfactory bulbectomy (*see* Chapter 9). However, it might be predicted that such analogs should either result in no change in the NMDA receptor complex or in adaptation that opposes those induced by chronic antidepressant administration. Inasmuch as the forced-swim test was designed as a preclinical screen for antidepressant activity *(30)*, the increased potency of glycine to displace [3H]5,7-DCKA binding, observed in cortical homogenates from rats exposed to this paradigm, is consistent with this prediction.

When subjected to a chronic, variable regimen of mild stressors, rats display a progressive reduction in responsiveness to rewarding stimuli, such as the consumption of sweetened solutions *(31;* and *see* Fig. 2), place-preference conditioning *(32–34)*, and intracranial self-stimulation *(35)*. This reduction has been termed "anhedonia" and is reversible by chronic, but not acute, treatment with a variety of antidepressants (reviewed in ref. *36; see* Chapter 12). Papp and Moryl *(9,10)* have demonstrated that, like clinically active antidepressants, chronic administration of antagonists of the glutamate recognition site of the NMDA receptor complex also reverses CMS-induced reductions in sucrose solution consumption. The activity of NMDA receptor antagonists in this model suggested to us that CMS might induce antidepressant-reversible adaptation of the NMDA receptor complex.

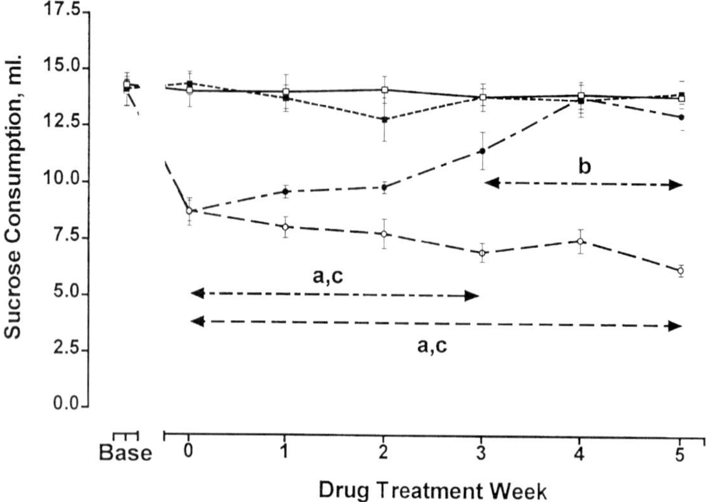


**Fig. 2.** Chronic imipramine treatment reverses chronic mild stress-induced reductions in sucrose drinking. Rats were treated as described. Data represent the mean ± SEM of 10 rats/group. Filled symbols = imipramine-treated. Squares = Sham-treated. Circles = CMS-treated. Base = Baseline sucrose consumption. (a) $p < 0.01$ vs baseline sucrose consumption (Univariate F-test of group contrasts). (b) $p < 0.01$ vs vehicle controls (Univariate F-test). (c) $p < 0.01$ vs sham controls (Univariate F-test).

In preliminary experiments, male Wistar rats were subjected to either sham or CMS treatment, as described *(10)*, for 7 wk. After the third week of CMS treatment, animals were divided into groups matched for sucrose consumption, and were treated with either 10 mg/kg/d imipramine or vehicle (ip) for 4 wk. Twenty-four h after the last treatment, animals were decapitated and cortical samples dissected, as described in Nowak et al. *(14)*. Glycine displacement of [$^3$H]5,7-DCKA binding was assessed using eight concentrations of glycine ($10^{-9}$–$10^{-3}M$) *(14,17)*. Data were analyzed for IC$_{50}$ values using an iterative curve-fitting routine (InPlot 4.03), and were subsequently analyzed using a two-factor analysis of variance, followed by Fisher's LSD test.

As previously reported, CMS treatment over a period of 3 wk reduced sucrose drinking in male Wistar rats (Fig. 2). This reduction was reversed following co-treatment with imipramine (10 mg/kg/d ip) for 4 wk. In cortical samples, chronic imipramine treatment resulted in a 109 ± 19% increase in the IC$_{50}$ of glycine to displace [$^3$H]5,7-dichlorokynurenic acid (DCKA) binding to neocortical homogenates (Fig. 3). In contrast, CMS treatment resulted in a 45 ± 13% reduction in the IC$_{50}$ of glycine to displace [$^3$H]5,7-DCKA binding (Fig. 3). Moreover, imipramine treatment (10 mg/kg ip) reversed the effects of CMS to decrease the IC$_{50}$ of glycine, returning the values to control levels (93 ± 19%; *see* Fig. 3). In contrast to our previous results in cortical tissue from Sprague-Dawley rats, all three treatments (imipramine alone, CMS alone, and imipramine + CMS) resulted in 30–41% decreases in the specific binding of [$^3$H]5,7-DCKA (Fig. 4). These data are consistent with our previous observations that some antidepressant treatments can reduce specific [$^3$H]5,7-DCKA binding *(15,16)*, as well as our more recent unpublished findings, which suggest that most, if not all, antidepres-

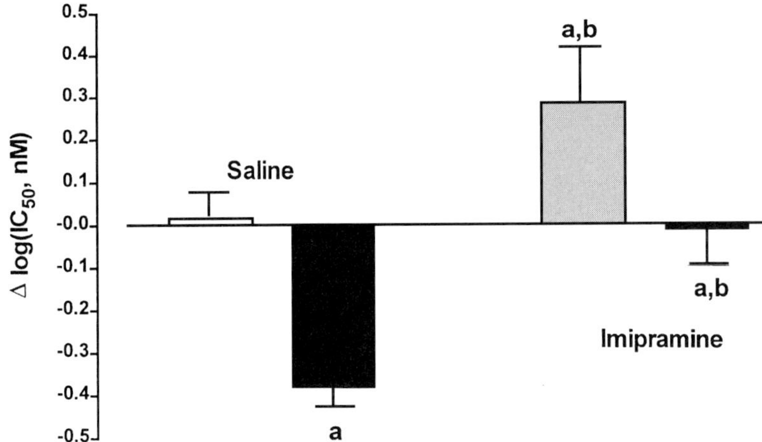


**Fig. 3.** Chronic mild stress increases the potency of glycine to displace [$^3$H]5,7-DCKA in an imipramine-reversible manner. Rats were treated as described. Data are the mean ± SEM of the difference in log[IC$_{50}$] relative to sham/vehicle controls of 7–10 subjects/group. Open bars = sham-treated. Filled bars = CMS-treated. (a) $p < 0.05$ vs vehicle-treated. (b) $p < 0.05$ vs sham-treated, Fisher's LSD test.

sants can produce this effect is administered for sufficient time. Specific binding at a single ligand concentration is a composite measure reflecting both the affinity of [$^3$H]5,7-DCKA for the receptor and the density of [$^3$H]5,7-DCKA binding sites. Consequently, these data must be interpreted cautiously, since these treatments may reduce specific binding by divergent mechanisms. For example, the reduction in specific binding produced by long-term imipramine treatment in these rats may result from a reduction in the density of [$^3$H]5,7-DCKA binding sites similar to that previously observed with electroconvulsive shock *(15)*. In contrast, the reduction in specific binding produced by CMS treatment might be the result of a reduction in receptor affinity. However, the fact that CMS produced an effect on specific binding comparable to that of imipramine may not fit the simplest version of the above working hypothesis.

These data demonstrated that the inhibition of sucrose consumption in Wistar rats by exposure to CMS is accompanied by an increase in the potency of glycine to displace [$^3$H]5,7-DCKA from cortical homogenates comparable to that previously observed in rats processed in the forced-swim procedure. Likewise, as previously demonstrated, chronic treatment with imipramine (10 mg/kg/d) reduced the potency of glycine to displace [$^3$H]5,7-DCKA. Moreover, the potency of glycine to displace [$^3$H]5,7-DCKA in rats exposed to CMS and imipramine co-treatment did not differ from that of unstressed, vehicle-treated controls. In contrast, CMS exposure did not significantly affect either glycine-displaceable [$^3$H]CGP-39653 binding or basal-, glycine-, or glutamate-stimulated [$^3$H]dizocilpine binding in these rats (data not shown). Chronic imipramine treatment, either alone or combined with CMS, also did not affect these recognition sites on the NMDA receptor complex. The dose of imipramine used in this study was lower than previously employed, and we have shown the effects of imipramine (as well as citalopram) on radioligand binding to NMDA receptors are dose-dependent. In addition, this may reflect a strain-difference between Wistar and

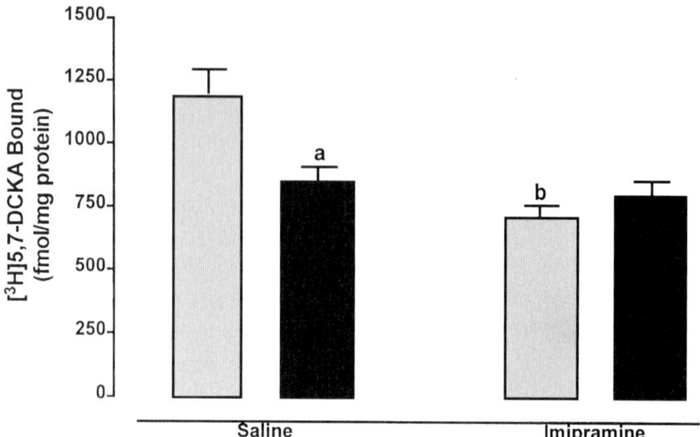

**Fig. 4.** Both chronic mild stress and imipramine treatment reduce the specific binding of [³H]5,7-DCKA (20 n*M*). Rats were treated as described for Fig. 3. Data are presented as the mean ± SEM of 7–10 subjects/group. Open bars = sham-treated. Filled bars = CMS-treated. (a) $p < 0.05$ vs vehicle-treated. (b) $p < 0.05$ vs sham-treated, Fisher's LSD test.

Sprague-Dawley rats. The contrast in effect between these measures underscores the lack of stoichiometry between them. This lack of stoichiometry was also observed following chronic antidepressant treatments, most notably with citalopram. Thus, chronic citalopram treatment produces only a modest increase in the $IC_{50}$ of glycine to inhibit [³H]5,7-DCKA binding, while essentially abolishing the high-affinity component of glycine-displaceable [³H]CGP-39653 binding (*see* Chapter 7). In sum, the data from these experiments suggests that a behavioral stressor can selectively and differentially affect recognition sites of the NMDA receptor complex.

### 3.3. Olfactory Bulbectomy

The olfactory bulbectomized rat analog of depression possesses significant predictive validity for detecting antidepressants after chronic, but not acute, drug administration (reviewed in refs. *37* and *38*). In addition, this analog possesses significant face validity as a model of agitated forms of major depressive disorder *(38,39)*. Bulb ablation has also been found to affect excitatory amino acid transmitters in CNS. In the olfactory cortex, Sandberg et al. *(40)* have reported decreased levels of aspartate and glutamate following olfactory bulbectomy (OB), whereas Harvey et al. *(41)* reported increased levels of glycine. Notwithstanding its limited theoretical validity *(42)*, we have employed this analog to compare with our results, using the CMS model. Moreover, examination of this analog has permitted us to determine whether an analog of agitated depression differs from one of the anhedonic features of depression in its effects on the NMDA receptor complex.

Male Sprague-Dawley rats (175–200 g on arrival, 6–7 wk old) were housed in groups of 4/cage under standard laboratory conditions, as described above for the CMS paradigm. After acclimating to the colony for 2 wk, animals were anesthetized and a midline sagittal incision of the scalp and periostium was then made at least 1 cm rostral to bregman. Two burr holes (2 mm diameter) were made through the

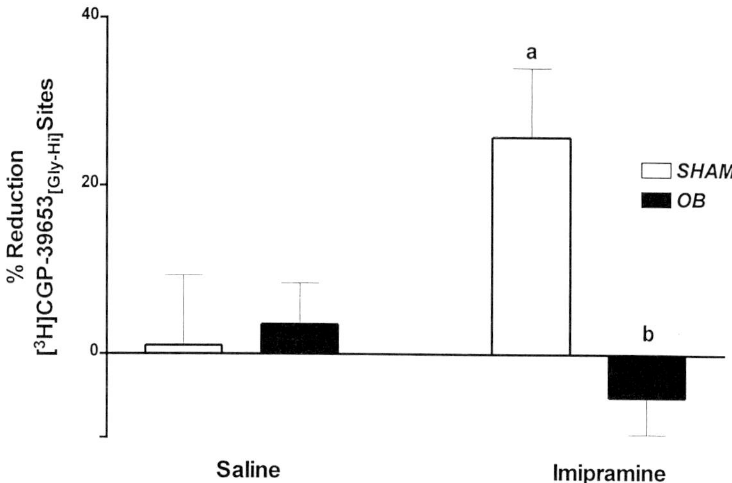



**Fig. 5.** Olfactory bulbectomy blocks the effects of chronic imipramine treatment on the proportion of high-affinity, glycine-displaceable [³H]CGP-39653 binding sites. Rats were treated as described. Data are presented as the mean ± SEM of 8–11 subjects per group. Open bars = vehicle-treated. Filled bars = imipramine-treated. (a) $p < 0.05$ vs vehicle-treated. (b) $p < 0.05$ vs sham-treated, Fisher's LSD test.

skull 5 mm rostral to bregma and 2 mm lateral to the midsagittal suture. Olfactory bulbectomy was performed by carefully aspirating the olfactory bulbs using a 1.9-mm cannula connected to the house vacuum system, bleeding controlled using Gelfoam plugs (Upjohn, Kalamazoo, MI), and the wound closed. For sham-treated animals, the dura was penetrated without disturbing underlying vasculature or forebrain. After a 2-wk recovery period, animals were randomly assigned to groups of 8–10 and treated with imipramine (15 mg/kg) or vehicle for 28 d. On d 21 of drug treatment (6 wk postsurgery), locomotor activity was assessed for all animals for 10 min, using a $100 \times 50$ cm testing arena equipped with an overhead-horizontal tracking system (Videomex-V, Columbus Instruments, Columbus, OH). Twenty-four hours after the last drug treatment, all animals were sacrificed and the tissue dissected for frontal cortex and frozen until assayed. The tissue was then assayed for glycine displacement of [³H]5,7-DCKA and [³H]CGP-39653 binding, as described previously *(14)*.

As previously reported, OB increased locomotor activity in rats and this effect was antagonized by imipramine cotreatment (data not shown). Bulbectomy alone was without effect on glycine displacement of either [³H]5,7-DCKA (Table 2) or [³H]CGP-39653 binding (Fig. 5). However, OB completely antagonized the imipramine-induced reduction in the proportion of high-affinity, glycine-displaceable [³H]CGP-39653 binding sites (Fig. 5). In contrast, OB did not alter the imipramine-induced reduction in the potency of glycine to displace [³H]5,7-DCKA (Table 2).

These data demonstrate that, like CMS, OB can antagonize the effects of imipramine on radioligand binding to the NMDA receptor complex. However, in contrast to CMS, bulbectomy itself does not appear to significantly affect either the glycine or glutamate recognition sites of this receptor. Moreover, the ability of OB to antagonize

Table 2
**Effect of Imipramine Treatment in Olfactory Bulbectomized Rats on Glycine Displacement of [3H]5,7-DCKA Binding**

| Treatment | Specific binding, fmol/mg protein | $\log(IC_{50})$, $M$ |
|---|---|---|
| Sham + Saline | $439 \pm 96$ | $-6.74 \pm 0.12$ |
| Sham + Imipramine | $578 \pm 65$ | $-6.33 \pm 0.12^{a}$ |
| OB + Saline | $502 \pm 65$ | $-6.12 \pm 0.07^{a}$ |
| OB + Imipramine | $530 \pm 69$ | $-6.29 \pm 0.12^{a}$ |

Data represent the mean ± SEM of 5–8 animals per group. Analysis of variance revealed a significant main effect of olfactory bulbectomy on $\log(IC_{50})$ in frontal cortex (F[1,26] = 9.709, $p$ = 0.004), as well as an interaction between olfactory bulbectomy and imipramine treatment (F[1,26] = 7.375, $p$ = 0.01).

[a]$p < 0.05$ vs sham + saline (Fisher's LSD).

imipramine's effects on the NMDA receptor appears to be confined to the allosteric interaction between the glycine- and glutamate-recognition sites. These data demonstrate that animal models of depression affect the NMDA receptor complex and/or interact with antidepressant-induced adaptation of the complex. However, the differences observed between the two models in the specific site and nature of the interaction indicate that the glycine and glutamate recognition sites of the NMDA receptor complex are independently regulated. These differences are not entirely surprising in view of the divergence of the methods employed in generating these two models. It remains to be determined whether these differences between the CMS and OB models relate to differences in the depressive symptomatology that they model.

## 4. DISCUSSION

An increase in the potency of glycine to displace frontal cortical [3H]5,7-DCKA binding in chronic mild stress-treated rats is consistent with predictions based on the antidepressant-induced reduction in the potency of glycine to displace [3H]5,7-DCKA binding *(14–16)*. Since chronic antidepressant treatments may also reduce specific [3H]5,7-DCKA binding *(16)*, it might also be predicted that the CMS model would increase specific binding of [3H]5,7-DCKA. However, in contrast to our predictions, CMS reduces the specific binding of [3H]5,7-DCKA to the same extent as imipramine, and has no effect on the displacement of [3H]CGP-39653 by glycine.

Similarly, based on our earlier studies with chronic antidepressant administration to naïve rodents, it might be predicted that suicide victims would display an increased proportion of high-affinity, glycine-displaceable [3H]CGP-39653 binding sites, compared to normal sudden-death controls. Moreover, our previous studies would suggest that the potency of glycine to displace [3H]5,7-DCKA binding should have been increased in suicide victims, compared to normal controls. In fact, we observed a reduction in the proportion of high-affinity, glycine-displaceable [3H]CGP-39653 binding sites in suicide victims, and no change in that of the potency of glycine to displace [3H]5,7-DCKA binding. Together with our data from the CMS model of depression in rats, these data suggest that the relationship of NMDA receptor activity to depression and/or suicide is a complex phenomena, likely to involve multiple regulatory sites and neurotransmitter systems.

The functional significance of both the observations in suicide victims and our previous observations of antidepressant-induced adaptation of the NMDA receptor complex are unclear. Several interpretations are possible, including a shift in the equilibrium between agonist and antagonist preferring conformations or populations of NMDA receptors, alterations in subunit expression, and posttranslational modifications of the NMDA receptor complex. These possibilities are currently under investigation. Certainly, the similarities between our results in human suicide victims and our previous results in rodents following chronic antidepressant administration are puzzling. If, as we have argued, adaptation of the NMDA receptor complex is critical to the action of antidepressants and is involved in the pathophysiology of depression, it would seem unlikely that both antidepressant treatments and the psychopathological states resulting in suicide would result in identical effects on the NMDA receptor complex. Therefore, either the NMDA receptor complex is unrelated to the pathophysiology of depression and antidepressant actions, or subtle differences between the effects observed following antidepressant treatments and the behavior of the complex in suicide victims must result in significant differences in the function of the NMDA receptor complex. In this light, it is worth noting that significant differences were observed between the recent studies of suicide victims and animal analogs of major depression and our previous studies with antidepressant treatments in naïve rodents. For example, in the study of human suicide victims, we observed that both the specific binding of [$^3$H]CGP-39653 and the proportion of high-affinity, glycine-displaceable [$^3$H]CGP-39653 sites were reduced in the frontal cortex of suicide victims, compared to control subjects. However, in contrast to the results obtained in antidepressant-treated rodents, we observed no differences between suicide victims and control subjects in either the potency of glycine to displace [$^3$H]5,7-DCKA or in basal nonequilibrium [$^3$H]dizocilpine binding. Likewise, although CMS exerts profound effects on the potency of glycine to displace [$^3$H]5,7-DCKA, it is without effect on glycine-displaceable [$^3$H]CGP-39653 binding. Conversely, OB itself has no effect on glycine-displaceable [$^3$H]CGP-39653 binding, but it blocks the effects of chronic imipramine treatment on this site. Moreover, the effects of OB on glycine-displacement of [$^3$H]5,7-DCKA parallel those of chronic imipramine treatment.

Thus, the behavioral, pharmacological, and, now, human postmortem data accumulated to date would argue that, although similarities exist between the behavior of the NMDA receptor complex in suicide victims and in animal analogs of major depressive symptoms and following antidepressant treatments, significant differences are also observed in the behavior of this complex, which may result in profound differences in NMDA receptor function. However, until the functional implications of these radioligand-binding differences are ascertained, it remains unclear how or if the radioligand-binding characteristics observed in rodent analogs of major depressive symptoms and human suicide victims are related.

Regardless of the functional interpretation, these data lend significant support to the hypothesis that disturbance of the NMDA receptor complex accompanies the symptoms of major depressive disorder in both rodents and humans. Moreover, these data represent the first direct demonstration that changes in the NMDA receptor complex in frontal cortex attend behaviors linked to major depression. The present data, coupled with previous reports that functional NMDA antagonists are active in the forced-swim test *(5–8)*, learned helplessness and CMS models of depression *(9,11)*, and reduce cortical β adrenoceptor density *(12,13)*, suggest the NMDA receptor complex may be a common pathway for antidepressant action and depressive symptomatology. Thus, these observations provide both new avenues for the identification and development of novel antidepressant therapies and a novel heuristic model for the study of depressive disorders.

## REFERENCES


1. Watkins, J. C. and Olverman, H. J. (1987) Agonists and antagonists for excitatory amino acid receptors. *Trends. Neurosci.* **10**, 265–272.
2. Marvizon, J. C. G. and Skolnick, P. (1988) [$^3$H]glycine binding is modulated by $Mg^{+2}$ and other ligands of the. *Eur. J. Pharmacol.* **151**, 157–158
3. Watson, G. and Lanthorn, T. H. (1990) Pharmacological characteristics of cyclic homologues of glycine at the. *Neuropharmacology* **29**, 727–730.
4. Wong, E. H. F., Kemp, J. A., Priestley, T., Knight, A. R., Woodruff, G. N., and Iversen, L. L. (1986) The anticonvulsant MK-801 is a potent N-methyl-D-aspartate antagonist. *Proc. Natl. Acad. Sci.* USA **83**, 7104–7108.
5. Trullas, R. and Skolnick, P. (1990) Functional antagonists at the NMDA receptor complex exhibit antidepressant actions. *Eur. J. Pharmacol.* **185**, 1–10.
6. Trullas, R., Folio, T., Young, A., Miller, R., Boje, K. and Skolnick, P. (1991) 1-aminocyclopropanecarboxylates exhibit antidepressant and anxiolytic actions in animal models. *Eur. J. Pharmacol.* **203**, 379–385.
7. Maj, J., Rogóz, Z., Skuza, G., and Sowinska, H. (1992) The effects of MK-801 and antidepressant drugs in the forced swimming test in rats. *Eur. J. Neuropsychopharmacol.* **2**, 37–41.
8. Skolnick, P., Miller, R., Young, A., Boje, K., and Trullas, R. (1992) Chronic treatment with 1-aminocyclopropanecarboxylic acid desensitizes behavioral responses to compounds acting at the N-methyl-D-aspartate receptor complex. *Psychopharmacology* (Berlin) **107**, 489–496.
9. Papp, M. and Moryl, E. (1993) Similar effect of chronic treatment with imipramine and the NMDA antagonists CGP-37849 and MK-801 in a chronic mild stress model of depression in rats. *Eur. J. Neuropsychopharmacol.* **3**, 348–349.
10. Papp, M. and Moryl, E. (1994) Antidepressant activity of non-competitive and competitive NMDA antagonists in a chronic mild stress model of depression. *Eur. J. Pharmacol.*, submitted.
11. Meloni, D., Gambarana, C., Graziella de Montis, M., Dal Prá, T., and Tagliamonte, A. (1993) Dizocilpine antagonizes the effect of chronic imipramine on learned helplessness in rats. *Pharmacol. Bioc. Behav.* **46**, 423–426.
12. Paul, I. A., Trullas, R., Skolnick, P. and Nowak, G. (1992) Down-regulation of cortical beta-adrenoceptors by chronic treatment with functional NMDA antagonists. *Psychopharmacology* (Berlin) **106**, 285–287.
13. Klimek, V. and Papp, M. (1994) The effect of MK-801 and imipramine on beta-adrenergic and 5-HT$_2$ receptors in the chronic mild stress model of depression in rats. *Pol. J. Pharmacol.* **46**, 67–69.
14. Nowak, G., Trullas, R., Layer, R. T., Skolnick, P., and Paul, I. A. (1993) Adaptive changes in the N-methyl-D-aspartate receptor complex after chronic treatment with imipramine and 1-aminocyclopropanecarboxylic acid. *J. Pharmacol. Exp. Ther.* **265**, 1380–1386.

15. Paul, I. A., Layer, R. T., Skolnick, P., and Nowak, G. (1993) Adaptation of the NMDA receptor in rat cortex following chronic electroconvulsive shock or imipramine. *Eur. J. Pharmacol.-Mol. Pharm.* **247,** 305–311.

16. Paul, I. A., Nowak, G., Layer, R. T., Popik, P. and Skolnick, P. (1994) Adaptation of the N-methyl-D-aspartate receptor complex following chronic antidepressant treatments. *J. Pharmacol. Exp. Ther.* **269,** 95–102.

17. Baron, B. M., Siegel, B. W., Sloan, A. L., Harrison, B. L., Palfreyman, M. G., and Hurt, S. D. (1991) [$^3$H]5,7-Diclorokynurenic acid, a novel radioligand labels NMDA receptor. *Eur. J. Pharmacol.* **206,** 149–154.

18. Sills, M. G., Fagg, G., Pozza, M., Angst, C., Brundish, D., Hurt, S. D., Wilusz, E. J., and Williams, M. (1991) [$^3$H]CGP 39653; a new N-methyl-D-aspartate antagonist radioligand with low nanomolar affinity in rat brain. *Eur. J. Pharmacol.* **192,** 19–24.

19. Baraclough, B., Bunch, J., Nelson, B., and Sainsbury, P. (1974) A hundred cases of suicide: clinical aspects, *Br. J. Psychiatry* **125,** 355–373.

20. Rich, C. L., Young, D., and Fowler, R. C. (1986) San Diego suicide study. I. Young vs old subjects, *Arch. Gen. Psychiatry* **43,** 577–582.

21. Nowak, G., Ordway, G. A., and Paul, I. A. (1995) Alterations in the N-methyl-D-aspartate (NMDA) receptor complex in the frontal cortex of suicide victims. *Brain Res.* **675,** 157–164.

22. Mugnaini, M., Giberti, A., Ratti, E., and van Amsterdam, F. T. M. (1993) Allosteric modulation of [$^3$H]CGP-39653 binding by glycine in rat brain. *J. Neurochem.* **61,** 1492–1497.

23. Palmer, A. M., Burns, M. A., Arango, V., and Mann, J. J. (1994) Similar effects of glycine, zinc and an oxidizing agent on [$^3$H]dizocilpine binding to the N-methyl-D-aspartate receptor in neocortical tissue from suicide victims and controls. *J. Neural Transm.-Gen. Sect.* **96,** 1–8.

24. Holemans, S., De Paermentier, F., Horton, R. W., Crompton, M. R., Katona, C. L. E., and Maloteaux, J.-M. (1993) NMDA glutamatergic receptors, labelled with [$^3$H]MK-801, in brain samples from drug-free depressed suicides. *Brain Res.* **616,** 138–143.

25. Arato, M., Demeter, E., Rihmer, Z., and Somogyi, E., (1988) Retrospective psychiatric assessment of 200 suicides in Budapest. *Acta Psychiatry Scand.* **77,** 454–456.

26. Gilad, G. M., Gilad, V. H., Wyatt, R. J., and Tizabi, Y. (1990) Region-selective stress-induced increase of glutamate uptake and release in rat. *Brain Res.* **525,** 335–338.

27. Moghaddam, B. (1993) Stress preferentially increases extraneuronal levels of excitatory amino acids in the prefrontal cortex: comparison to hippocampus and basal ganglia. *J. Neurochem.* **60,** 1650–1657.

28. Shors T., Seib T., Levine S., and Thompson R. (1989) Inescapable versus escapable shock modulates long-term potentiation in the rat. *Science* **244,** 224–226.

29. Nowak, G., Redmond, A., McNamara, M., and Paul, I. A. (1995) Swim stress increases the potency of glycine at the N-methyl-D-aspartate receptor complex. *J. Neurochem.* **64,** 925–927.

29a. Paul, I. A., Duncan, G. E., Mueller, R. A., Hong, J.-S., and Breese, G. R. (1990) Assessment of the effects of time course, stress, rat strain, and uptake blockade on the regionally selective rapid adaptation of beta-adrenergic and 5HT$_2$ receptors in imipramine-treated rats processed in the forced swim test. *J. Pharmacol. Exp. Ther.* **252,** 997–1005.

30. Porsolt R., Anton G., Blavet N., and Jalfre M. (1978) Behavioral despair in rats: a new model sensitive to antidepressant treatments. *Eur. J. Pharmacol.* **47,** 379–391.

31. Willner, P., Towell, A., Sampson, D., Sophokleous, S., and Muscat, R. (1987) Reduction of sucrose preference by chronic unpredictable mild stress, and its restoration by a tricyclic antidepressant. *Psychopharmacology* (Berlin) **93,** 358–364.

32. Papp, M., Willner, P. and Muscat, R. (1991) An animal model of anhedonia: attenuation of sucrose consumption and place preference conditioning by chronic unpredictable mild stress. *Psychopharmacology* **104,** 255–229.
33. Muscat, R., Papp, M., and Willner, P. (1994) Reversal of stress-induced anhedonia by the atypical antidepressants, fluoxetine and maprotiline. *Psychopharmacology* **109,** 433–438.
34. Papp, M., Lappas, S., Muscat, R., and Willner, P. (1992) Attenuation of place preference conditioning but not place aversion conditioning by chronic mild stress. *J. Psychopharmacol.* **6,** 352–356.
35. Moreau, J.-L., Jenck, F., Martin, J. R., Mortas, P., and Haefely, W. E. (1992) Antidepressant treatment prevents chronic unpredictable mild stress-induced anhedonia as assessed by ventral tegmental self-stimulation in rats. *Eur. J. Neuropsychopharmacol.* **2,** 43–49.
36. Willner, P., Muscat, R., and Papp, M. (1992) Chronic mild stress-induced anhedonia: a realistic animal model of depression. *Neurosci. Biobehav. Rev.* **16,** 525–534.
37. Leonard, B. E. and Tuite, M. (1981) Anatomical, physiological and behavioural aspects of olfactory bulbectomy in the rat. *Rev. Neurobiol.* **22,** 251–286.
38. Jancsar, S. and Leonard, B. E. (1983) The olfactory bulbectomized rat as a model of depression, in: *Frontiers in Neuropsychiatric Research* (Usdin, E., Goldstein, M., Friedhoff, A. J. and Georgotas, A., eds.). Macmillan, New York, pp. 357–372.
39. Lumia, A. R., Teicher, M. H., Salchli, F., Ayers, E., and Possidente, B. (1992) Olfactory bulbectomy as a model of hyposerotonergic depression. *Brain Res.* **587,** 181–185.
40. Sandberg, M., Bradford, H. F., and Richards C. D. (1984) Effects of lesions of the olfactory bulb on the levels of amino acids and related enzymes in the olfactory cortex of guinea pig. *J. Neurochem.* **43,** 276–279.
41. Harvey, J. A., Scholfield, C. N., Graham, L. T., and Aprison, M. H., (1975) Putative transmitters in denervated olfactory cortex. *J. Neurochem.* **24,** 445–449.
42. Willner, P. (1984) The Validity of animal models of depression. *Psychopharmacology* **83,** 1–16.

# 9

# The Potential Contribution of Sigma Receptors to Antidepressant Actions

Brian E. Leonard

## 1. INTRODUCTION

The origin of the concept of the σ-receptor is closely linked with the unraveling of the complexity of the opiate receptor. The potent effects of the naturally occurring opiates have been recorded since antiquity, but it was only in the late nineteenth century that the scientific basis for action of the morphine-like opioids was laid. There were several reasons for this development, but undoubtedly the need to produce effective antitussive agents to facilitate the healing of tubercular pulmonary lesions motivated such changes. In addition, the increase in opiate abuse, particulary in the United States, emphasized the need to understand the pharmacological basis of opiate abuse and thereby provide a rational basis for its treatment. An outcome from such studies was the synthesis of several opiate receptor antagonists. One of the first of these was nalorphine, which Lasagna and Beecher *(1)* demonstrated to have analgesic activity, but which would also precipitate abstinence symptoms in morphine-dependent subjects. Further development of nalorphine as an analgesic was prevented by the dysphoric effects it caused in many subjects. The search for drugs that combined the potent analgesic effect of morphine with a negligible abuse potential eventually led to the synthesis of pentazocine and cyclazocine *(2)*. Again, it was found that the analgesic effects of these drugs were combined with dysphoric effects that were dose-related, the morphine-like effects being most pronounced at low doses and the dysphoric-hallucinogenic effects at high doses. Furthermore, it was found that pentazocine, like nalophine, would precipitate withdrawal in morphine-dependent subjects.

A major advance in differentiating the hallucinogenic effects of pentazocine and related opioids from their analgesic effects came with the studies of Martin and co-workers *(3)* on the effects of various opioids on dogs. The benzomorphan *N*-allyl normetazocine (SKF-10047) was found to cause a characteristic canine delirium that was the equivalent of hallucinations found in man. Further studies on the enantiomers of SKF-10047 showed that the (–) enantiomer bound predominantly to the μ and κ

*From:* Antidepressants: *New Pharmacological Strategies*
*Edited by: P. Skolnick, Humana Press Inc., Totowa, NJ*

opioid receptors *(4)*, but the (+) enantiomer bound to the same receptor as that occupied by the potent hallucinogen, phencyclidine (PCP) *(5)*. The question arose regarding the explanation of the hallunicogenic effects of SKF-10047 in terms of its affinity for the PCP receptor. Quirion and coworkers *(6)* showed that SKF-10047 not only bound to PCP receptors, but also to a unique class of receptors termed sigma receptors. The use of nonselective ligands to characterize sigma receptors ultimately resulted in over a decade of confusing and conflicting data, which slowed progress in understanding the importance of this receptor.

Considerable progress has, however, been made in recent years, with the discovery of novel ligands with high affinity and selectivity for the sigma receptors. Furthermore, sigma receptors have been implicated in several behavioral and biochemical effects, which should enable a functional role for these receptors to be delineated. For example, it has been shown that the sigma receptors are involved in the motor side effects of neuroleptics *(7)*, have been implicated in psychotic disorders *(8)*, and may play a role in neuroprotective mechanisms *(9)*. This subject has been extensively reviewed by Walker et al. *(10)*.

Based on specific ligand and biochemical studies, Quirion et al. *(11)* have classified the σ-receptors into two main types, σ-1 and σ-2. Currently, the affinities of specific ligands are used to identify these receptor subtypes. For example, the benzomorphan (+) pentazocine is highly selective for the σ-1 subtype; ditolyl guanidine (DTG) is more selective for the σ-2 receptor.

It would appear that the σ-receptors are unique among receptors found in the brain with respect to the high affinities of a very diverse range of naturally occurring compounds, including steroids and neuropeptides, such as neuropeptides Y and PYY. Which, if any, of these compounds act as endogenous ligands for the σ-receptors is presently uncertain.

The σ-receptors are also unique in that drugs from different therapeutic groups bind to these sites with high affinities. For example, the neuroleptics haloperidol and chlorpromazine bind to σ-receptors, but this is not a common property of most neuroleptics *(12)*. Similarly, cytochrome P450 inhibitors, such as SKF-525A (2-(diethyllamino)-ethyl-,-2-diphenyl valerate) and lobeline, are potent inhibitors of σ-receptors, but this is not a common property of all cytochrome P450 inhibitors *(13)*. In addition to the neuroleptics, other psychotropic drugs that bind to σ-receptors include some antidepressants *(14,15)*, including the monoamine oxidase inhibitors *(16)*. Some antihistamines have also been shown to bind with a high affinity to these receptors *(17)*.

It is possible that, because such diverse chemical structures possess high affinities for σ-receptors, it has been difficult to obtain highly selective ligands. However, several ligands showing a high selective affinity for σ-receptors without having an affinity for common neurotransmitter receptors have been synthesized. Some of these are shown in Table 1.

Regarding the subcellular distribution of σ-receptors, it has been shown that these receptors are twice as enriched in the microsomal fraction than in the synaptosomal fraction of brain *(18)*. These receptors appear to be particularly enriched in the plasma membranes, which suggests that they reside in nonsynaptic regions of these membranes. Table 2 summarizes the main properties of the σ-1 and σ-2 receptors.

**Table 1**
**Selective Ligands for Sigma Receptors**

| Compound | σ-receptor affinity, n$M$ | Ref. |
| --- | --- | --- |
| BD 737 | 1.3 | (73) |
| DUP 734 | 10.0 | (74) |
| NPC 16377 | 36.0 | (75) |
| JO 1784 | 39.0 | (76) |
| PRE-084 | 44.0 | (77) |
| Remoxipride | 60.0 | (78) |
| BMY-14802 | 74.0 | (78) |
| Rimcazole | 1460 | (78) |

None of these compounds have any affinity for the following receptors at concentrations less than 1500 n$M$: PCP, $D_2$, 5-$HT_{1A}$, 5-$HT_2$, α- and β-adrenergic, or muscarinic receptors. Remoxipride is an exception, in that it has an affinity of 950 n$M$ for Dopamine receptors.

## 2. ENDOGENOUS LIGANDS FOR DOPAMINE RECEPTORS

There have been several preliminary reports of the isolation of endogenous compounds that inhibit the binding of selective ligands to σ sites. The first of these were two substances with mol wt of approx 4000 and 10,000, which were isolated from guinea pig brain by Su et al. *(19)* and called sigmaphins. This was followed by a report of Contreras and coworkers *(20)* on the partial purification of a peptide from pig brain; Zhang et al. *(21)* isolated a peptide-like substance from human brain with a mol wt of approx ~700 Dalton and an affinity for both σ and PCP receptor sites. Unfortunately, the precise identity of these various materials isolated from mammalian brains remains unknown and no further publications from these groups have arisen.

Another approach that has been used to identify putative endogenous ligands for σ-receptors assessed the binding of various neurotransmitters, neuropeptides, and steroids to σ sites in vitro. There is no evidence that the biogenic amine transmitters histamine and GABA, or the excitatory amino acid transmitters, have any action on σ-receptors (*see* ref. *22*). Similarly, the peptide neurotransmitters and neuromodulators, the dynorphins, enkephalins, endorphins, substance P, and tachykins, CCK, melatonin, and the prostaglandins were also found to lack affinity for σ-receptors *(22)*. Neuropeptide Y (NPY) and peptide PPY were also found to lack affinity for σ sites. However, there is experimental evidence to show that these peptides can modulate the functional activity of σ-receptors in the gastrointestinal tract *(23)*; Monnet et al. *(24)* have shown that NPY stimulates *N*-methyl-D-aspartate (NMDA) receptor-induced firing of CA3 pyramidal cells of the hippocampus. These findings raise the possibility that NPY-like peptides could act as endogenous modulators of σ sites in the brain.

Su and colleagues *(25)* have identified several steroids that inhibit the binding of [$^3$H](+)SKF-10047 and [$^3$H]haloperidol to the σ-receptor. However, the affinities of these steroids (which was found to vary between 268 and 4000 n$M$, respectively, for the steroids progesterone, testosterone, deoxycorticosteorne, pregnenalone, and corticosterone) make it unlikely that they occupy a significant percentage of the σ-receptors under physiological conditions (*see* ref. *26*). Thus, it is unlikely that even progesterone,

**Table 2**
**Summary of the Biochemical Properties and Distribution of Sigma Receptors**

| σ-Receptor subtype | Putative signal transduction pathway | Representative potential selective agonist |
| --- | --- | --- |
| σ-1 | G protein linked negatively linked to phosopholipase C | (+) pentazone<br>(+) 3PPP<br>    dextromethorphan<br>(+) *N*-allylnormethazocine |
| σ-2 | K+ channel linked | Haloperidol (+)<br>    pentazocine DTG |
| | **Anatomical distribution**[a]<br>(high density of receptors) | **Possible therapeutic application** (limited evidence) |
| σ-1 | Hypothalamus<br>Thalamus<br>Mid-brain<br>Pons/medulla | Antidepressant<br>Antipsychotic<br>Drugs (stimulants) abuse<br>Stroke/Antidementia |
| σ-2 | Hippocampus<br>Cerebellum<br>Cortex<br>Red nucleus<br>Pontine and cranial nerve nuclei | Motor disorders |
| **Representative potential selective antagonist** | **Proposed functional assay** | |
| Haloperidol | NMDA-potentiation in CA region of rat hippocampus<br><br>Dystonia following injection into red nucleus of rats | |
| **Endogenous ligands** | | |
| ? NPY<br>? Steroids<br>? Unknown peptides<br>? $Zn^{2+}$ | | |

[a]Receptor distribution varies between mammalian species. For distribution in rat brain, *see* ref. *79*. Table modified from ref. *80*.

the most potent of the steroids, is a candidate for the endogenous ligand. Nevertheless, Monnet and coworkers *(27)* and Bergeron et al. *(28)* have shown that several sulfated forms of these neurosteroids differentially affect the NMDA receptor-mediated release of [3H]noradrenaline from preloaded hippocampal slices, and suggest that this modulatory affect occurs via σ-receptors. Whether such changes are relevant in vivo is presently uncertain.

Together with the various proteins, peptides, and steroids that have been suggested as candidates for the endogenous σ-ligands, convincing evidence has been produced by Connor and Chavkin *(29)* that $Zn^{2+}$ can competitively displace [3H]DTG from hippocampal slices with an $IC_{50}$ value of 130 $\mu M$. These studies showed that $Zn^{2+}$ is 50-fold more selective for the $\sigma$-2 than the $\sigma$-1 sites, suggesting that this ion may be the endogenous modulator of $\sigma$-2 sites. More recently, Patterson et al. *(22)* have reported that two $\sigma$-1 selective ligands occur in acid extracts of mammalian brain. These appear to be a peptide and a low mol wt, nonpeptide substance, the precise identity of which are unknown.

## 3. NPY, DEPRESSION, AND σ-RECEPTORS

Recent experimental studies have shown that NPY may play a role in the mechanism of action of antidepressant drugs and in the pathology of mood disorders. For example, when rats are chronically treated with antidepressants or subjected to repeated electro-convulsive shock treatments, there is an increase in NPY immuno reactivity in several brain regions *(30–32)*. Conversely, NPY immunoreactivity is decreased in the brains of rats that have been subjected to bilateral olfactory bulbectomy (OB), a procedure that has been used in the development of a rodent model of depression *(see* ref. *33)*. NPY is known to modulate the release of noradrenaline and serotonergic neurons *(34,35)*. Such findings provide a tangible link among the brain biogenic amine systems, depression, and the mode of action of antidepressant drugs. Several clinical studies have also implicated a dysfunctional NPY system in the etiology of depression. Thus, the concentration of NPY in the cerebrospinal fluid of depressed patients has been found to be decreased in some *(36,37)*, but not all *(38)*, studies. Widdowson et al. *(39)* reported that the NPY content of the frontal cortex and caudate nucleus of suicide victims was lower than normal, although a subsequent study by these investigators failed to replicate these findings *(40)*.

The high concentration of NPY and NPY receptor sites in limbic regions of the mammalian brain, together with the co-localization of NPY with noradrenaline and serotonin, suggest that this peptide may be involved in the regulation of cortical and limbic activity. In a series of studies on the OB rat model of depression, we have shown that many of the changes in behavioral, endocrine, immune, and neurotransmitter functions are similar to those occurring in depressed patients *(33,41–43)*. These findings, together with the observation by Widerlov et al. *(37)* that the concentration of NPY is reduced in the rat brain following OB, suggest that there may be a causal relationship between a defect in NPY function and the biological changes observed in the bulbectomized rat and in depressed patients. Furthermore, the functional modulation of $\sigma$-receptors by NPY *(24)* suggests that these receptors may play a crucial role in the pathogenesis of depression. Song et al. *(44)* have shown that the icv administration of NPY (0.5 nmol) daily for 7 d to OB rats attenuated the hyperactivity and increased rearing behavior that the animals exhibit when placed in the novel, stressful environment of the open-field apparatus. Such changes are characteristic of all chronically administered antidepressants *(33)*. In addition, changes in the turnover of noradrenaline, serotonin, and dopamine in limbic regions of the brain that occurred following OB were largely reversed by the subchronic administration of NPY. In the immune system, the suppression of lymphocyte proliferation that occurred following OB was reversed in those animals treated with NPY, as was the reduction in the serum corticosterone concentration. The results of this study suggest that NPY produces its effects by modulating the biogenic amine neurotransmitters and the pituitary adrenal axis. Whether these are direct effects on neurotransmitter function, or indirect effects via the action of NPY on σ-receptors or corticotropin releasing factor (CRF) release, is uncertain. It has been shown that NPY blocks CRF-induced stress in colonic motor activity *(23)* and CRF-induced gastric acid secretion in rats *(45)*. Because we have recently shown that a selective σ-ligand such as igmesine (JO-1784) can attenuate many of the behavioral, neurotransmitter, and immune changes initiated by the intracerebroventricular administration of CRF in the rat *(81)*, the possibility remains that the action of NPY on σ-receptor function could account for the reversal of any of the changes seen in the OB rat model of depression. Table 3 gives a summary of some of the changes observed in this model and compares the qualitative effects of NPY and antidepressants on these changes.

## 4. THE EFFECT OF ANTIDEPRESSANTS ON σ AND NMDA RECEPTORS IN THE MAMMALIAN BRAIN

Despite the proven therapeutic efficacy of antidepressant drugs for over 30 yr, their mechanism of action remains an enigma. The delay in the onset of action of the therapeutic response to an antidepressant has been variously explained in terms of adaptational changes in different neurotransmitter receptors (*see* refs. *46* and *47*). Particular attention has been paid to the adaptive changes in β-adrenoceptors and 5-HT$_{2A}$ receptors that may be dysfunctional in the depressed patient *(48,49)*. Chronic administration of many antidepressant drugs causes a downregulation (decrease in receptor density) of β-receptors and desensitization of β-receptor mediated increase in cyclic adenosine monophosphate accumulation *(50–52)*. Similarly, different types of antidepressants have been shown to downregulate 5-HT$_{2A}$ receptors and desensitize 5-HT$_{2A}$-mediated phosphotidyl inositol turnover *(53)*. However, it now seems unlikely that these changes are sufficient to explain the efficacy of antidepressant treatments. Thus, many effective antidepressants do not affect the density of either β- or 5-HT$_{2A}$ receptors in cortical regions of the rat brain. Furthermore, many subcortical regions of the rat brain show no evidence of β- or 5-HT$_{2A}$ receptor adaptation following prolonged treatment with antidepressant drugs *(54,55)*. Such findings have led to the search for other mechanisms whereby antidepressants, irrespective of their structure or acute actions or biogenic amine neurotransmitters, bring about their therapeutic effects.

Several groups of investigators have reported that tricyclic antidepressants, as well as such atypical antidepressants as iprindole and the monoamine oxidase A inhibitor, clorgyline, inhibit the binding of selective σ-receptor ligands to brain tissue in vitro *(56,57)*. It has also been shown that sertraline, a selective serotonin reuptake inhibitor (SSRI) with little affinity for neurotransmitter receptors in the brain, potently displaces [$^3$H](+)3-(3-hydroxyphenyl-*N*-1(1-propyl)piperidine (3PPP) binding to σ-receptors in rat brain *(14)*. These findings led Paul et al. (I. Paul, personal communication) to examine the inhibition by various classes of antidepressants of [$^3$H](+) pentazocine to σ-1 receptor sites in mouse brain in vitro. The result of these studies showed that the majority of the 22 clinically active antidepressants inhibited the bindings of (+) pentazocine to the σ-1 site in mouse forebrain homogenates with potencies ranging from 0.1 to 10 μ*M*. The majority of these antidepressants displayed noncompetitive inhibitory characteristics and Paul et al. (I. Paul, personal communication) speculate that the action of antidepressants at the haloperidol-sensitive σ-1 site might represent an initial step in the

**Table 3**
**Effect of NPY and Antidepressants on Some Changes in the OB Rat Model of Depression**

| | Antidepressants[a] | NPY[b] |
|---|---|---|
| 1. Hyperactivity in the open field apparatus | Normalized | Normalized |
| 2. Hyposecretion of corticosterone (a.m.) | Normalized | Normalized |
| 3. Increased sertotonin turnover in limbic regions | Normalized | Normalized |
| 4. Reduction in the concentration of noradrenaline in the amygdala | Normalized | Normalized |
| 5. Reduced mitogen stimulated lymphocyte proliferation | Normalized | Normalized |
| 6. Increased percentage of neutrophils and reduced percentages of lymphocytes and monocytes in different white blood cell count | Normalized | Partially reversed |

[a]Changes characteristic of most classes of antidepressants following their chronic (14 d) administration.
[b]Changes reported following 7 d icv administration to OB rats.
(*See* refs. *33* and *44.*)

therapeutic action of antidepressant drugs. However, the relevance of such a study to the chronic effects of these antidepressants in vivo is uncertain. In a preliminary study *(58)*, it has been shown that several SSRIs and desipramine decrease the density of σ-1 receptors in several regions of the limbic system following the chronic administration of the antidepressants, as shown by a decrease in the binding of [3H](+)pentazocine. By contrast, acute administration of these same antidepressants increased the density of these receptors in limbic regions of the rat brain. Thus, there is some circumstantial evidence to suggest that chronic antidepressant treatment downregulates σ-1 receptors in the rodent brain.

It has been estimated that excitatory amino acids modulate the activity of approx 50% of synapses in the mammalian brain. Because glutamate is the most important of the excitatory amino acid transmitters, it follows that changes in the activity of this transmitter will have a profound impact on the functional activity of many brain regions. It is known that, although many antidepressant drugs block the reuptake of biogenic amines throughout the brain, many brain regions that contain noradrenergic and serotonergic terminals show no evidence of amine receptor adaptation following prolonged antidepressant treatment *(54)*. As antidepressants effectively correct most of the psychiatric and physical symptoms of depression, it is surmised that they do so by correcting the malfunctioning of those brain regions that are responsible for most of the symptoms of depression. Thus, the possibility arises that antidepressants act by modulating the activity of neurotransmitters, such as glutamate, that play a crucial role in central neurotransmission. What is the evidence that antidepressants can modulate such glutamate receptors as the NMDA receptor in vivo? There is no evidence that these drugs affect the binding of [3H]MK801 or [3H]TCP to the NMDA receptor-associated ionophore at concentrations that are near the therapeutic range *(59)*. However, it has been demonstrated that chronic antidepressant treatment results in a dose-dependent adaptation of the NMDA receptor complex in the mouse brain *(60)*. Furthermore, Paul et al. *(61)* have shown that the ability of such antidepressants to modulate NMDA receptor function is highly correlated with their therapeutic efficacy. Such finding may provide an explanation for the mechanism of action in the diverse range of drugs that act as antidepressants.

The question now arises regarding the relationship between the action of antidepressants on σ-1 receptors and their ability to functionally modulate NMDA receptors for which they appear to have no affinity in vitro, at least at relevant therapeutic concentrations *(59)*. Bergeron and coworkers *(62)* have shown by their electrophysiological studies that both sertraline and clorgyline selectively potentiate the effect of NMDA on pyramidal neurons in the CA3 region of the rat hippocampus. This potentiation was reversed by the σ-receptor ligands haloperidol and BMY-14802. However, other studies have shown that tricyclic antidepressants, such as opipramol, can provide significant neuronal protection against ischemia-induced neuronal loss in the hippocampus of the Mongolian gerbil *(63)*. This effect was believed to be caused by a modulation of the strychnine-insensitive glycine site on the NMDA receptor complex. Thus, it would appear that antidepressants, which modulate NMDA receptor function indirectly via σ-1 receptors, potentiate NMDA function. Because there is evidence that glycine, which potentiates NMDA receptor function by stimulating the strychnine-insensitive glycine site on the receptor complex, has antidepressant-like activity in the OB rat model of depression *(64)*, it may be speculated that antidepressants owe their efficacy to such a mechanism of action. By contrast, psychotropic drugs that reduce NMDA receptor function are more likely to have an anti-ischemic action.

The mechanism whereby σ-1 receptors modulate the glycine site on the NMDA receptor complex is still uncertain. Yamamoto et al. *(65)* have investigated a series of six σ-1 receptor ligands for their ability to influence the binding of [$^3$H]TCP to cultured neuronal cells prepared from fetal rat telencephalon. The $IC_{50}$ values of the six σ-ligands for the [$^3$H]TCP binding site in the NMDA receptor ionophore were closely correlated with their $K_i$ values for the DTG site in the guinea pig brain. These investigators concluded that σ-ligands indirectly modulate the NMDA receptor ion channel both in vitro and in vivo. Although it would appear that antidepressants modulate the NMDA complex by acting on the glycine site, other σ-1 ligands appear to do so by inhibition of phosphorylation through a mechanism in which phospholipase C is involved *(66)*. Thus, there appear to be different mechanisms whereby antidepressants and σ-1 receptor ligands can alter NMDA receptor function. This suggests that the action of antidepressants on σ-1 receptors may be a necessary, but not a sufficient, action to account for their antidepressant activity.

## 5. CAN THE LOCOMOTOR SIDE EFFECTS OF THE SSRIs BE ATTRIBUTED TO A FUNCTIONAL CHANGE IN σ-2 RECEPTORS?

Matsumoto and Walker *(67)* have shown that σ-2 ligands in the red nucleus, when applied microontophoretically, inhibit the firing in rubral neurons in a dose-dependent manner. These experimental studies show that σ-ligands, acting on σ-2 receptors, alter the spontaneous activity in two brain regions that play a significant role in motor behavior and possess high densities of σ-receptors.

Recently, we have studied *(68)* the effects of a single injection of the SSRIs sertraline, paroxetine, citalopram, fluoxetine, and fluvoxamine into the red nucleus of the rat and compared the resulting motor disturbances with those elicited by known σ-ligands (DTG, [+]-pentazocine, haloperidol, and [3PPP]). The known σ-ligands predictably caused an acute dystonic reaction and torticollis lasting for 5 min following the injection. Of the SSRIs investigated, however, only fluvoxamine and fluoxetine caused moderate dystonia. Such dystonic behavior caused by these two SSRIs would suggest that they produce their effects by acting at σ-2 sites in the rubrocerebellar circuit. Somewhat similar results have recently been found by us following chronic administration of these SSRIs *(68a)*.

In another series of studies, we investigated the effects of the chronic administration of fluoxetine, sertraline, and paroxetine on the density of σ-receptors in some 32 discrete regions of rat brain (Ivory et al., in preparation). The results of these studies showed that the density of the σ-receptors (mainly σ-1 type) was increased in the cortex, caudate hippocampus, and the entromedial hypothalamus, but decreased in the amygdala, nucleus, accumbens, arcuate nucleus, septum, thalamus, and olfactory tubercle.

σ-Ligands have also been shown to enhance dopamine-mediated activity in the nigrostriatal pathway of the rat, an effect that has also implicated that involvement of σ-2 receptors *(69)*. Thus, DTG and (+)-pentazocine have been shown to increase the concentration of extracellular dopamine in the striatum *(70)*. However, the effects of σ-ligands on the functional activity of dopaminergic neurons in the striatum is far from clear, largely because of the relative lack of specificity of the ligands used to assess their electrophysiological effects on the nigrostriatal system and the complication of the general anesthetic used to maintain the anesthesia during the experiments *(71)*. It is also possible that the effects of some of the σ-ligands used could be indirect, because of an interaction with NMDA receptors that are known to play an important role in the nigrostriatal system *(72)*. Thus, it may be concluded that, although the SSRIs may produce their main pharmacological effects by facilitating central serotonergic transmission, they also have actions on other central neurotransmitter processes that could account for some of their therapeutic and adverse effects.

## 6. CONCLUSION

Although the σ-receptor was once thought to be a subtype of the opioid receptor family, in the past few years it has been established that not only are there at least two distinct types of σ-receptor, but also that they are quite distinct from all known neurotransmitter receptors. The σ-receptors are widely distributed both centrally and in peripheral tissues, including immune cells. Since the σ-receptors exhibit a high affinity for diverse types of psychotropic drugs, and have been postulated to be involved in various disorders of the central nervous system, researchers have speculated how these receptors may be involved in the actions of antipsychotic, antidepressant, and neuroprotective agents. There is already substantial evidence that antidepressants may act by modulating σ- (particularly σ-1) receptor function in limbic regions of the brain. Whether such an action is sufficient to explain the therapeutic action of these drugs is presently uncertain, but the consequent modulation of the glutamate–NMDA receptor complex by such an action on σ-1 receptors may open up new possibilities, not only in our understanding of the mechanism of action of antidepressants, but also in the development of novel therapeutic agents.

## REFERENCES


1. Lasagna, L. and Beecher, H. K. (1954) The analgesic effectiveness of nalorphine and nalorphine-morphine combinations in man. *J. Pharmacol. Exp. Ther.* **112,** 356–363.
2. Harris, L. S. and Pierson A. K. (1964) Some narcotic antagonists in the benzomorphan series. *J. Pharmacol. Exp. Ther.* **143,** 141–148.
3. Martin, W. R., Eades, C. G., Thompson, W. O., Thompson, J. A., Huppler, R. E., and Gilbert, P. E. (1976) The effect of morphine and nalophrine-like drugs in the non dependent and morphine dependent chronic spinal dog. *J. Pharmacol. Exp. Ther.* **197,** 517–532.
4. Magnan, J., Patterson, S. J., Tavani, A., and Zosterlitz, H. W. (1982) The binding specturm of narcotic analgesic drugs with different agonists and antagonists properties. *Naunyn-Schmeideberg's Arch. Pharmacol.* **319,** 197–319.
5. Zukin, S. R., Tempel, A., Gardner, E. L., and Zukein, R. S. (1986) Interaction of 3H (−) SKF 10,047 with brain σ-receptors: characterization and autoradiogrphic visualization. *J. Neurosci.* **46,** 1032–1041.
6. Quirion, R., Chicheportiche, R., Contreras, P. C., Johnson, K. M., and Lodge, D. (1987). Classification and nomenclature of phencyclidine and σ-receptor sites. *Trends Neurosci.* **10,** 444–446.
7. Walker, J. M., Matsumato, R. R., Bowen, W. D., Gans, D. L., Jones, K. D., Walker, F. O. (1988) Evidence for a role of haloperidol-sensitive σ 'opiate' receptors in the motor effects of antipsychotic drugs. *Neurology* **38,** 961–965.
8. Deutsch S. I., Weizman, A., Goldman, M. E., and Morihisa, J. M. (1988) The σ-receptor: a novel site implicated in psychosis and antipsychotic drug efficacy. *Clin. Neuropharmacol.* **11,** 105–119.
9. Contreras, P. C., Ragan, D. M., Bremer, M. E., Lanthorn, T. H., Gay, N. M., Iyengar, S., et al. (1991) Evaluation of U50, 488H analogues anti-ischaemic activity in the gerbil. *Brain Res.* **546,** 798–782.
10. Walker, J. M., Bowen, W. D., Walker, F. O., Matsumoto, R. R., de Costa, B., and Rice, K. C. (1990) σ-receptors: biology and function. *Pharmacol. Rev.* **42,** 356–401.
11. Quirion, R., Bowen, W. D., Itzhak, Y., Junien, J. L., Musacchio, R. B., et al. (1992) A proposal for the classification if σ binding sites. *Trends Pharmacol. Sci.* **13,** 85–87.
12. Tam, S. W. and Cook, L. (1984) σ opiates and certain antipsychotic drugs mutually inhibit (+) 3H SKF 10,047 and 3H-haloperidol binding in guinea pig brain membranes. *Proc. Natl. Acad. Sci. USA* **81,** 5618–5621.
13. Klein, M., Carroll, P. D., and Musacchio, J. M. (1991) SKF S25-A and cytochrome P450 ligands inhibit with high affinity the binding of 3H-dextromethorphan and ligands to guinea pig brain. *Life Sci.* **48,** 543–550.
14. Schmidt, A., Lebel, L., Koe, B. K., Seeger, T., and Heym, J. (1989) Sertraline potently displaces (+) 3H-3 PPP binding to sites in rat brain. *Eur. J. Pharmacol.* **165,** 335–336.
15. Ferris, C. D., Hirsch, D. J., Brooks, B. P., and Snyder, S. H. (1991) σ-receptors: from molecule to man. *J Neurochem.* **57,** 729–737.
16. Itzhak, Y., Mash, D., Zhang, S. H., and Stin, I. (1991) Characterization of MPTP binding sites in C57BL/6 mouse brain; mutual effects of monoamine oxidase inhibitors and σ ligands on MPTP and σ binding sites. *Mol. Pharmacol.* **39,** 385–393.
17. Gray, N. M., Contreras, P. C., Allen, S. E., and Taylor, D. P. (1990) $H_1$ antihistamines interact with central σ-receptors. *Life Sci.* **47,** 175–180.
18. McCann, D. J. and Su, T. P. (1990) Haloperidol-sensitive (+) 3H-SKF 10,047 binding sites (σ sites) exhibit a unique distribution in rat brain subcellular fractions. *Eur. J. Pharmacol. (Mol. Pharmacol.)* **188,** 211–218.

19. Su, T.-P., Weissman, A. D., and Yeh, S.-Y. (1986) Endogenous ligands for σ opioid receptors in the brain ("σphin"): evidence from binding assays. *Life Sci.* **38,** 2199–2210.

20. Contreras, P. C., Di Maggio, D. A., and O'Donohue, T. L. (1987) An endogenous ligand for the σ opioid binding site. *Synapse* **1,** 57–61.

21. Zhang, A. Z., Mitchell, K. N., Cook, L., and Tam, S. W. (1988) Human endogenous brain ligands for σ and phencyclidine receptors, in: *σ and Phencyclidine-like Compounds as Molecular Probes in Biology* (Domino, E. F. and Kameuka, J. M., eds.). NPP Books, Ann Arbor, MI, pp. 335–343.

22. Patterson, T. A., Connor, M., and Chavkin, C. (1994) Recent evidence for endogneous substances for σ-receptors, in: *σ-receptors* (Itzhak, Y., ed.). Academic, New York, pp. 171–189.

23. Junien, J. L., Gue, M., and Bueno, L. (1991) Neuropeptide-Y and σ ligand JO 1784 act through a Gi protein to block the psychological stress and CRF induced colonic motor activation in rats. *Neuropharmacology* **30,** 119–124.

24. Monnet, F. P., Debonnel, G., and de Montigny, C. (1990) Neuropeptide-Y selectively potentiates NMDA induced neuronal activation. *Eur. J. Pharmacol.* **182,** 207–208.

25. Su, T.-P., London, E. D., and Jaffe, J. H. (1988) Steroid binding at σ-receptors suggests a link between endocrine, necrosis and immune systems. *Science* **240,** 219–221.

26. Schwartz, S., Pohl, P., and Zhou., G. Z. (1990) Steroid binding at σ 'opioid' receptors. *Science* **246,** 1635–1637.

27. Monnet, F. P., Mahe, V., Robel, P., and Beulien, E. D. (1995) Neurosteroids, via σ-receptors, modulate the 3H-norepinephrine release evoked by N-methyl-D-aspartate in the rat hippocampus. *Proc. Natl. Acad. Sci. USA* **92,** 3774–3778.

28. Bergeron, R., de Montigny, C., and Debonnel, G. (1996) Biphasic effects of σ ligands on the neuronal response to N-methyl-D-aspartate. *Naunyn-Schmiedeberg's Arch. Pharmacol.* **351,** 252–260.

29. Connor, M. A. and Chavkin, C. (1992) Ionic zinc may function as a endogenous ligand for the haloperidol sensitive σ-2 receptor in rat brain. *Mol. Pharmacol.* **42,** 471–479.

30. Heilig, M., Wahlestedt, C., Ekman, R., and Widerlov, E. (1988) Antidepressant drugs increase the concentration of neuropeptide-Y like immuno reactivity in the rat brain. *Eur. J. Pharmacol.* **147,** 465–467.

31. Wahstedt, C., Blendy, J. A., Kellar, K. T., Heilig, M., Widerlöv, E., and Ekman, R. (1990) Electroconvulsive shocks increase the concentration of neocortical and hippocampal NPY-like immunoreactivity in the rat. *Brain Res.* **507,** 65–68.

32. Widdowson, P. S. and Halaris, A. E. (1994) Increased levels of NPY immunoreactivity in rat brain limbic structures following antidepressant treatment. (Abst), *J. Neurochem.* **52 (Suppl.)** S 77.

33. van Riezen, H. and Leonard, B. E. (1991) Effects of psychotropic drugs on the behaviour and neurochemistry of olfactory bulbectomized rats. *Pharmacol. Ther.* **47,** 21–34.

34. Halliday, G. M., Li, Y. W., Oliver, J. R., Jh, T. J., Cotton, R. G., et al. (1988) The distribution of NPY-like immunoreactive neurons in the human medulla oblongata. *Neuroscience* **26,** 179–191.

35. Finta, E. P., Regenold, J. J., and Illes, P. (1992) Depression by NPY of noradrenergic inhibitory postsynaptic potentials of the locus coeruleus neurones. *Naunyn-Schmiedeberg's Arch. Pharmacol.* **346,** 472–474.

36. Berrettini, W. H., Doran, A. R., Kelsoe, J. Y., Roy, A., and Pickar, D. (1987) Cerebrospinal fluid NPY in depression and schizophrenia. *Neuropsychopharmacology* **1,** 81–83.

37. Widerlöv, E., Lindstrom, L. H., Wahlested, C., and Ekman, R. (1988) NPY and PYY as possible cerebrospinal fluid markers for major depression and schizophrenia respectively. *J. Psychiatr. Res.* **22,** 69–79.

38. Gjerris, A., Widerlov, E., Werdelin, L., and Ekman, R. (1992) Cerebrospinal fluid concentrations of NPY in depressed patients and controls. *J. Psychiatr. Neurosci.* **17,** 23–27.

39. Widdowson, P. S., Ordway, G. A., and Halaris, A. E. (1992) Reduced neuropeptide Y concentrations in suicide brains. *J. Neurochem.* **59,** 730–780.

40. Ordway, G. A., Stockmeier, C. A., Meltzer, H. Y., Overholser, J. C., et al. (1995) NPY in frontal cortex is not altered in major depression. *J. Neurochem.* **65,** 1646–1650.

41. O'Neill, B., O'Conner, W. T., and Leonard, B. E. (1987) Depressed neutrophil phagocytosis in the rat following olfactory bulbectomy reversed by chronic desipramine treatment. *Med. Sci. Res.* **15,** 267, 268.

42. Song, C. and Leonard, B. E. (1994) The effects of chronic lithium chloride administration on some behavioral and immunological changes in the bilaterally olfactory bulbectomized rat. *J. Psychopharmacol.* **8,** 40–47.

43. Song, C. and Leonard, B. E. (1995) The effect of olfactory bulbectomy in the rat, alone or in combination with antidepressants and endogenous factors, an immune function. *Human Psychopharmacol.* **10,** 7–18.

44. Song, C., Earley, B., and Leonard, B. E. (1996) The effects of the central administration of NPY on behaviour, neurotransmitter and immune functions in the olfactory bulbectomized rat model of depression. *Brain Behav. Immun.* **10,** 1–16.

45. Gue, M., Yoneda, H., Monnikes, H., Junien, J. L., and Tache, Y. (1992) Central NPY and σ ligand JO 1784 reverse CRF induced inhibition of gastric acid secretion in rats. *Br. J. Pharmacol.* **107,** 642–647.

46. Leonard, B. E. (1994) Effect of antidepressants on specific neurotransmitters: are such effects relevant to their specific action? in: *Handbook of Depression and Anxiety—a Biological Approach* (van den Boer, J. A. and Sitsen, J. M. A., eds.). Marcel Dekker, New York, pp. 379–404.

47. Leonard, B. E. (1995) Mechanisms of action of antidepressants. *CNS Drugs* **4 (Suppl 1),** 1–12.

48. Healy, D., Carney, P. A., and Leonard, B. E. (1983) Monoamine related markers of depression: changes following treatment. *J. Psychiatr. Res.* **17,** 251–260.

49. Healy, D., Carney P. A., O'Halloran, A., and Leonard, B. E. (1995) Peripheral adrenoceptors and serotonin receptors in depression. Changes associated with response to treatment with trazodone or amitriptyline. *J. Affect. Dis.* **9,** 285–296.

50. Vetulani, J., Stawarz, R. J., Dingell, J. V., and Sulser, F. (1976) A possible common mechanism of action of antidepressant treatments. *Naunyn-Schmiedeberg's Arch. Pharmac.* **293,** 109–114.

51. Sarai, K., Frazer, A., Brunswick, D., and Mendels, D. (1978) Desipramine induced decrease in beta adrenergic receptor binding in rat cerebral cortex. *Neuropharmacology* **27,** 2179–2187.

52. Wolfe, B. B., Harden, T. K., Sporn, J. R., and Molinoff, P. B. (1978) Presynaptic modulation of beta adrenergic receptors in rat cerebral cortex after treatment with antidepressants. *J. Pharmacol. Exp. Ther.* **207,** 446–457.

53. Stockmeier, C. A. and Kellar, K. T. (1989) Serotonin depletion unmarks serotonergic component of 3H-dihydroalprenolol binding in rat brain. *Mol. Pharmacol.* **36,** 903–911.

54. Paul, I. A., Duncan, G. E., Powell, K. R., Mueller, R. A, Hong, J.-S., and Breese, G. R. (1988) Regionally specific neuronal adaptation of beta adrenergic and 5HT2 receptors after antidepressant administration in the forced swim test and after chronic antidepressant drug treatment. *J. Pharmacol. Exp. Ther.* **246,** 956–962.

55. Paul, I. A., Duncan, G. E., Mueller, R. A., Hong, J. S., and Breese, G. R. (1991) Adaptation in response to chronic imipramine and electroconvulsive shock: evidence for separate mechanisms. *Eur. J. Pharmacol.* **205,** 135–143.

56. Largent, B. L., Wikstrom, H., Gunlach, A. L., and Snyder, S. H. (1987) Structural determinants of σ-receptor affinity. *Mol. Pharmacol.* **32,** 772–784.

57. Itzhak, Y. and Kassim, C. O. (1990) Clorgyline displays high affinity for σ sites in C57-BL/6 mouse brain. *Eur. J. Pharmacol.* **176,** 107–108.

58. Ivory, J. D. (1994) The effects of putative antipsychotic drugs and antidepressants on σ-receptors in the rat brain. M.Sc. Thesis, National University of Ireland.

59. Kitamura, Y., Zhao, X.-H., Takei, M., et al. (1991) Effects of antidepressants on the glutaminergic system in mouse brain. *Neurochem. Int.* **3,** 247–253.

60. Nowak, G., Trullas, R., Layer, R. T., and Skolnick, P. (1993) Adaptative changes in the NMDA receptor complex after chronic treatment with imipramine and 1-aminocyclopyropanecarboxylase acid. *J. Pharmacol. Exp. Ther.* **265,** 1380–1386.

61. Paul, I. A., Nowak, G., Layer, R. T., Popik, P., and Skolnick, P. (1994) Adaptation of the NMDA receptor complex following chronic antidepressant treatments. *J. Pharmacol. Exp. Ther.* **269,** 95–102.

62. Bergeron, R., Debonnel, G., and de Montigny, C. (1993) Modification of the NMDA response by antidepressant σ-receptor ligands. *Eur. J. Pharmacol.* **240,** 319–323.

63. Rao, T. S., Cler, J. A., Mick, S. J., Ragan, D. M., et al. (1990) Opipramol, a potent σ ligand, is an antiischaemic agent. *Neuropharmacology* **29,** 1199–1204.

64. Redmond, A., Kelly, J. P., and Leonard, B. E. (1996) Behavioural effects of chronic glycine in the olfactory bulbectomized rat in the 'open field' test and on phencyclidine induced hyperactivity. *Med. Sci. Res.* **24,** 55–56.

65. Yamamoto, H., Yamamoto, Y., Sagi, N., Klenerova, V., Goji. K., et al. (1995) σ ligands indirectly modulate the NMDA receptor ion channels complex on intact neuronal cells via σ-1 sites. *J. Neurosci.* **15,** 731–736.

66. Bowen, W. D., Tolentino, P., and Varghese, P. (1989) Investigation of the mechanism by which σ ligands inhibit stimulation of phosphoinositide metabolism by muscarinic cholinergic agonists. *Prog. Clin. Biol. Res.* **328,** 21–24.

67. Matsumoto, R. R. and Walker, J. M. (1992) Ionophoretic effects of putative σ ligands on rubral neurons in the rat. *Brain Res. Bull.* **29,** 419–425.

68. Leonard, B. E. and Faherty C. (1996) SSRIs and movement disorders: is serotonin the culprit? *Human Psychopharmacol.* **11 (Suppl. 2),** S75–S82.

68a. Faherty, C. J., Harkin, A. J., and Leonard, B. E. (1997) The functional modulation of σ-receptors following chronic SSRI treatment. *Eur. J. Pharmacol.,* in press.

69. Walker, J. M., Bowen, W. D., Patrick, S. L., Williams, W. E., and Marsearella, S. W. (1993) A comparison of (−) deoxybenzomorphans devoid of opiate activity with their dextro rotatory phenolic counterparts suggest a role for σ type receptors for psychtomimetic opiaties and antipsychotic drugs. *Proc. Natl. Acad. Sci. USA* **83,** 8784–8788.

70. Patrick, S. L., Walker, J. M., Pekel, J. M., Lockwood, M., and Patrick, R. O. (1993) Increases in rat striatal extracellular dopamine and vacuous chewing produced by two σ ligands. *Eur. J. Pharmacol.* **231,** 243–249.

71. Steinfels, G. F., Tam, W. T., and Cook, L. (1989) Electrophysiological effects of selective σ-receptor agonists, antagonists and the selective phencyclidine receptor agonist MK 801, on mid brain dopamine. *Neuropsychopharmacology* **2,** 201–208.

72. Iyengar, S., Dilworth, V., Mick, S. J., and Contrecras, P. C. (1990) σ-receptors modulate both A9 and A10 dopaminergic neurons in the rat brain: functional interactions with NMDA receptors. *Brain Res.* **523,** 322–326.

73. Bowen, W. D., Einstein, G., and Walfe, S. A. (1992) σ-1 and σ-2 binding sites of rat kidney. *Soc. Neurosci. Abstracts* **18,** 456.

74. Tam, S. W., Steinfels, G. F., Gilligan, P. J., Schmidt, W. K., and Cook, L. (1992) DUP 734, a σ and 5HT2 receptor antagonist: receptor binding, electrophysiological and neuropharmacological profiles. *J. Pharmacol. Exp. Ther.* **263,** 1167–1174.

75. Karbon, E. W., Bailey, M., Borowsky, S., Abreu, M., Martin, L., et al. (1991) In vitro and in vivo binding properties of NPC 16377, a potent and selective ligand for σ binding sites. *Soc. Neurosci. Abstracts* **17,** 334.

76. Roman, F. J., Pascaud, X., Martin, B., Vanche, D., and Junien, J. L. (1990) JO 1784, a potent and selective ligand for rat and mouse σ sites. *J. Pharm. Pharmacol.* **42,** 439–440.

77. Su, T.-P., Wu, X.-Z., Cone, E. J., Shulcla, K., Dodge, A. L., and Parish, D. W., (1991) σ compounds derived from phencyclidine. Identification of PRE-O84 a new selective σ ligand. *J. Pharmacol. Exp. Ther.* **259,** 543–550.

78. Largent, B. L., Wikstrom, H., Snowman, A. H., and Snyder, S. H. (1988) Novel antipsychotic drugs share high affinity for σ-receptors. *Eur. J. Pharmacol.* **155,** 345–347.

79. McCann, D. J., Weissman, A. D., and Su, J.-P. (1994) σ-1 and σ-2 sites in rat brain: comparison of regional, ontogenetic and subcellular patterns. *Synapse* **17,** 182–189.

80. Leonard, B. E. and Nicholson, C. D. (1994) σ ligands as potential psychotropic drugs. *J. Psychopharmacol.* **8,** 64–65.

81. Song, C., Earley, B., and Leonard, B. E. (1997) Effect of chronic pretreatment with the sigma ligand Jo 1784 on CRF-induced changes in behavior, neurotransmitter and immunological function in the rat. *Neuropsychobiol.*, in press.

**10**

# A Role for CREB in Antidepressant Action



## Ronald S. Duman, Masashi Nibuya, and Vidita A. Vaidya



## 1. INTRODUCTION

A role for monoamine neurotransmitter systems, particularly norepinephrine (NE) and serotonin (5-HT), in the actions of antidepressant treatments is supported by preclinical and clinical studies *(1–4)*. Many antidepressant treatments block the reuptake or metabolism of NE and 5-HT and acutely increase synaptic levels of these monoamines. However, the therapeutic action of antidepressants is dependent on chronic treatment, leading to the hypothesis that adaptations to the acute elevation of NE and 5-HT underlie the therapeutic effects of antidepressants. Early studies focused on adaptations of NE and 5-HT receptors and the second messenger systems regulated by these receptors *(5,6)*. Advances in molecular and cellular neurobiology have demonstrated that the receptor-coupled second messenger systems represent the first step in a cascade of intracellular messengers and regulatory proteins. These intracellular cascades control virtually every aspect of neuronal function, including the regulation of gene transcription. It is reasonable to hypothesize that adaptations of these intracellular sites are responsible for the therapeutic action of antidepressant treatments *(7–11)*.

Early studies have implicated the cAMP second messenger cascade in the action of antidepressant treatments. One of the most consistent actions of antidepressant treatments is downregulation of $\beta_1$-adrenergic receptor ($\beta_1$AR)-stimulated cAMP production in limbic brain regions *(12,13)*. More recent studies have examined the role of postreceptor components of the cAMP cascade in the action of antidepressant treatments. In contrast to downregulation, the results of these studies demonstrate that antidepressant treatments upregulate the cAMP system at several levels, including expression of the cAMP response-element binding protein (CREB) *(14)*. Regulation of CREB also indicates that there are certain target genes that are regulated by antidepressant treatments *(14,15)*. The possibility that the actions of antidepressant treatments are mediated via upregulation of CREB and certain target genes is critically examined in this chapter.



*From:* Antidepressants: *New Pharmacological Strategies*
*Edited by: P. Skolnick, Humana Press Inc., Totowa, NJ*



*173*

## 2. MONOAMINE RECEPTOR-COUPLED SECOND MESSENGER PATHWAYS AND REGULATION OF GENE TRANSCRIPTION FACTORS

The influence of NE and 5-HT receptor activation on neuronal function is mediated by coupling, via G proteins, to intracellular second messenger cascades, as well as to ion channels. Only the cAMP and phosphatidylinositol systems, which are reported to be influenced by antidepressant treatments, will be discussed in the present review. These and other second messenger pathways have been reviewed in detail elsewhere (for reviews, *see* refs. *16–18*).

### 2.1. NE and 5-HT Receptor Regulation of the cAMP Cascade

There are several monoamine receptor subtypes that either stimulate (e.g., $\beta_1$AR and 5-HT$_{4,6,7}$ receptors) or inhibit (e.g., $\alpha_2$-adrenergic and 5-HT$_{1A}$ receptors) adenylyl cyclase via coupling to the G-protein subtypes Gs$\alpha$ or Gi$\alpha$, respectively *(17,18)*. Activation of adenylyl cyclase and increased formation of cAMP levels lead to activation of cAMP-dependent protein kinase (PKA) (Fig. 1); cAMP binds to regulatory subunits of PKA and causes their dissociation from the catalytic subunits. Once dissociated, the catalytic subunits phosphorylate substrate proteins and thereby influence the function of many different classes of cellular proteins. One phosphorylation substrate of PKA is the transcription factor CREB (Fig. 1). In the dephosphorylated form, CREB can bind to the promoter region of genes and is capable of constitutively regulating transcription. Phosphorylation of CREB dramatically increases its ability to regulate gene transcription. The consensus DNA binding site for CREB, the cAMP response element (CRE), is a canonical sequence TGACGTCA, but CREB is also reported to recognize cryptic CRE sites that contain a portion of this consensus sequence.

### 2.2. NE and 5-HT Receptor Regulation of the Phosphatidylinositol Cascade

The phosphatidylinositol system is activated by both NE and 5-HT receptor subtypes (e.g., $\alpha_1$-adrenergic and 5-HT$_{2A}$ receptors) *(16–18)*. Activation of these receptors leads to stimulation of phospholipase C via coupling to Gq, or related G-protein subtypes. Activation of phospholipase C catalyzes the breakdown of phosphatidylinositol 4,5-bisphosphate, resulting in the formation of the second messengers inositol triphosphate (IP$_3$) and diacylglycerol. IP$_3$ acts to release Ca$^{2+}$ from intracellular stores. Elevated intracellular Ca$^{2+}$ activates Ca$^{2+}$/calmodulin-dependent protein kinase and, in conjunction with diacylglycerol, stimulates protein kinase C (PKC) (Fig. 1). One of the best-studied transcription factors regulated by PKC is the immediate early gene c-*fos*. This occurs via phosphorylation of the serum response factor (SRF), which binds to the serum response element within the promoter of the c-*fos* gene. Fos heterodimerizes with another transcription factor, Jun, to form an AP-1 DNA binding complex. In addition, the function of the AP-1 complex can be influenced by dimerization of Fos with other Jun-like or Fos-like transcription factors.

### 2.3. Mechanisms For Regulation of Gene Transcription Factors

The regulation of CREB and c-*fos* serve as examples of the two major mechanisms for regulation of transcription factors *(19–22)*. The function of a transcription factor such as CREB is activated primarily by phosphorylation; the function of imme-

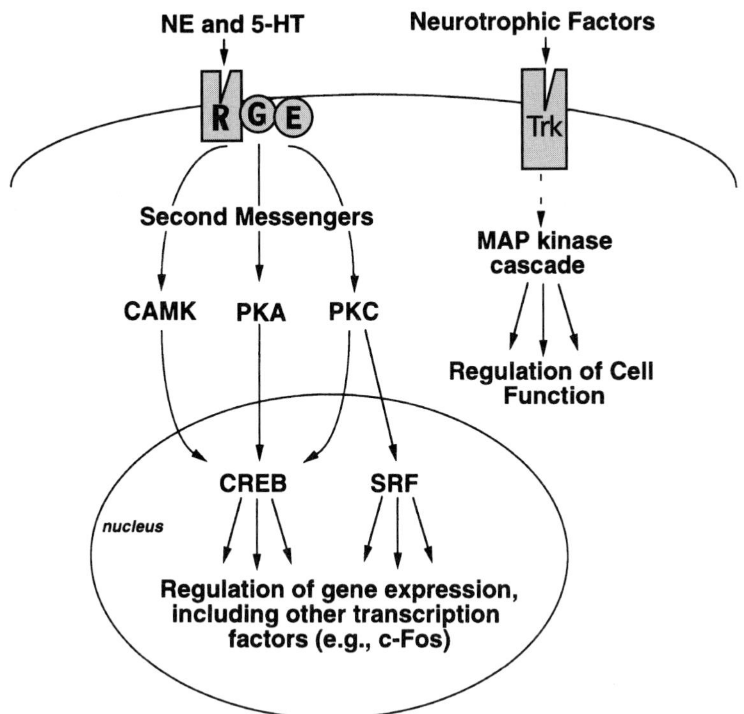



**Fig. 1.** A general model for regulation of transcription factors by receptor-coupled signal transduction pathways. NE and 5-HT receptor subtypes regulate second messenger pathways via coupling to G proteins (G) and activation of effectors (E), such as adenylyl cyclase and phospholipase C. Some common second messenger cascades influenced by NE and 5-HT receptors are cAMP, inositol triphosphate, diacylglycerol, and $Ca^{2+}$. These second messengers lead to regulation of second messenger-dependent protein serine-threonine kinases, such as PKA, PKC, and $Ca^{2+}$/calmodulin kinase (CAMK). Many cellular proteins are substrates for these kinases, including gene transcription factors. CREB is activated by PKA, but can also be regulated by PKC and CAMK. PKC also regulates the immediate early gene transcription factor, c-*fos*, via phosphorylation of the serum response factor (SRF). In addition, the c-*fos* gene contains a CRE and can be regulated by CREB. These mechanisms for regulation of CREB and c-*fos* demonstrate the potential for regulation of transcription factors by multiple second messenger-dependent protein kinases. Neurotrophic factors act through different types of intracellular pathways, via receptors (Trk) that have intrinsic tyrosine kinase activity on the intracellular aspect of the receptors. Activation of these tyrosine kinases leads to autophosphorylation of the receptor, which promotes its association with, and regulation of, other cellular proteins. One of the most notable pathways activated by neurotrophic factors is the MAP kinase cascade.

diate early genes, like c-*fos*, are regulated primarily by induction of the total amount of transcription-factor protein. However, these two mechanisms are not necessarily exclusive of each other. For example, CREB is also active, albeit to a much lower level, in the dephosphorylated state, and Fos is regulated by its phosphorylation state. Another mechanism for regulation of transcription factors is that utilized by intracellular glucocorticoid receptors. Upon binding of ligand, the glucocorticoid receptors are translocated to the nucleus, where they act on glucocorticoid response elements

(GRE). Translocation of cytoplasmic transcription factors to the nucleus also occurs in response to tyrosine phosphorylation by growth factor or cytokine-activated pathways *(23,24)*.

### 2.4. Transcription Factors Can Be Regulated by Multiple Receptor-Coupled Intracellular Pathways

The function of transcription factors may be regulated by a single, primary second messenger pathway, but may also be influenced by other second messenger pathways. For example, CREB is phosphorylated and activated not only by PKA, but also by PKC and $Ca^{2+}$/calmodulin-dependent protein kinase (Fig. 1) *(21,22,25)*. Many promoter elements regulated by agents that increase $Ca^{2+}$, referred to as $Ca^{2+}$ response elements (CaRE), are regulated in this way by CREB. As discussed, the expression of c-*fos* and other immediate early gene transcription factors is induced by activation of PKC and the serum response factor, but also by cAMP and $Ca^{2+}$/calmodulin-dependent protein kinase. In addition, the function of Fos can be regulated by phosphorylation, in some cases by these same protein kinases.

In this way, transcription factors may serve as common intracellular targets for different receptor-coupled second messenger systems. This could explain how different classes of antidepressant drugs, which influence NE and 5-HT, and possibly other neurotransmitter systems, may have similar effects on neuronal gene expression. This will be discussed in greater detail below.

### 2.5. Receptor Sensitivity and the Function of Intracellular Pathways

Early studies of antidepressants focused on the regulation of NE and 5-HT receptors. The results of these studies demonstrated that the density and function of NE and 5-HT receptors are decreased by antidepressant treatments, leading to a receptor subsensitivity hypothesis of antidepressant action. In addition to downregulation of the $\beta_1$AR system, antidepressant treatments are also reported to decrease levels of 5-HT$_2$ receptors in limbic brain regions *(26)*. The receptor subsensitivity hypotheses postulated that decreased function of these receptor systems contributed to the therapeutic action of antidepressant treatments *(5,6)*. Recent studies have not provided support for these receptor sensitivity hypotheses. For example, the time-course for desensitization of $\beta_1$AR and 5-HT$_2$ receptors is not in agreement with the time-course for the therapeutic action of antidepressant treatments *(see* refs. *27* and *28)*. In addition, downregulation of these $\beta_1$AR and 5-HT$_2$ receptors is not observed in response to all classes of antidepressants, although it is possible that there is more than a single mechanism of action for antidepressants.

Although receptor subsensitivity may not underlie the therapeutic actions of antidepressants, downregulation of $\beta_1$AR and 5-HT$_2$ receptors may provide information about the state of receptor–second messenger function. This is illustrated for NE and the $\beta_1$AR in Fig. 2. In the absence of antidepressant, the levels of NE are relatively low and the level of receptor activation is correspondingly low. Acute antidepressant treatment increases synaptic levels of NE and leads to activation of the $\beta_1$AR-coupled cAMP system. This level of activation is relatively high, because there is a full compliment of receptors in the acute state. After chronic antidepressant treatment, the level of $\beta_1$ARs is downregulated by the sustained elevation of NE levels. However, in the presence of

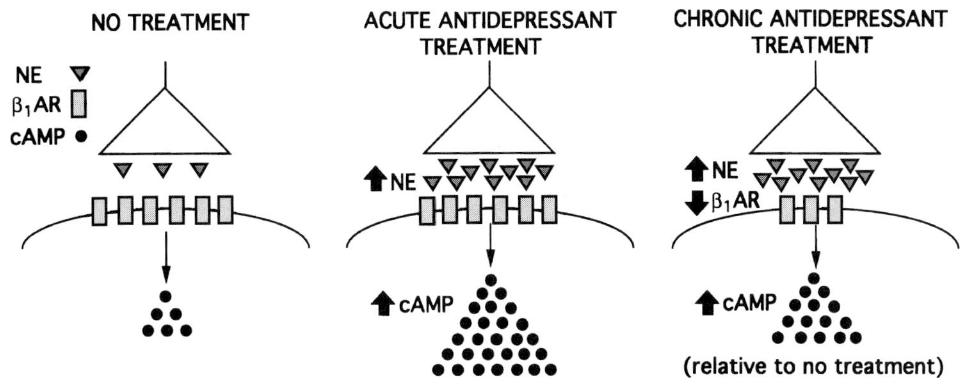


**Fig. 2.** A model demonstrating antidepressant regulation of β₁AR and cAMP levels in brain. This model explains how chronic antidepressant treatments could activate the cAMP pathway, even though levels of β₁AR are decreased. In the no-treatment condition, there is a low level of cAMP production in response to basal levels of NE acting on the normal compliment of β₁AR. In the acute antidepressant treatment condition, levels of NE are increased; this results in activation of β₁AR and increased levels of cAMP formation. In the chronic treatment condition, the levels of β₁AR are downregulated by antidepressant treatments, resulting in reduced levels of cAMP production. However, with synaptic levels of NE still elevated as a result of continued antidepressant treatment, the level of cAMP production is increased relative to the no-treatment condition.

higher NE levels, the level of β₁AR-stimulated cAMP production would be increased relative to the untreated condition. Although this model is hypothetical, it can be argued that the continued downregulation of β₁ARs by antidepressant treatments indicates that these receptors remain activated. This model suggests that the response to chronic antidepressant treatment is an increase, not decrease, in cellular function.

The apparent contradiction of this model with the receptor subsensitivity hypothesis has resulted from the way antidepressant experiments are normally conducted. In most studies, the responses to exogenous NE in the untreated control and antidepressant-treated condition are compared. Although this provides a measure of maximal receptor response, it is not a reflection of the level of in vivo receptor activation in response to endogenous NE levels. The appropriate comparison would be to determine the level of in vivo receptor activation in the untreated condition and compare it to the antidepressant-treated condition. This could be determined by measuring the influence of a β₁AR antagonist on the levels of cAMP in vivo or on the physiological activity of neurons known to be regulated by the β₁AR. Advances in techniques that enable the in vivo measurement of cAMP by microdialysis make it possible to test this hypothesis.

## 3. REGULATION OF THE cAMP CASCADE BY ANTIDEPRESSANT TREATMENT

Although early studies demonstrated that the β₁AR and the activation of cAMP production are downregulated by antidepressant treatments, recent studies have demonstrated that certain postreceptor components of the cAMP cascade are upregulated by chronic antidepressant treatments. The regulation of adenylyl cyclase and PKA by antidepressant treatments is discussed in this section.

### 3.1. Regulation of Adenylyl Cyclase

The possibility that antidepressant treatments increase, not decrease, the cAMP second messenger cascade is supported by studies of the postreceptor components of this system (Fig. 3). The level of adenylyl cyclase enzyme activity is reported to be increased in response to several different types of antidepressant treatments *(29,30)*. Increased adenylyl cyclase activity in response to chronic antidepressant treatment appears to result from enhanced coupling of Gs to the adenylyl cyclase catalytic subunit, and not increased expression of these proteins. The expression of adenylyl cyclase types I and II is reported to be increased by chronic administration of lithium *(31)*. It is possible that antidepressant treatments influence the expression of one of the many adenylyl cyclase isozymes expressed in brain. There are at least nine different adenylyl cyclase isozymes that differ in their ability to be regulated by $Ca^{2+}$ and calmodulin, as well as by activated $Gs\alpha$- and free $\beta\gamma$-subunit *(17,32)*. The availability of specific cDNA probes and antibodies makes it possible to begin to study the influence of antidepressant treatments on the expression of the multiple forms of adenylyl cyclase in brain.

In addition to being regulated by the rate of production, intracellular levels of cAMP are highly regulated by the rate of metabolism. The metabolism of cAMP is catalyzed by a group of enzymes known as phosphodiesterases (PDE) *(17,33)*. There are five major types of PDE, each with multiple subtypes, and three of the these (I, II, and IV) are found in brain. Type I, or $Ca^{2+}$/calmodulin, PDE is stimulated by $Ca^{2+}$/calmodulin and has a relatively low affinity for cAMP. Type II, or cGMP, PDE is activated by cGMP and has a low affinity for cAMP. The physiological regulators of type IV, or cAMP specific PDE, are not known, but this form of PDE has a high affinity for cAMP and is thought to play an important role in cAMP metabolism in brain. The function of the PDEs are highly regulated by phosphorylation and by changes in levels of expression *(34)*. For example, the expression of certain PDE type IV isozymes is increased by activation of the cAMP pathway. This is thought to act as a feedback mechanism for regulation of cAMP levels. The influence of antidepressant treatments on the expression and function of the PDE subtypes is an area of interest for future studies.

### 3.2. Regulation of PKA

PKA is also reported to be influenced by antidepressant treatments (Fig. 3) *(35,36)*. These studies have demonstrated that chronic, but not acute, antidepressant treatments increase levels of PKA enzyme activity in particulate fractions of cerebral cortex. One study has reported that this effect occurs in the microtubule compartment of the particulate fraction and that this leads to increased phosphorylation of microtubule-associated protein *(36)*. Nestler and colleagues *(35)* have reported that levels of PKA are increased in a crude nuclear fraction, but decreased in the cytosolic fraction, of cerebral cortex, suggesting that PKA is translocated to the nucleus in response to antidepressant treatments. The discrepancy between these studies may result from the technical problems in obtaining pure subcellular fractions of brain tissue. Although additional studies are required to firmly identify the cellular localization of these effects, the results indicate that PKA can be regulated by antidepressant treatments.

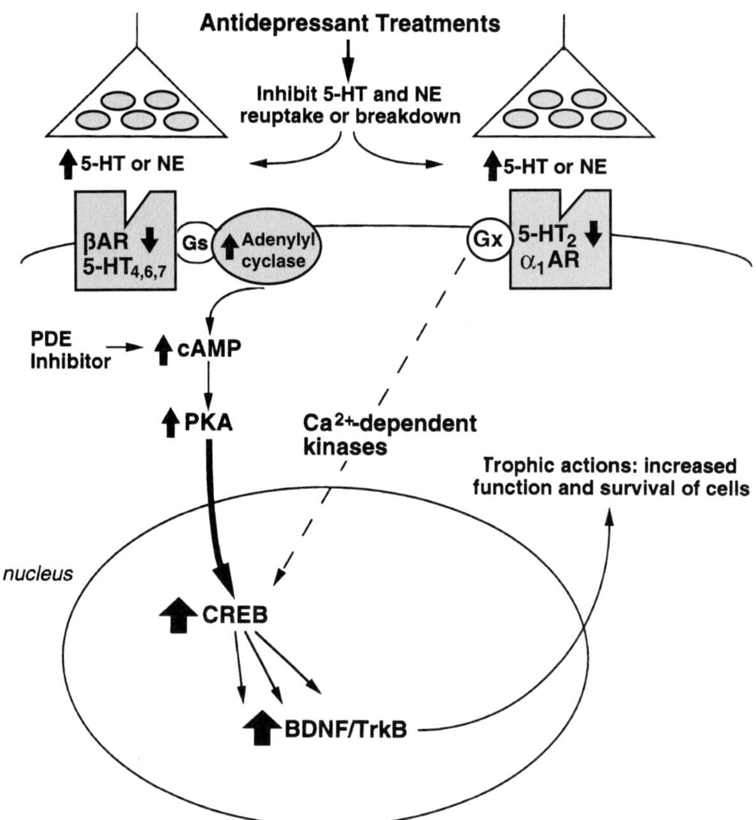


**Fig. 3.** A cellular model for the action of chronic antidepressant treatments. Antidepressant treatment increases levels of 5-HT and NE by inhibiting the reuptake or metabolism of these monoamines. Chronic antidepressant treatment decreases the function and expression of certain 5-HT and NE receptors (e.g., βAR and 5-HT$_2$). However, analysis of postreceptor components of the cAMP pathway demonstrate that this second messenger system is increased, not decreased, by chronic antidepressant treatments. This includes increased levels of adenylyl cyclase and PKA, and possibly translocation of PKA to the cell nucleus. In addition, the expression of CREB is increased by antidepressant treatments. CREB could serve as a common postreceptor target for monoamine receptors that stimulate the cAMP-PKA cascade (5-HT$_{4,6,7}$ and βAR) or receptors that lead to activation of Ca$^{2+}$-dependent kinases (e.g., 5-HT$_2$ and α$_1$AR). Upregulation of CREB could result in regulation of specific target genes, such as BDNF and TrkB, which are also increased by antidepressant treatment. BDNF and TrkB could influence the function and survival of hippocampal neurons or the neurons innervating this brain region, such as 5-HT and NE neurons.

## 4. REGULATION OF CREB BY ANTIDEPRESSANT TREATMENTS

Reports that antidepressant treatments increase levels of adenylyl cyclase activity, and the possibility that PKA is translocated to the nucleus, suggest that CREB is also regulated by these treatments. Regulation of CREB could mediate the action of antidepressant treatments on gene expression and thereby result in long-term changes in neuronal function. In addition, CREB may serve as a common target for antidepressant treatments that influence different receptor-coupled second messenger pathways. In addition to the possible regulation of CREB by $\beta_1$AR and 5-HT$_{4,6,7}$ receptor subtypes, CREB may be regulated by $\alpha_1$AR and 5-HT$_2$ receptors via Ca$^{2+}$-activated protein kinases (Figs. 1 and 3).

### 4.1. In Vivo Regulation of CREB by Antidepressants

Recent studies provide direct evidence for the regulation of CREB by antidepressant treatments *(14)*. We have reported that chronic, but not acute, administration of several different classes of antidepressant treatment increases the expression of CREB mRNA in rat hippocampus (Fig. 3). The antidepressants tested included selective NE and 5-HT reuptake inhibitors, a nonselective tricyclic monoamine reuptake inhibitor, a monoamine oxidase inhibitor, and electroconvulsive seizure. In contrast, several nonantidepressant psychotropic drugs, including cocaine, morphine, and haloperidol, do not influence the level of CREB mRNA in hippocampus, demonstrating the pharmacological specificity of this effect. However, this does not rule out the possibility that CREB is influenced by these nonantidepressant drugs in other brain regions. Indeed, cocaine and morphine are reported to regulate the expression and/or phosphorylation of CREB in the striatum and locus ceruleus *(37,38)*, demonstrating that CREB is regulated in a region-specific manner by different classes of psychotropic drugs.

Within the hippocampus, the expression of CREB mRNA is increased in the dentate gyrus granule cell layer and the CA1 and CA3 pyramidal cell layers in response to antidepressant treatments (Fig. 4) *(14)*. In addition, the level of CREB immunoreactivity is increased in the same hippocampal cell layers (not shown), indicating that the elevated expression of mRNA leads to increased levels of CREB protein. This conclusion is supported by studies demonstrating that the level of CRE binding in extracts of hippocampus is increased by chronic antidepressant treatment (Fig. 5). The presence of CREB in the CRE binding complex was confirmed by demonstrating that anti-CREB antisera disrupts the complex, but antisera to a related transcription factor has no effect. A recent report from another laboratory has also reported that antidepressant treatments increase the level of CRE binding in hippocampus and frontal cortex *(39)*. These results indicate that antidepressant induction of CREB mRNA leads to the formation of functional CRE binding.

The results of these studies indicate that antidepressant treatments increase the function and expression of CREB, but have not provided information on the phosphorylation state of this transcription factor. This is a critical point, since the level of CREB phosphorylation is a major determinant of the overall function of this transcription factor. Our attempts to determine the level of CREB phosphorylation using phospho-CREB-specific antisera have been unsuccessful, because of the variability and inconsistency of the immunostaining. We are currently testing new lots of phospho-CREB antisera in an attempt to answer this important question.

### 4.2. Mechanisms Underlying the Regulation of CREB Expression

The mechanisms underlying the regulation of CREB may be explained by upregulation of the cAMP system by antidepressant treatments. In Sertoli cells, activation of the cAMP pathway is reported to increase the expression of CREB mRNA *(40–42)*, and we have found that stimulation of cAMP production in C6 glioma cells has a similar

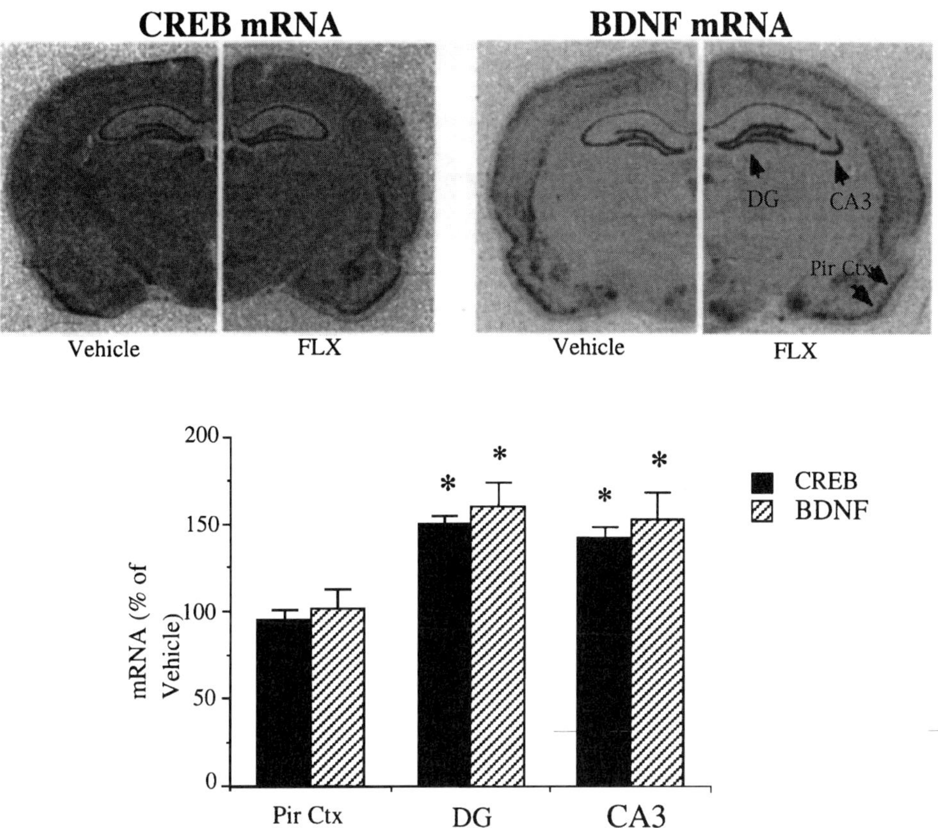


**Fig. 4.** Chronic antidepressant treatment increases the expression of CREB and BDNF mRNA in rat hippocampus. The influence of chronic (21 d) fluoxetine (FLX) administration on levels of CREB and BDNF mRNA was determined by *in situ* hybridization analysis. Levels of CREB and BDNF mRNA were increased in the major subfields of the hippocampus, including CA3 pyramidal and dentate gyrus (DG) granule cell layers. In contrast, levels of CREB and BDNF mRNA were not influenced in the piriform cortex (Pir Ctx). These results are consistent with the hypothesis that regulation of CREB results in increased expression of BDNF. The levels of CREB and BDNF mRNA were quantified by densitometry and the results are expressed as percent of vehicle (mean ± SEM). Adapted from ref. *14*.

effect (unpublished observations). The possibility that the cAMP system mediates the induction of CREB in hippocampus is supported by the finding that chronic administration of a PDE inhibitor also increases the expression of CREB mRNA in this brain region (*see below*). The induction of CREB in response to cAMP could be mediated by a CRE in the 5'-flanking region found in the CREB gene *(25,41)*. However, another study in a neuronal cell line has demonstrated that activation of cAMP decreases expression of CREB *(37)*, indicating that cAMP regulation of CREB is cell-type specific. The type of regulation in a cell may depend on the compliment of other transcription factors that act as repressors of CREB and may be regulated by the cAMP cascade. For example, tissue-specific expression of the cAMP response modulator (CREMs) proteins, which also bind to CRE sites, but act as transcriptional repressors *(43)*, may lead to a different response to cAMP activation.

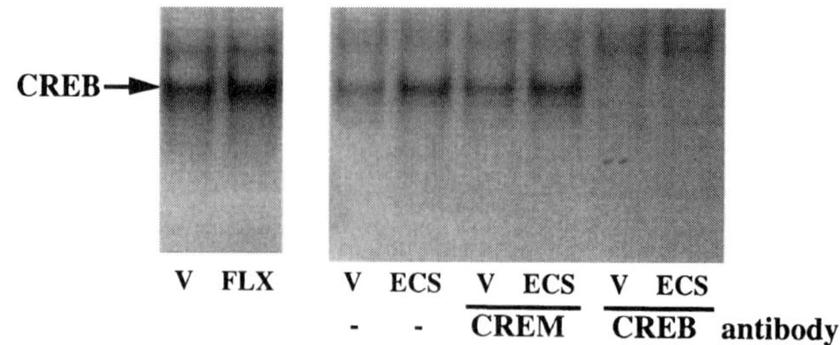



**Fig. 5.** Chronic antidepressant treatment increases CREB binding in hippocampus. Chronic administration of fluoxetine (FLX, 21 d) or ECS (10 d) increased the level of CREB binding, which was determined by gel-shift analysis. For this assay, homogenates of hippocampus are incubated with a radiolabeled, synthetic fragment of DNA that contains a consensus CRE. Binding of CREB shifts the migration of the radiolabeled DNA through the gel. The identity of CREB is confirmed by supershift studies, which demonstrate that a specific antibody to CREB, but not CREM, disrupts the CREB binding band. For supershift analysis, the homogenates are preincubated with antibody before adding the radiolabeled DNA. Adapted from ref. *14*.

### 4.3. Antidepressant Regulation of CREB in Cultured Cells

The upregulation of CREB in hippocampus in response to antidepressant treatments appears to contradict another report that antidepressants inhibit CREB activity *(44)*. This study was conducted in cultured PC12 cells, an adrenal chromaffin cell line, and examined the influence of acute antidepressant incubation on levels of CRE-mediated transcription. The influence of chronic antidepressant treatment on CREB in these cells was not examined. In PC12 cells, antidepressants inhibited the activation of CREB in response to KCl-induced depolarization, but not in response to cAMP. This effect was found to be caused by antidepressant blockade of L-type voltage-dependent $Ca^{2+}$ channels, which open in response to depolarization and result in activation of CREB via the $Ca^{2+}$-activated protein kinases (*see* Fig. 3). The contribution of $Ca^{2+}$ channel blockade to the actions of antidepressant treatments in vivo is unclear. Although antidepressant treatments may reduce CREB activity in response to activation of $Ca^{2+}$ channels, they increase CREB activity via activation of the cAMP and phosphatydylinositol pathways in brain. Determination of the influence of chronic antidepressant treatments on levels of phospho-CREB may provide information to address this question.

### 4.4. Identification of Target Genes Regulated by CREB

These results suggest that chronic antidepressant treatment, via increased expression of CREB, regulates the expression of target genes that contain CRE sites. Although this may suggest that antidepressant treatment has a very broad effect on gene expression, it is likely that there are additional levels of control that determine which genes are activated by these treatments. For example, upregulation of the cAMP system and CREB by antidepressant treatment is region-specific. In addition, the regulation of a specific gene by CREB may be dependent on the expression of other regulatory elements in that gene. A major goal of ongoing studies is to identify the genes regulated by antidepressant treatments and CREB, and to determine the relevance of these genes to the action of antidepressant treatment. Recent studies have provided evidence that brain-derived neurotrophic factor (BDNF) is one target of antidepressant treatments.

## 5. REGULATION OF BDNF BY CREB AND ANTIDEPRESSANT TREATMENTS

BDNF is a member of the nerve growth factor family, which also includes nerve growth factor (NGF), neurotrophin-3 (NT-3), and neurotrophin-4/5 (NT-4/5). These neurotrophic factors were first characterized for their important role in the differentiation and development of neurons, but it is now known that they also play a critical role in the maintenance and survival of neurons in adult brain *(45–47)*. In addition to maintaining the overall health of neurons, neurotrophic factors are capable of influencing synaptic strength of neurons. This has been demonstrated by reports that exogenous BDNF can enhance long-term potentiation (LTP), and that endogenous BDNF is involved in the formation of LTP in hippocampal neurons *(48–52)*. Neurotrophic factors influence cellular function via binding to specific receptors, knows as Trks (*see* Fig. 1). Activation of the Trk receptor leads to activation of its intracellular tyrosine kinase domain and regulation of several intracellular pathways, most notably the MAP kinase pathway. The primary neurotrophic factor receptor subtype for BDNF is TrkB.

### 5.1. Antidepressant Regulation of BDNF and TrkB

The possibility that BDNF is a target of antidepressant treatments is supported by recent studies from our laboratory. We have found that chronic, but not acute, administration of several different classes of antidepressant increases the expression of BDNF and TrkB in rat hippocampus *(14,15)*. This includes the same classes of antidepressant treatments found to increase the expression of CREB. In addition, upregulation of BDNF and TrkB mRNA is observed in the same cell layers of hippocampus as CREB: the dentate gyrus granule cell layer and CA1 and CA3 pyramidal cell layers (Fig. 4).

These studies provide correlative evidence for a functional link between increased expression of CREB and BDNF in hippocampus. Further evidence that the cAMP system mediates the induction of BDNF is provided by our studies with PDE inhibitors *(14)*. We have found that chronic administration of rolipram, a selective inhibitor of PDE IV, or papaverine, a nonselective PDE inhibitor, increases the expression of BDNF, as well as CREB, in hippocampus (Fig. 6). In addition, coadministration of rolipram with imipramine results in more rapid induction of BDNF, as well as CREB (Fig. 6). Studies in cultured cells have demonstrated that activation of cAMP or $Ca^{2+}$ pathways increases the expression of BDNF and TrkB *(53,54)*. Finally, using an antisense oligonucleotide strategy (*see* refs. *38 and 55*), we have found that local infusion of CREB antisense, but not sense, oligonucleotide decreases levels of CREB immunoreactivity and reduces basal and electroconvulsive therapy (ECT)-induction of BDNF mRNA in the hippocampus *(55)*. Taken together, the results demonstrate that expression of BDNF is positively regulated by cAMP and activation of CREB.

### 5.2. A Role for BDNF in Antidepressant Action

In addition to antidepressant regulation, there are several additional lines of evidence that suggest a role for neurotrophic factors in the pathophysiology, as well as treatment, of depression. First, infusion of BDNF into the midbrain is reported to have anti-

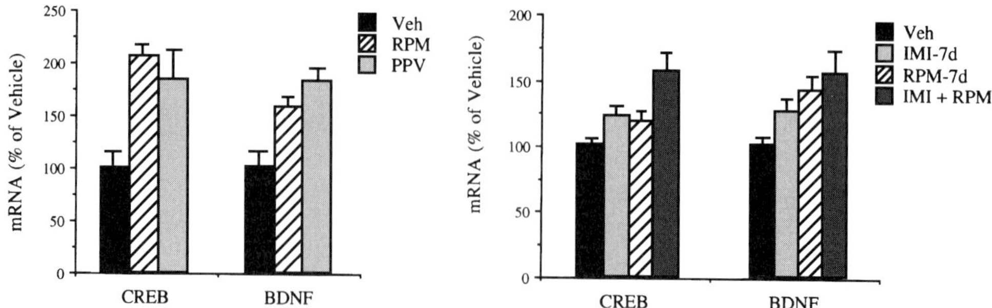

**Fig. 6.** Chronic administration of a PDE inhibitor increases the expression of CREB and BDNF in hippocampus and enhances the response to another antidepressant. Chronic administration of rolipram (RPM) or papaverine (PPV) (21 d) significantly increased levels of CREB and BDNF mRNA (left), which was determined by Northern blot analysis. Coadministration of rolipram and imipramine (IMI) for 7 d resulted in a more rapid upregulation of CREB and BDNF mRNA in hippocampus than observed with either treatment alone (right). The results are expressed as percent of vehicle and are the mean ± SEM. Adapted from ref. *14*.

depressant effects in two behavioral models of depression, the forced-swim and the learned-helplessness paradigms *(57)*. Second, BDNF is reported to enhance the growth of 5-HT and NE neurons and to protect these neurons from neurotoxic damage, suggesting that dysfunction of this neurotrophic factor could adversely influence monoamine neurotransmission *(58,59)*. Third, immobilization stress is reported to decrease the expression of BDNF in hippocampus *(60)*, and we have found that chronic antidepressant pretreatment blocks this stress-induced downregulation of BDNF *(15)*. Fourth, chronic physical or psychosocial stress, as well as glucocorticoid treatment, can cause atrophy or death of vulnerable neurons in the hippocampus of rats and non-human primates *(61–68)*. Decreased expression of BDNF may contribute to the atrophy of hippocampal neurons, although the influence of stress paradigms that are reported to induce atrophy of these neurons (i.e., restraint or psychosocial stress) on expression of BDNF has not been reported.

Finally, the results of clinical imaging studies demonstrate that the volume of the hippocampus is decreased in patients with depression or posttraumatic stress disorder (PTSD) *(69–71)*, suggesting that atrophy of hippocampal neurons may also occur in association with these illnesses. Atrophy and/or dysfunction of hippocampal neurons could underlie many of the neuroendocrine, emotional, and cognitive abnormalities observed in depression. For example, hippocampal feedback inhibition of the hypothalamic pituitary–adrenal axis is decreased in depression *(72)*. However, the relationship between the decreased hippocampal volume and the pathophysiology of depression, as well as PTSD, remains to be determined.

### 5.3. Cellular Consequences of Antidepressant Treatments

The induction of BDNF and TrkB suggests that antidepressant treatments may also influence the synaptic strength or the morphology of hippocampal neurons. The possibility that synaptic strength is altered has been examined by studies on antidepressant treatments and LTP. The results of these studies have been mixed, with reports that antidepressant treatments either augment or reduce LTP in hippocampus *(73,74)*. In addition, uncontrollable stress is reported to reduce the formation of LTP in hippocampus *(75)*. There is also behavioral evidence that antidepressants can enhance memory and cognitive function in rats and depressed patients *(76–78)*, although these effects are complicated by other pharmacological actions of antidepressants (e.g., anticholinergic activity). Additional preclinical and clinical studies using more selective antidepressant drugs (e.g., 5-HT selective reuptake inhibitors) are required to further examine the influence of antidepressants on LTP, memory, and cognitive performance tasks.

There is less known about the influence of antidepressant treatment on the morphology of neurons. There is one report that antidepressant treatment induces the regeneration of catecholamine neurons in the cerebral cortex *(79)*. There is also one study that examines the influence of antidepressant treatment on the atrophy of hippocampal neurons in response to chronic stress *(80)*. In this study, the number and length of the CA3 pyramidal cell dendritic branch points were determined in Golgi-stained sections of hippocampus. Chronic administration of tianeptine, an atypical antidepressant, blocked the stress-induced atrophy of CA3 pyramidal neurons, but alone did not influence the branching of apical dendrites. This study indicates that certain antidepressants may reverse the morphological effects of stress on CA3 pyramidal neurons, but alone have no effects. However, it is possible that these treatments have more subtle effects on hippocampal neurons that cannot be observed with the approach used for this study.

Another approach used for analysis of neuronal morphology is the Timm histochemical technique. The Timm silver sulfide protocol selectively stains zinc-containing dentate gyrus granule cells and synaptic boutons in hippocampus. This method has been used to demonstrate sprouting of granule cells during development, in models of kindling, and after excitotoxin-induced damage *(81–84)*. We have found that chronic ECT treatment (10 d) dramatically increases the level of Timm granules in the supragranular cell layer and inner molecular layer of dentate gyrus (Fig. 7). This effect was observed 12 d after the last ECT treatment. Unlike the sprouting observed in response to kindling or excitotoxins, there was no obvious cell damage after chronic ECT. In addition, the influence of ECT on sprouting appears to be reversible over time, another difference from the kindling- and excitotoxin-induced sprouting. Finally, preliminary studies demonstrate that the sprouting of hippocampal neurons is diminished in BDNF heterozygous knock-out mice, indicating that induction of high levels of BDNF is necessary for the observed sprouting in these animals.

These results provide evidence that ECT treatment causes sprouting of hippocampal neurons. However, preliminary studies indicate that antidepressant drug treatments do not significantly influence the level of Timm staining. Alternative approaches that measure more subtle changes in synaptic architecture are required to determine if antidepressant drugs alter neuronal morphology. However, it is also possible that regulation of BDNF and TrkB influence neuronal function without altering the morphology of these neurons.

### 5.4. Hypothesis for Enhanced Vulnerability to Depression

The potential relevance of CREB and regulation of BDNF and TrkB in the therapeutic actions of antidepressant treatments is supported by preclinical studies demonstrating that stress causes atrophy and death of vulnerable neurons in the hippocampus, and clinical studies demonstrating that the volume of hippocampus is decreased in

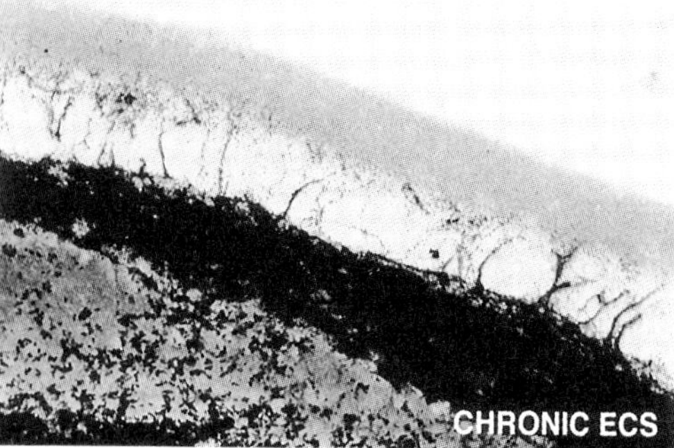


**Fig. 7.** Chronic ECT induces sprouting of dentate gyrus granule cells in the hippocampus. The induction of BDNF by antidepressants raises the possibility that the morphology of hippocampal neurons is also altered by these treatments. We have found that chronic ECT increases levels of dentate gyrus granule cell sprouting, which was determined using the Timm histochemical technique. For this study rats were administered ECT for 10 and 12 d after the last treatment the brains were analyzed. The additional time is necessary for the development of sprouting. The sprouting scores for the control and chronic ECT treated sections were 0.71 ± 0.11 and 3.07 ± 0.12, respectively (mean ± SEM, $n = 4$).

depression and stress-related psychiatric disorders. This has led to the hypothesis that one action of antidepressant treatment, through activation of CREB and increased expression of BDNF and TrkB, is to reverse the damaging effects of stress or protect neurons from further damage. This hypothesis could also explain why stress causes depression in certain individuals, but not in others (*see* ref. *11*). It is possible that vulnerable individuals have been exposed to prior episodes of stress or other neuronal insults, such as hypoxia-ischemia, hypoglycemia, neurotoxins, or viral infection, or that there is a genetic vulnerability. This could result in a certain level of damage that alone is not sufficient to produce abnormal behavioral or physiological effects, but could leave the neurons in a compromised or vulnerable state. Upon subsequent exposure to another episode of stress, the vulnerable neurons are further damaged, to the degree that depressive behavior is observed. This hypothesis provides a framework for future studies to characterize the pathophysiology and genetic basis of depression and other stress-related disorders.

## 6. DEVELOPMENT OF NOVEL THERAPEUTIC AGENTS

The finding that antidepressant treatment upregulates the cAMP system and expression of BDNF and TrkB makes it possible to rationally design novel therapeutic agents for the treatment of depression. The most obvious strategy would be to develop or test the efficacy of agents that activate the cAMP pathway or the specific components of this second messenger cascade. PKA and CREB are interesting molecular targets, but must await the development of better pharmacological tools. Another possibility is to develop synthetic BDNF agonists for the TrkB receptor, although efforts to date in this area have been unsuccessful. The best possibilities are agents that influence cellular levels of cAMP, including agonists for NE and 5-HT receptors that are positively coupled to the cAMP pathway, or to increase the level of cAMP with a PDE inhibitor.

### 6.1. Neurotransmitter Receptor Regulation of CREB

There are several NE and 5-HT receptor subtypes that are positively coupled to the cAMP system. In the hippocampus, this includes the $\beta_1 AR$ and $5\text{-}HT_{4,7}$ receptor subtypes. In addition, there are neuropeptide receptors (e.g., VIP) known to stimulate cAMP production that may also be useful targets for antidepressant treatments. The $5\text{-}HT_{4,7}$ receptors may be of particular interest, since they could mediate the actions of 5-HT selective reuptake inhibitors on CREB and BDNF/TrkB. A potential problem with this approach is that direct-acting receptor agonists may induce desensitization of the receptors that is even greater than that observed with the antidepressants that block reuptake or metabolism of NE and 5-HT. However, the development of subtype selective agonists that are able to cross the blood–brain barrier would provide useful pharmacological tools to test the therapeutic potential of such NE and 5-HT receptor agonists.

The function of CREB can also be influenced by receptors that regulate the $Ca^{2+}$-activated protein kinases. This includes the $\alpha_1$-adrenergic and $5\text{-}HT_{2A/2C}$ receptor subtypes that are coupled to the phosphatidylinositol second messenger system. We have found that $5\text{-}HT_{2A}$ receptor agonists increase levels of BDNF in neocortical regions. However, administration of a $5\text{-}HT_{2A}$ receptor agonist also decreases levels of BDNF in hippocampus. In the hippocampus, $5\text{-}HT_{2A}$ agonists may activate GABA-ergic interneurons that inhibit dentate gyrus granule cells, resulting in downregulation of BDNF expression. This illustrates how receptor agonists may influence neuronal function via polysynaptic mechanisms, and not via a single postsynaptic action. In addition, $5\text{-}HT_{2A}$ receptor agonists, such as LSD, are hallucinogenic and would not be useful therapeutic agents. It is also possible that other neurotransmitter receptors coupled to the phosphatidylinositol system, or agents that influence ligand-gated $Ca^{2+}$ channels, could be possible targets for antidepressant treatments.

### 6.2. Activation of the cAMP Pathway with PDE Inhibitors

Another strategy is to activate the cAMP pathway at sites located past the level of the receptors. This strategy bypasses the problem of receptor desensitization and could potentially result in a greater activation of the cAMP pathway than possible with direct-acting receptor agonists. On the other hand, regulators of postreceptor sites could influence the cAMP system in cells located throughout the brain, as well as peripheral tissues, and may have unwanted side effects. With these advantages and limitations in mind, we have examined the possibility that inhibitors of PDE may have antidepressant effects. This possibility was raised several years ago, based on the hypothesis that elevation of cAMP levels would enhance the adaptive downregulation of the $\beta AR$-coupled cAMP system *(86,87)*. At that time, the neurochemical model of depression was that the $\beta AR$-coupled cAMP system was elevated and that the therapeutic action of antidepressant treatment involved downregulation of this system *(5,6)*. We have found that PDE inhibitors, when administered alone, increase CREB and expression of BDNF in hippocampus *(5,14)*. In addition, we have found that co-administration of a PDE inhibitor with a conventional antidepressant drug results in a more rapid induction of CREB and BDNF than observed with either treatment alone. These results suggest that the PDE inhibitors may be effective antidepressants and that they may enhance the response to other antidepressant agents.

The potential clinical efficacy of the PDE inhibitors for the treatment of depression has been supported by early clinical trials *(87)*. These studies reported that rolipram had a good to very good antidepressant effect in approx two-thirds of the patients tested, which included many who were resistant to conventional antidepressants. However, the lack of further clinical reports on the use of PDE inhibitors for the treatment of depression raises doubts about the therapeutic potential of these agents. One possible explanation is that the PDE inhibitors have side effects, such as nausea, that are not well tolerated at the doses required for antidepressant effects. With this in mind, we have initiated clinical trials to further examine the therapeutic potential PDE inhibitors for the treatment of depression. For these studies, we have used a nonselective PDE inhibitor, papaverine. Papaverine was administered to a patient who had been depressed for years and resistant to several types of antidepressants. At the time, the patient was being treated with venlafaxine, a NE/5-HT reuptake inhibitor, and administration of this drug was maintained during the papaverine treatment. The mood of the patient started to improve within 2 wk and the patient has since remained free of depressive symptoms *(88)*. Although preliminary, these results suggest that the potential of PDE inhibitors for the treatment of depression, possibly of resistant patients, should be reconsidered. We are currently extending these studies to more fully evaluate the clinical efficacy of PDE inhibitors for the treatment of depression.

We hypothesize that combination therapy may be the most useful strategy for the PDE inhibitors, for several reasons. First, the combination of two agents that have different primary actions may result in a more effective antidepressant therapy than possible with either agent alone. This possibility requires further testing at the preclinical level to determine if coadministration of a PDE inhibitor with an antidepressant for longer periods of time (e.g., 3–6 wk) results in a greater response than observed with either treatment alone. Second, combination of a PDE inhibitor with an antidepressant may be faster-acting, as suggested by the rapid upregulation of CREB and BDNF in rats by this treatment (*see* Fig. 5). Finally, coadministration of a PDE inhibitor and antidepressant drug may provide the regional specificity that would not be possible with the PDE inhibitor alone. This could occur by using a dose of the PDE inhibitor that alone has little effect on levels of cAMP, but that is able to augment the response to an antidepressant that activates NE and/or 5-HT receptor-stimulated cAMP production.

## 7. CONCLUSIONS

The studies discussed in this chapter suggest that chronic antidepressant treatments lead to an upregulation, not downregulation, of the cAMP intracellular signal transduction cascade. The results also indicate that increased expression of BDNF and TrkB is one functional consequence of the upregulated cAMP system and CREB, although there may be other relevant targets of CREB that are regulated by antidepressant treatments. This work and extensive studies of stress have led to the hypothesis that the regulation of BDNF could contribute to the therapeutic action of antidepressants by reversing or preventing the damaging effects of stress on vulnerable hippocampal neurons.

The challenge for research efforts is to test this hypothesis at both the preclinical and clinical levels. At the preclinical level, this can be accomplished by a combination of pharmacological and molecular genetic approaches to define the influence of the cAMP and BDNF systems on neuronal function and relate these effects to cell morphology and animal behavior. Selective pharmacological tools can be used to determine the role of specific NE and 5-HT receptor subtypes and the cAMP system in the action of antidepressant treatments. Molecular studies can provide mice that have a genetically engineered mutation or overexpressed gene (e.g., CREB, BDNF, TrkB) and the phenotype of these mice can be determined at the cellular and behavioral levels. For example, our hypothesis suggests that the heterozygous BDNF knock-out mice could be a useful animal model for stress-related disorders. These mice express approx one-half the normal level of BDNF, similar to that observed in rats after immobilization stress, and we have found that the neuronal plasticity (i.e., sprouting of hippocampal granule cells) of these mice is reduced. Behavioral studies of these mice in models of depression are currently being conducted. These studies could lead to novel therapeutic agents and, at the very least, will provide further tests of our hypothesis.

At the clinical level, the pharmacological tools identified in the basic research studies can be used to determine the therapeutic relevance of certain NE and 5-HT receptor subtypes and the cAMP system in depression. Additional brain-imaging studies are needed to determine if the altered hippocampal volume is a result or a cause of depression. Postmortem studies of CREB and BDNF levels in the brains of depressed patients are needed to relate the hippocampal volume changes to altered function of these proteins.

This combination of preclinical and clinical approaches will help to define the role of CREB and BDNF in the therapeutic action of antidepressant treatments. However, it is premature to conclude that these are the only intracellular pathways involved. Rather, it is possible and likely that there are other intracellular cascades, transcription factors, and regulatory proteins that play a role in the treatment and pathophysiology of depression, and it is critical that we continue to investigate these possibilities.

## REFERENCES


1. Schildkraut, J. J. (1965) The catecholamine hypothesis of affective disorders: a review of supporting evidence. *Am. J. Psychiatry* **122**, 509–522.
2. Bunney, W. E. and Davis, J. (1965) Norepinephrine in depressive reactions: a review. *Arch. Gen. Psychiatry* **13**, 483–494.
3. Heninger, G. R. and Charney, D. S. (1987) Mechanisms of action of antidepressant treatments: implications for the etiology and treatment of depressive disorders, in: *Psychopharmacology: The Third Generation of Progress* (Meltzer, H.Y., ed.). Raven, New York, pp. 535–544.
4. Delgado, P. L., Price, L. H., Miller, H. L., Salomon, R. M., Aghajanian, G., Heniner, G. R., Charney, D. (1994) Serotonin and the neurobiology of depression. *Arch. Gen. Psychiatry* **51**, 865–874.
5. Sulser, F., Vetulani, J., and Mobley, P. (1978) Mode of action of antidepressant drugs. *Biochem. Pharmacol.* **27**, 257–261.
6. Charney, D. S., Menkes, D. B., Heninger, G. R. (1981) Receptor sensitivity and the mechanism of action of antidepressant treatment. *Arch. Gen. Psychiatry* **38**, 1160–1173.
7. Sulser, F. (1989) New perspectives on the molecular pharmacology of affective disorders. *Eur. Arch. Psych. Neurol.* **238**, 231–239.
8. Hudson, C. J., Young, L. T., Li, P. P., and Warsh, J. J. (1993) CNS signal transduction in the patho-physiology and pharmacotherapy of affective disorders and schizophrenia. *Synapse* **13**, 278–293.

9. Duman, R. S., Heninger, G. R., and Nestler, E. J. (1994) Adaptations of receptor-coupled signal transduction pathways underlying stress- and drug-induced neural plasticity. *J. Nerv. Ment. Disease* **182,** 692–700.

10. Manji, H. K., Potter, W. C., and Lenox, R. H. (1995) Signal transduction pathways. *Arch. Gen. Psychiatry* **52,** 531–543.

11. Duman, R. S., Heninger, G. R., and Nestler, E. J. (1996) A molecular and cellular theory of depression. *Arch. Gen. Psychiatry,* in press.

12. Vetulani, J. and Sulser, F. (1975) Action of various antidepressant treatments reduces reactivity of noradrenergic cAMP generating system in limbic forebrain. *Nature* **257,** 495–496.

13. Banerjee, S. P., Kung, L. S., Riggi, S. J., and Chanda, S. K. (1977) Development of β-adrenergic receptor subsensitivity by antidepressants. *Nature* **268,** 455–456.

14. Nibuya, M., Nestler, E. J., and Duman, R. S. (1996) Chronic antidepressant administration increases the expression of cAMP response element binding protein (CREB) in rat hippocampus. *J. Neurosci.* **16,** 2365–2372.

15. Nibuya, M., Morinobu, S., and Duman, R. S. (1995) Regulation of BDNF and trkB mRNA chronic electroconvulsive seizure and antidepressant drug treatments. *J. Neurosci.* **15,** 7539–7547.

16. Berridge, M. J. (1993) Inositol triphosphate and calcium signaling. *Nature* ***361,*** 315–325.

17. Nestler, E. J. and Duman, R. S. (1994) G-proteins and cyclic nucleotides in the nervous system, in: *Basic Neurochemistry,* 5th ed., (Siegel, G., Agranoff, B., Albers, R. W., Molinoff, P., eds.). Raven, New York, pp. 429–448.

18. Duman, R. S. and Nestler, E. J. (1995) Signal transduction pathways for catecholamine receptors, in: *Psychopharmacology: The Fourth Generation* (Meltzer, H., ed.), Raven, New York, pp. 303–320.

19. Montminy, M. R., Gonzalez, G. A., and Yamamoto, K. K. (1990) Regulation of cyclic AMP inducible genes by CREB. *TINS* **13,** 184–188.

20. Morgan, J. I. and Curran, T. (1991) Stimulus-transcription coupling in the nervous system: involvement of the inducible proto-oncogenes for and jun. *Ann. Rev. Neurosci.* **14,** 421–451.

21. Armstrong, R. C. and Montminy, M. R. (1993) Transsynaptic control of gene expression. *Ann. Rev. Neurosci.* **16,** 17–29.

22. Ghosh, A. and Greenberg, M. E. (1995) Calcium signaling in neurons: molecular mechanisms and cellular consequences. *Science* **268,** 239–247.

23. Blenis, J. (1993) Signal transduction via the MAP kinases: proceed at your own RSK. *Proc. Natl. Acad. Sci. USA* **90,** 5889–5892.

24. Darnell, J. E., Kerr, I. M., and Stark, G. R. (1994) Jak-STAT pathways and transcriptional activation in response to IFNs and other extracellular signaling proteins. *Science* **264,** 1415–1421.

25. Meyer, T. E. and Habener, J. F. (1993) Cyclic adenosine 3',5'-monophosphate response element binding protein (CREB) and related transcription-activating deoxyribonucleic acid-binding proteins. *Endocrine Rev.* **14,** 269–290.

26. Peroutka, S. L. and Snyder, S. H. (1980) Long-term antidepressant treatment decreases spiroperidol labeled serotonin receptor binding. *Science* **210,** 88–90.

27. Riva, M. A. and Creese, I. (1989) Reevaluation of the regulation of b-adrenergic receptor binding by desipramine treatment. *Mol. Pharmacol.* **36,** 211–218.

28. Butler, M. O., Morinobu, S., and Duman, R.S. (1993) Chronic electroconvulsive seizures increase the expression of serotonin2 receptor mRNA in rat frontal cortex. *J. Neurochem.* **61,** 1270–1276.

29. Menkes, D. B., Rasenick, M. M., Wheeler, M. A., and Bitensky, M. W. (1983) Guanosine triphosphate activation of brain adenylate cyclase: enhancement by long-term antidepressant treatment. *Science* **129,** 65–67.

30. Ozawa, H. and Rasenick, M. M. (1991) Chronic electroconvulsive treatment augments coupling of the GTP-binding protein Gs to the catalytic moiety of adenylyl cyclase in a manner similar to that seen with chronic antidepressant drugs. *J. Neurochem.* **56,** 330–338.
31. Colin, S. F., Chang, H.-C., Mollner, S., Pfeuffer, T., Reed, R. R., Duman, R. S., and Nestler, E. J. (1991) Chronic lithium regulates the expression of adenylate cyclase and Gia in rat cerebral cortex. *Proc. Natl. Acad. Sci. USA* **88,** 10,634–10,637.
32. Sunahara, R. K., Dessauer, C. W., and Gilman, A. G. (1996) Complexity and diversity of mammalian adenylyl cyclases. *Annu. Rev. Pharma. Toxicol.* **36,** 461–480.
33. Beavo, J. A. and Reifsnyder, D. H. (1990) Primary sequence of cyclic nucleotide phosphodiesterase isozymes and the design of selective inhibitors. *TIPS* **11,** 150–155.
34. Conti, M., Catherine Jin, S.-L., Monaco, L., Repaske, D. R., and Swinnen, J. V. (1991) Hormonal regulation of cyclic nucleotide phosphodiesterases. *Endocrine Rev.* **12,** 218–234.
35. Nestler, E. J., Terwilliger, R. Z., and Duman, R. S. (1989) Chronic antidepressant administration alters the subcellular distribution of cyclic AMP-dependent protein kinase in rat frontal cortex. *J. Neurochem.* **53,** 1644–1647.
36. Perez, J., Tinelli, D., Brunello, N., and Racagni, G. (1989) cAMP-dependent phosphorylation of soluble and crude microtubule fractions of rat cerebral cortex after prolonged desmethylimipramine treatment. *Eur. J. Pharmacol.* **172,** 305–316.
37. Widnell, K. L., Russell, D. S., and Nestler, E. J. (1994) Regulation of expression of cAMP response element-binding protein in the locus coeruleus in vivo and in a locus coeruleus-like cell line in vitro. *Proc. Natl. Acad. Sci. USA* **91,** 10,947–10,951.
38. Konradi, C., Cole, R. L., Heckers, S., and Hyman, S. E. (1994) Amphetamine regulates gene expression in rat striatum via transcription factor CREB. *J. Neurosci.* **14,** 5623–5634.
39. Frechilla, D., Otano, A., and Del Rio, J. (1996) Effect of chronic antidepressant treatment on cre- and sp1-binding activity in rat hippocampus and frontal cortex. *Soc. Neurosci. Abst.* 26.
40. Waeber, G., Meyer, T. E., LeSieur, M., Herman, H. L., Gerard, N., and Haebner, J. F. (1991) Developmental stage-specific expression of cyclic adenosine 3',5'-monophosphate response element-binding protein CREB during spermatogenesis involves alternative exon splicing. *Mol. Endocrinol.* **5,** 1418–1430.
41. Ruppert, S., Cole, T. J., Boshart, M., Schmid, W., and Schutz, G. (1992) Multiple mRNA isoforms of the transcription factor CREB: generation by alternative splicing and specific expression in primary spermatocytes. *EMBO J.* **11,** 1503–1512.
42. Walker, W. H., Fucci, L., and Habener, J. F. (1995) Expression of the gene encoding transcription factor cyclic adenosine 3',5'-monophosphate (cAMP) response element-binding protein (CREB): regulation by follicle stimulating hormone-induced cAMP signalling in primary rat sertoli cells. *Endocrinology* **136,** 3534–3545.
43. Molina, C. A., Foulkes, N. S., Lalli, E., and Sassone-Corsi, P. (1993) Inducibility and negative autoregulation of CREM: an alternative promoter directs the expression of ICER, an early response repressor. *Cell* **75,** 875–886.
44. Schwaninger, M., Schoel, C., Blume, R., Rossig, L., and Knepel, W. (1995) Inhibition by antidepressant drugs of cyclic AMP response element-binding protein/cyclic AMP response element-directed gene transcription. *Mol. Pharmacol.* **47,** 1112–1118.
45. Lindsay, R. M., Wiegand, S. J., Anthony Altar, C. A., and DiStefano, P. S. (1994) Neurotrophic factors: from molecule to man. *TINS* **17,** 182–190.
46. Lindvall, O., Kokaia, Z., Bengzon, J., Elmer, E., and Kokaia, M. (1994) Neurotrophins and insults. *TINS* **17,** 490–496.
47. Thoenen, H. (1995) Neurotrophins and neuronal plasticity. *Science* **270,** 593–598.
48. Kang, H. and Schuman, E. M. (1995) Long-lasting neurotrophin-induced enhancement of synaptic transmission in the adult hippocampus. *Science* **267,** 1658–1662.

49. Levine, E. S., Dreyfus, C. F., Black, I. B., and Plummer, M. R. (1995) Brain derived neurotrophic factor rapidly enhances transmission in hippocampal neurons via postsynpatic tyrosine kinase receptors. *Proc. Natl. Acad. Sci. USA* **92**, 8074–8077.

50. Korte, M., Carroll, P., Wolf, E., Brem, G., Thoenen, H., and Bonhoeffer, T. (1995) Hippocampal long-term potentiation is impaired in mice lacking brain-derived neurotrophic factor. *Proc. Natl. Acad. Sci. USA* **92**, 8865–8860.

51. Patterson, S. L., Abel, T., Deuel, T. A., Martin, K. C., Rose, J. C., and Kandel, E. R. (1996) Recombinant BDNF rescues deficits in basal synaptic transmission and hippocampal LTP in BDNF knockout mice. *Neuron* **16**, 1137–1145.

52. Figurov, A., Pozzo-Miller, L. D., Olafsson, P., Wang, T., and Lu, B. (1996) Regulation of synaptic responses to high-frequency stimulation and LTP by neurotrophins in the hippocampus. *Nature* **381**, 706–709.

53. Condorelli, D. F., Dell'Albani, P., Mudo, G., Timmusk, T., and Belluardo, N. (1994) Expression of neurotrophins and their receptors in primary astroglial cultures: induction by cAMP elevating agents. *J. Neurochem.* **63**, 509–516.

54. Ghosh, A., Carnahan, J., and Greenberg, M. E. (1994) Requirement for BDNF in activity-dependent survival of cortical neurons. *Science* **263**, 1618–1623.

55. Widnell, K. L., Self, D. W., Lane, S. B., Russell, D. S., Vaidya, V. A., Miserendino, M. J. D., Rubin, C. R., Duman, R. S., and Nestler, E. J. (1996) Regulation of CREB expression: in vivo evidence for a functional role in morphine action in the nucleus accumbens. *J. Pharmacol. Exp. Ther.*, in press.

56. Duman, R. S., Vaidya, V. A., Nibuya, M., Morinobu, S., and Rydelek Fitzgerald, L. (1995) Stress, antidepressant treatments, and neurotrophic factors: molecular and cellular mechanisms. *Neuroscientist* **1**, 351–360.

57. Siuciak, J. A., Lewis, D. R., Wiegand, S. J., and Lindsay, R. M. (1996) Antidepressant-like effect of brain derived neurotrophic factor (BDNF). *Pharmacol. Biochem. Beh.* **56**, 131–137.

58. Mamounas, L. A., Blue, M. E., Siuciak, J. A., and Anthony Altar, C. (1995) BDNF promotes survival and sprouting of serotonergic axons in the rat brain. *J. Neurosci.* **15**, 7929–7939.

59. Sklair-Tavron, L. and Nestler, E. J. (1995) Opposing effects of morphine and the neurotrophins, NT-3, NT-4, and BDNF, on locus coeruleus neurons in vitro. *Brain Res.* **702**, 117–125.

60. Smith, M. A., Makino, S., Kvetnansky, R., and Post, R. M. (1995) Stress alters the express of derived neurotrophic factor and neurotrophin-3 mRNAs in the hippocampus. *J. Neurosci.* **15**, 1768–1777.

61. Sapolsky, R. M., Krey, L. C., and McEwen, B. S. (1985) Prolonged glucocorticoid exposure reduces hippocampal neuron number: implications for aging. *J. Neurosci.* **5**, 1222–1227.

62. Uno, H., Tarara, R., Else, J. G., Suleman, M. A., and Sapolsky, R. M. (1989) Hippocampal damage associated with prolonged and fatal stress in primates. *J. Neurosci.* **9**, 1705–1711.

63. Sapolsky, R. M., Uno, H., Robert, C. S., and Finsh, C. E. (1990) Hippocampal damage associated with prolonged glucocorticoid exposure in primates. *J. Neurosci.* **10**, 2897–2902.

64. Wooley, C. S., Gould, E., and McEwen, B. S. (1990) Exposure to excess glucocorticoids alters dendritic morphology of adult hippocampal pyramidal neurons. *Brain Res.* **531**, 225–231.

65. Watanabe, Y., Gould, E., and McEwen, B. S. (1992) Stress induces atrophy of apical dendrites of hippocampal CA3 pyramidal neurons. *Brain Res.* **588**, 341–345.

66. Stein-Behrens, B., Mattson, M. P., Chang, I., Yeh, M., and Sapolsky, R. (1994) Stress exacerbates neuron loss and cytoskeletal pathology in the hippocampus. *J. Neurosci.* **14**, 5373–5380.

67. Magarinos, A. M., McEwen, B.S., Flugge, G., and Fuchs, E. (1996) Chronic psychosocial stress causes apical dendritic atrophy of hippocampal CA3 pyramidal neurons in subordinate tree shrews. *J. Neurosci.* **16,** 3534–3540.

68. Sapolsky, R. (1990) Glucocorticoids, hippocampus, and the glutamatergic synapse. *Prog. Brain Res.* **86,** 13–23.

69. Bremner, J. D., Randall, P., Scott, T. M., Bronen, R. A., Seibyl, J. P., Southwick, S. M., Delaney, R. C., McCarthy, G., Charney, D. S., and Innis, R. B. (1995) MRI-based measurement of hippocampal volume in patients with combat-related posttraumatic stress disorder. *Am. J. Psychiatry* **152,** 973–981.

70. Sheline, Y. I., Wany, P., Gado, M. H., Csernansky, J. G., and Vannier, M. W. (1996) Hippocampal atrophy in recurrent major depression. *Proc. Natl. Acad. Sci. USA* **93.**

71. Sapolsky, R. M. (1996) Glucocorticoids and atrophy of the human hippocampus. *Science* **273,** 749,750.

72. Young, E. A., Haskett, R. F., Murphy-Weinberg, V., Watson, S. J., and Akil, H. (1991) Loss of glucocorticoid fast feedback in depression. *Arch. Gen. Psychiatry* **48,** 693–699.

73. Birnstiel, S. and Hass, H. L. (1991) Acute effects of antidepressant drugs on long-potentiation (LTP) in rat hippocampal slices. *Naunyn-Schmiedeberg's Arch. Pharmacol.* **344,** 79–83.

74. Massicotte, G., Bernard, J., and Ohayon, M. (1993) Chronic effects of trimipramine, an antidepressant, on hippocampal synaptic plasticity. *Behav. Neural. Biol.* **59,** 100–106.

75. Shors, T. J., Seib, T. B., Levine, S., and Thompson, R. F. (1989) Inescapable versus escapable shock modulates long-term potentiation in the rat hippocampus. *Science* **244,** 224–226.

76. Allain, H., Lieury, A., Brunet-Bourgin, F., Mirabaud, C., Trebon, P., Le Coz, F., and Gandon, J. M. (1992) Antidepressants and cognition: comparative effects of moclobemide, viloxazine and maprotiline. *Psychopharmacology* **106,** S56–S61.

77. Spring, B., Gelenberg, A. J., Garvin, R., and Thompson, S. (1992) Amitriptyline, clovoxamine and cognitive function: a placebo-controlled comparison in depressed outpatients. *Psychopharmacology* **108,** 327–332.

78. Yau, J. L. W., Olsson, T., Morris, R. G. M., Meaney, M. J., and Seckl, J. R. (1995) Glucocorticoids, hippocampal corticosteroid receptor gene expression and antidepressant treatment: relationship with spatial learning in young and aged rats. *Neuroscience* **66,** 571–581.

79. Nakamura, S. (1990) Antidepressants induce regeneration of catecholaminergic axon terminals in the rat cerebral cortex. *Neurosci. Lett.* **111,** 64–68.

80. Watanabe, Y., Gould, E., Daniels, D. C., Cameron, H., and McEwen, B. S. (1992) Tianeptine attenuates stress-induced morphological changes in the hippocampus. *Eur. J. Pharmacol.* **222,** 157–162.

81. Cronin, J. and Dudek, F. E. (1988) Chronic seizures and collateral sprouting of dentate fibers after kainic acid treatment in rats. *Brain Res.* **474,** 181–184.

82. Cavazos, J. E., Golarai, G., and Sutula, T. P. (1991) Mossy fiber synaptic reorganization induced by kindling: time course of development, progression, and permanence. *J. Neurosci.* **11,** 2795–2803.

83. Davenport, C. J., Jann Brown, W., and Babb, T. L. (1990) Sprouting of GABAergic and mossy fibers in dentate gyrus following intrahippocampal kainate in the rat. *Exp. Neurol.* **109,** 180–190.

84. Cavazos, J. E., Das, I., and Sutula, T. P. (1994) Neuronal loss induced in limbic pathways by kindling: evidence for induction of hippocampal sclerosis by repeated brief seizures. *J. Neurosci.* **14,** 3106–3121.

85. Vaidya, V. A., Marek, G. J., Aghajanian, G. K., and Duman, R. S. (1997) 5-HT$_{2A}$ receptor-mediated regulation of brain-derived neurotrophic factor mRNA in hippocampus and neocortex. *J. Neurosci.* **17,** 2785–2795.

86. Wachtel, H. (1983) Potential antidepressant activity of rolipram and other selective cyclic adenosine 3',5'-monophosphate phosphodiesterase inhibitors. *Neuropharmacology* **22,** 267–272.
87. Horowski, R. and Sastre-y-Hernandez, M. (1985) Clinical effects of the neurotrophic selective cAMP phosphodiesterase inhibitor rolipram in depressed patients: global evaluation of the preliminary reports. *Curr. Ther. Res.* **38,** 23–29.
88. Malison, R., Nestler, E. J., Price, L., Heninger, G. R., and Duman, R. S. (1996) Efficacy of papaverine addition in treatment-refractory major depression. *Am. J. Psychiatry*, in press.

# 11

# Antidepressants: *Beyond the Synapse*

## S. Paul Rossby and Fridolin Sulser

## 1. INTRODUCTION

In this chapter, we discuss the actions of antidepressants in the context of central nervous system (CNS) plasticity. The term "plasticity" implies that the CNS can adapt to conditions that threaten the physical and psychic/emotional well-being of the organism by altering programs of gene expression in specific neuronal and glial cell populations *(1–3)*. These changes affect the construction and pruning of synaptic connections, the concentrations of enzymes and ion channel proteins, the rates of synthesis, release, and degradation of neurotransmitters and neuromodulators, and the densities and activities of their receptors. Ultimately, changes in programs of gene expression determine the intensities of incoming signals, the sensitivities of neuronal systems to those signals, and the nature, amplitude, and duration of CNS responses.

The functional utility of inherent mechanisms of CNS plasticity depends on the efficacy of the changes produced, i.e., whether they actually protect the organism from harm, and upon the capacity of the system to remain responsive to changing conditions (i.e., its continued plasticity). Research focused on the nature and pathogenesis of affective disorders has identified what appear to be aberrant programs of gene expression *(4,5)*, suggesting that the CNS of an individual suffering from depression may be locked into programs that no longer respond appropriately to changing external circumstances.

The assertion that CNS plasticity involves changes in programs of gene expression is supported by considerable evidence, including: The redundancy of promoter sequences (e.g., CRE, AP-1, AP-2, and so on) in the genome strongly suggests that transcription factors activated by intracellular signal transduction pathways can regulate the expression of sets of genes *(6)*; the activation of immediate early gene expression (e.g., c-*fos* and c-*jun*) by pharmacological and environmental stimuli results in the production of transcription factors that apparently orchestrate long-term adaptations of neuronal function by controlling the expression of late response genes *(7,8)*; and steroid hormones modify the transcription rates of cell-specific gene networks *(9)*. The rhythmic nature of gene expression in the CNS is well established. It is reflected in the cir-


From: Antidepressants: *New Pharmacological Strategies*
Edited by: P. Skolnick, Humana Press Inc., Totowa, NJ

cadian rhythms of α- and β-adrenergic, serotonergic, cholinergic, dopaminergic, opiate, and benzodiazepine receptor numbers *(10)*; urinary metabolites of serotonin (5-HIAA), dopamine (HVA), and norepinephrine (MHPG) *(10)*; levels of melatonin, norepinephrine, thyroid stimulating hormone, ACTH, cortisol, corticosterone, growth hormone, and prolactin *(11–14)*; region-specific levels of serotonin and dopamine *(15)*; and messenger RNAs encoding tryptophan hydroxylase *(16)*, corticotropin releasing hormone (CRH) *(17)*, glucocorticoid, mineralocorticoid, and serotonin (5-HT$_{2C}$) receptors *(18)*, as well as immediate early genes (IEGs), such as c-*fos*, *NGFI-A*, *NGFI-B*, c-*jun*, *jun*B, and *jun*D *(19)*.

## 2. MECHANISMS OF ANTIDEPRESSANT THERAPY: POSTULATES

Collectively, our current research is based on the presumption that understanding the mechanisms of antidepressant therapy at the molecular level will lead to a more complete understanding of the nature and etiology of depression, the development of better antidepressants, and perhaps to a cure. We suggest two postulates that may guide us in this endeavor. The first is that antidepressants do not restore plasticity in the CNS; they compensate for its loss. The second is that increases and decreases in neurochemicals (transmitters, modulators, hormones, receptors, enzymes, transcription factors, and so on) induced by antidepressants probably represent changes in waveforms and/or phase positions, since the systems we are studying oscillate continuously from positive to negative, from maxima to minima.

### 2.1. Postulate I: Two Examples

The delayed desensitization of the norepinephrine (NE) β-adrenoceptor-coupled adenylate cyclase system by virtually all clinically effective treatments with a noradrenergic component, including electroconvulsive therapy (ECT) *(4)*, was perhaps the first indication that antidepressants can compensate for loss of plasticity at the level of gene expression. Implicit in this finding were the notions that at some time in the patient's history the amplitude of β-adrenergic receptor synthesis was increased in order to compensate for insufficient catecholamine levels, and that the CNS somehow lost the capacity to restore normal plasticity to this system, resulting in perpetual hypersensitivity. Parenthetically, it is possible that inescapable stress (e.g., prenatal) could have caused the initial state of catecholamine insufficiency (rate of release exceeding rate of synthesis) leading to β-adrenoceptor upregulation.

Glucocorticoid hypersecretion in major depression indicates a loss of plasticity in the hypothalamic-pituitary-adrenal system (HPA axis) and, consequently, in programs of gene expression that have evolved to protect the organism from overreactions to stress. These programs are orchestrated at the genomic level by glucocorticoid–receptor complexes, which bind to specific DNA sequences (glucocorticoid response elements or GREs) and enhance (or decrease) the transcription of cell-type-specific networks of genes *(9)*. The functional utility of the HPA axis, which regulates both the circadian and phasic release of adrenal glucocorticoids, depends on its capacity for autoregulation, i.e., its sensitivity to feedback control. It should be emphasized that loss of control in this system is extensively pathogenic, because of the wide distribution of glucocorticoid receptors throughout the CNS (including glial cells) and, by implication, the large number of genes involved.

Recently, tricyclic antidepressants (e.g., desipramine [DMI]) have been shown to increase brain glucocorticoid receptor type II (GRII) immunoreactivity *(20)*, density *(21)*, and mRNA in primary neuronal cultures *(22)* and in vivo (*see* Section 5.).

It has been suggested that the DMI-induced increase in GRII receptor density could restore feedback inhibition in the HPA system and thus be responsible for its normalization during successful antidepressant pharmacotherapy *(23)*. If this is indeed the case, it represents an important example of pharmacologic compensation for a loss of plasticity.

## 2.2. Postulate II

The hypothesis linking affective disorders with disturbances in circadian rhythms was originally based on four clinical symptoms of depression, i.e., early morning awakening, diurnal variation in symptom severity, seasonality, and cyclicity of the illness. The hypothesis states (in part) that long-term cycles of relapse and remission could occur if affective episodes resulted from an abnormal internal-phase relationship between two circadian rhythms, and if at least one of those rhythms escaped from entrainment to the day–night cycle and was free-running in and out of phase with the other rhythm.

In support of this hypothesis, the peak amplitude of cortisol secretion has been shown to occur earlier in its circadian rhythm (a phase-advance) in depressed patients, compared with normal controls *(24)*. This shift in timing was accompanied by a change in waveform and was confirmed by independent studies involving reasonably large numbers of patients *(25,26)*. In one of these studies *(25)*, the degree of phase-advance was correlated with the severity of the depression. Phase-advances in depressed patients have also been reported for circadian temperature rhythm, the onset of REM sleep, and for urinary metabolites of 5-HT, dopamine (DA), and morepinephrine (MHPG) *(27,28)*.

Of particular interest, regarding our second postulate, are the effects of chronic antidepressant treatment on circadian, ultradian (less than 24 h), and circannual rhythms. Replicable diurnal rhythms were initially demonstrated in densities (but not apparent affinities) of α- and β-adrenergic, cholinergic, dopaminergic, opiate, and benzodiazepine receptor numbers (not for apparent affinities). These rhythms were shown to persist in constant darkness, with a change in waveform, indicating that they are "endogenous, but are also subject to the direct (masking) effects of light" *(10)*. In response to chronic treatment with clorgyline or imipramine, all temporal characteristics of these receptor rhythms, i.e., waveform, amplitude, 24-h mean, and phase position, were changed. Peak α-adrenergic receptor binding occurred 8–12 h later than controls and phase delays in the other receptor rhythms ranged from 4–8 h *(10)*.

In human studies, Linkowski et al. *(12)* have reported the effects of antidepressant treatment on the circadian rhythms of plasma ACTH and cortisol and the timing of REM sleep in patients suffering from major depressive illness. During the acute phase of the illness, the patients presented with early onset of REM sleep, phase-advanced ACTH/cortisol rhythms, hypercortisolism, and shortened periods of quiescent nocturnal cortisol secretion. Following treatment with ECT or amytriptyline, the timing of the circadian rhythms of ACTH and cortisol and the duration of the quiescent period of cortisol secretion were normalized. In addition, REM latencies were increased and cortisol levels returned to normal because of a change in the amplitude of episodic pulses. The authors concluded that a disorder of circadian rhythmicity characterizes acute episodes of major depressive illness, and that this chronobiological abnormality, as well as the hypersecretion of ACTH and cortisol, are state, rather than trait, dependent.

## 3. INTRACELLULAR MECHANISMS OF COMPENSATION

We thus arrive at the critical question of how antidepressant therapies compensate for loss of plasticity in the CNS. In searching for answers to this question, we have begun perforce at the historical starting point. That is, the observation that, although all clinically effective antidepressants, regardless of chemical class, affect central noradrenergic and/or serotonergic neuronal systems at various levels of signal transduction (or are converted in vivo to metabolites, which, in concert with the parent drug, affect the synaptic availability of NE and/or 5-HT), these rapid actions (within minutes and hours) are obviously not ultimately responsible for the delayed therapeutic actions of these drugs in humans. Moreover, although significant changes in agonist–receptor interactions and the production of second messengers do occur after chronic, but not acute, administration of most (but not all) antidepressants, the search for common principles at these levels has been challenging. For example, the antidepressant-induced desensitization of the β-adrenoceptor-coupled adenylate cyclase system *(4,29–31)*, often linked to a downregulation of β-adrenoceptor density *(32,33)*, occurs after antidepressants with a strong NE component; however, selective 5-HT uptake inhibitors (SSRIs) either do or do not consistently downregulate β-adrenoceptors *(34)*. Furthermore, the density of 5-HT$_{2A}$ and/or 5-HT$_{1A}$ receptors is reportedly modified after chronic administration of antidepressants, but the data are inconsistent, particularly with regard to the SSRIs *(35–38)*. Finally, the upregulation of 5-HT$_{2A}$ receptor density following ECT *(39)*, the most effective antidepressant treatment, does not fit with the concept of 5-HT$_{2A}$ receptor downregulation as being a common, clinically relevant therapeutic action of antidepressant treatments. The increases in synaptic 5-HT, which result from the delayed desensitization of terminal 5-HT autoreceptors (5-HT$_{1B}$/5-HT$_{1D}$) by chronic administration of SSRIs (*see* Chapter 1) *(40)*, may provide at least a partial explanation for the latency of their antidepressant action in humans.

The antidepressant-induced desensitization of the β-adrenoceptor-coupled adenylate cyclase system is currently being viewed as a consequence of agonist-mediated receptor phosphorylation by protein kinase A (PKA), and possibly the second messenger-independent protein kinase BARK 1. This phosphorylation is apparently caused by the persistent occupancy of the β-adrenoceptor by NE, which occurs, for example, during chronic treatment with desmethylimipramine (DMI) *(41)*. Another recent study *(42)* has provided evidence that a cell line exhibiting a high level of constitutive 5-HT$_{2C}$ receptor activity undergoes agonist-mediated desensitization, which is consistent with current models of G protein-coupled receptor regulation *(43–47)*. Thus, after shifting our focus in the mid-1970s from acute presynaptic events to delayed changes in the sensitivities of postsynaptic β-adrenergic and 5-HT receptor subtypes, recent research mandates a further shift of attention to the convergence of aminergic signals beyond the receptors, which ultimately leads to changes in programs of gene expression (*see* Section 5.).

## 4. PUTATIVE TARGETS FOR ANTIDEPRESSANTS BEYOND THE SYNAPSE

Lipid soluble drugs, such as DMI, are taken up by cells in a dose- and time-dependent manner *(48)*, leading, after chronic administration, to intracellular drug concentrations that are high enough to exert biological actions on cytoplasmic and nuclear constituents. Putative intracellular targets for antidepressant drugs that affect cytoplasmic–nuclear signal transfer are listed in Table 1.

### *4.1. Heterotrimeric G Proteins as Targets for Antidepressants*

G proteins mediate the stimulation ($G_s$) or inhibition ($G_i$) of intracellular effector molecules (e.g., adenylate cyclase, phospholipase C, phospholipase A2, and various ion channels) via neurotransmitter receptor/complexes (e.g., NE, 5-HT), leading to the generation of second messengers (e.g., cyclic AMP, diacylglycerol [DAG], inositol triphosphate [IP3], and arachidonic acid), and to modifications of ion fluxes. Thus, the notion that G proteins may be intracellular targets for antidepressants (and other psychotropic drugs) is attractive. In support of this idea, chronic administration of antidepressant drugs has been reported to modify the levels of G proteins, their mRNAs, and function; however, the results of these studies have been rather equivocal and somewhat difficult to reconcile. For example, the chronic administration of imipramine has been reported to decrease the levels of $G_{s\alpha}$ immunolabeling, cholera toxin ADP ribosylation, and $G_{s\alpha}$ mRNA in rat brain *(49,50)*. However, other investigators did not find changes in $G_{s\alpha}$ or $G_{i\alpha}$ mRNAs or the levels of G proteins following chronic treatment with various antidepressants *(51–54)*. When Gs function was assayed, chronic treatment with antidepressants was found to facilitate the activation of adenylate cyclase by $G_{s\alpha}$ without changing the content of G proteins *(52,55)*. Since direct activation of adenylate cyclase by $Mn^{2+}$ was not altered after chronic treatment with antidepressants, a modification of the $G_s$ protein, rather than adenylate cyclase, appears to be responsible for the enhanced activation *(56)*. However, based on agonist-induced increases in guanine nucleotide binding, $G_s$ function is apparently attenuated by chronic treatment with antidepressants *(57)*. Lithium has also been reported to alter G-protein function, G-protein levels, and G-protein subunit mRNAs, but these effects also remain equivocal *(58)*. Thus, although G proteins and their subunits are exciting prospects for mediating the therapeutic effects of antidepressants beyond the receptors, unequivocal conclusions in this regard may have to await more powerful methodology. For example, recent crystallographic studies of G proteins and their subunits promise to reveal how G proteins actually function *(59)*. Such increased knowledge could eventually lead to the synthesis of drugs acting selectively on G-protein subunits.

### *4.2. The Convergence of Neurotransmitter Signals Beyond the Receptors*

Since antidepressants affect signal transduction cascades, which activate various protein kinases, i.e., protein kinase A (PKA), protein kinase C (PKC), and calcium-

**Table 1**
**Putative Targets for Antidepressants Beyond the Receptors**

G proteins: $G_s$, $G_i$, $G_o$
Second messengers: cyclic AMP, cyclic GMP, diacylglycerol, inositol trisphosphate ($IP_3$), arachidonic acid
Second messenger-dependent protein kinases:
  cyclic AMP-dependent protein kinase (PKA)
  cyclic GMP-dependent protein kinase (PKG)
  $Ca^+$ +/calmodulin-dependent protein kinases
  Protein Kinase C (PKC)
Second messenger-independent protein kinases:
  Family of G-protein receptor-coupled kinases (GRKs), e.g., β-adrenergic receptor kinase (BARK)
Isoforms of protein kinase inhibitor proteins
Transcription factors (*see* Table 2)

From ref. *41*.

calmodulin (CaCAM) dependent protein kinases (Fig. 1), crosstalk among these kinases may be relevant to the modes of action of antidepressants. For example, counterregulation of agonist-induced desensitization of the β-adrenoceptor-coupled adenylate cyclase system by various protein kinases has been demonstrated in various cell lines *(60)*. Oligodeoxynucleotides antisense to mRNAs encoding PKA and BARK 1 nearly abolished desensitization of the β-adrenoceptor-coupled adenylate cyclase system in $DDT_1$ MF-2 and A-431 cell lines, but oligodeoxynucleotides antisense to PKC mRNA enhanced the desensitization. More recently, Döbbeling and Berchtold *(61)* convincingly demonstrated in a fibroblast cell line that both intracellular calcium and PKC activation downregulate the PKA pathway, and, furthermore, that the inhibitory effect of PKC occurs upstream of PKA activation.

Although these results are difficult to extrapolate from one cell line to another (or to the CNS in vivo), they are instrumental for research currently focused on the dual uptake inhibitor venlafaxine. In contrast to DMI-like antidepressants, chronic venlafaxine treatment does not desensitize the β-adrenoceptor-coupled adenylate cyclase system, except in the absence of 5-HT *(41)*. It should be noted that midalcipran, another 5-HT/NE dual reuptake inhibitor, neither desensitizes the β-adrenoceptor-coupled adenylate cyclase system nor downregulates the density of β-adrenoceptors *(62)*.

Research on the (unexpected) pharmacology of venlafaxine has apparently revealed an in vivo mechanism of crosstalk, involving the counterregulation by PKC of PKA-mediated desensitization of the β-adrenoceptor system *(41)*. This crosstalk has been suggested to be the consequence of hyperactivation of the inositol pathway (and thus PKC) by increased synaptic 5-HT (because of reuptake inhibition).

A hypothetical model for the convergence of PKA and PKC pathways at the level of transcriptional activation, i.e., phosphorylation of transcription factors, has also been presented *(63)*. The PKC and calcium pathways nearly always synergize, since they are often triggered simultaneously by the same agonist receptor interactions (e.g., 5-HT via $5\text{-}HT_{2A}$ or $5\text{-}HT_{2C}$ receptors), but, depending on the cell type, they sometimes

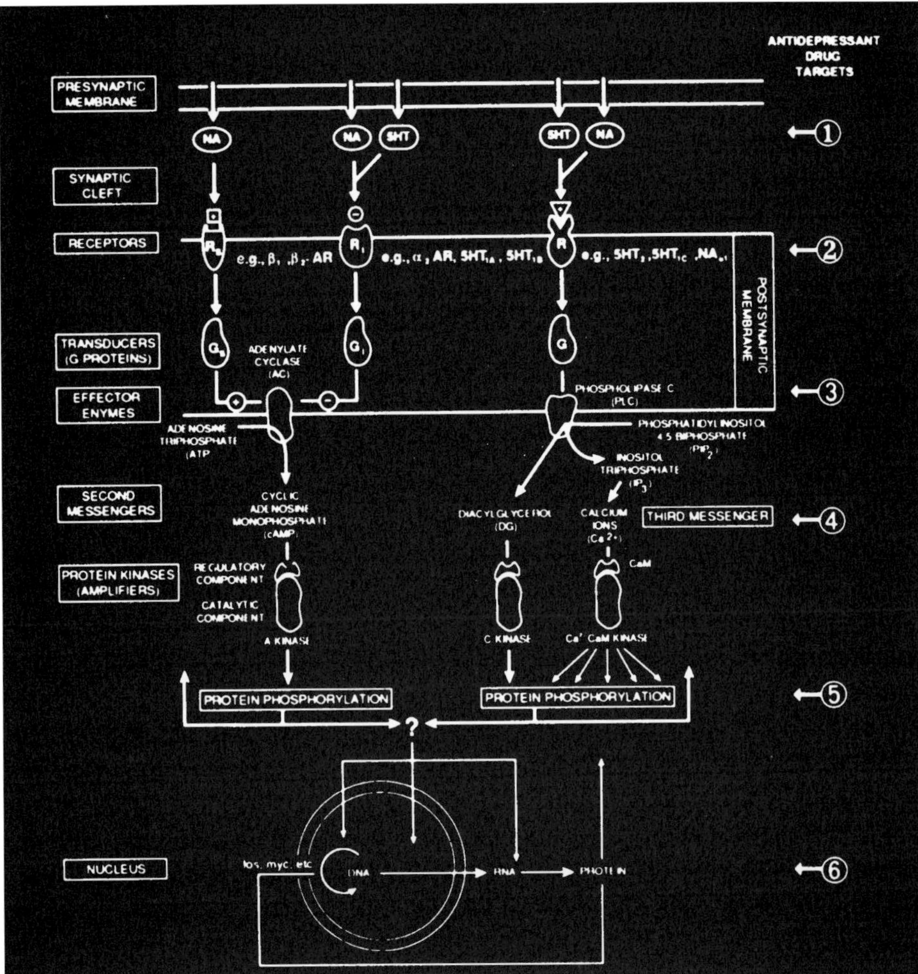


**Fig. 1.** Neurotransmitter signal transduction cascades as targets for antidepressant drugs. Antidepressant drugs can influence the information flow at various levels of the signal transduction cascades: (1) change in the synaptic availability of the primary signals NE and/or 5HT (MAO inhibitors; blockade of reuptake; autoreceptor subsensitivity); (2) change in the receptor number or sensitivity; (3) change in the function of G proteins; (4) change in the formation of second messengers; (5) change in the activity of protein kinases (amplifier function); (6) modification of nuclear events (e.g., gene transcription). From ref. *79.*

antagonize or synergize with the PKA pathway *(64).* Antidepressants acting through the NE and 5-HT signal transduction pathways can activate both PKA and PKC, and have also been shown to increase the density of GRII receptors (and presumably GR-mediated gene expression). Based on these data, it is tempting to suggest that the antidepressant-sensitive, 5-HT-linked, and glucocorticoid-responsive β-adrenoceptor signal transduction system in brain is involved in a much more general way as an amplification–adaptation system of stimulus-transcription coupling and the regulation of brain-specific gene expression (*see* Chapter 10).

### 4.3. Transcription Factors and Their Nuclear Import as Targets for Antidepressants

In Section 6., we advance the concept that the ultimate therapeutic effects of antidepressants are mediated via changes in the expression of subsets of genes. One of the events that links signals initiated at the cell surface with altered gene expression is the activation of cytoplasmic transcription factors and their translocation to the nucleus (nuclear import). Some of the eukaryotic transcription factors that are putative targets for antidepressants are listed in Table 2. It has been estimated that the human genome encodes as many as 3000 gene-specific transcription factors *(65)*. These transcription factors contain functionally distinct domains that predispose them to the effects of pharmaceutical agents. Antidepressant drugs could alter gene expression by altering the nuclear import of transcription factors, e.g., at the level of nuclear pore docking or translocation across the nuclear pore complex, by changing the formation of homo- and/or heterodimers between, e.g., CREB, CREM, *fos*B, *jun*B, and other transcription factors containing the leucine zipper motif, thus modifying transcription factor activity, and by altering the affinities of transcription factors to specific DNA promoter sequences (nuclear receptors). The processes of phosphorylation and dephosphorylation play major roles in altering the conformations of most (if not all) transcription-factor proteins to produce active/inactive DNA binding structures *(66)*. It follows that antidepressants could directly or indirectly modify transcription factor activity and DNA binding by affecting phosphorylation via PKA-, PKC-, or CaCAM-dependent protein kinases (target number 5 in Fig. 1). Additionally, cytoplasmic–nuclear trafficking of transcription factors is regulated by phosphorylation *(67)*. Examples of such regulation involve direct phosphorylation of the transporter protein, masking of the nuclear localization signals, modulation of the import machinery itself, and so on, all of which constitute an important regulatory checkpoint in the control of gene expression *(68)*. There are numerous possibilities for targeting transcription factors directly with drugs, including: The nuclear transport machinery can be altered so that it no longer recognizes a specific transcription factor; the binding affinities of nuclear localization signals for their receptors at the nuclear membrane can be modified; the translocation of transcription factors to the nucleus can be modified, and/or their affinities for responsive promoter elements can be altered. The demonstration that the coupling of hormonal stimulation and transcription via the transcription factor CREB is rate-limited by the nuclear entry of PKA *(69)*, and the reported translocation of PKA from the cytoplasmic to the particulate (nuclear) fraction in the frontal cortex in rats after chronic treatment with various antidepressants *(70)*, are pertinent findings. CREB is a member of the ATF1 family of transcription factors *(71)*, and phosphorylation at serine-133 leads to its dimerization and ultimately to the transcriptional activation of genes with CRE-containing promoters.

The translocation of the glucocorticoid receptor complex to the nucleus is of psychopharmacological interest, because antidepressants have been shown to increase glucocorticoid receptor density. In the absence of the hormone, the glucocorticoid receptor (GR) protein exists in the cytoplasm complexed with a mol wt of 90,000-$M_r$ heat-inducible protein (hsp 90). Occupancy of the GR by glucocorticoid hormone causes the dissociation of hsp 90, followed by translocation of the receptor complex to the nucleus, where it binds to GRE, enhancing or inhibiting the transcription of gene networks.

**Table 2**
**Transcription Factors in Brain as Putative Targets of Antidepressants**

Leucine zipper proteins
   CREB/ATF family
   Fos/Jun family
   Fos related antigens (FRAs) Jun B, Jun D
Zinc finger proteins (e.g., Zif 268)
Steroid hormone receptors (e.g., Glucocorticoid receptors, 1,25[OH]$_2$ D$_3$ receptors)
Thyroid hormone receptors
Retinoic acid family

From ref. *91.*

## 5. ANTIDEPRESSANTS AND PROGRAMS OF GENE EXPRESSION

Since the transcriptional activities of DNA-binding proteins are regulated by the convergent activities of various protein kinases *(63)*, and antidepressants activate aminergic signal transduction cascades leading to activation of various protein kinases (Fig. 1), it is not surprising that this class of psychotropic drugs has been shown to cause changes in gene expression *(72–82)*. Indeed, the view has been advanced that the clinically relevant delayed therapeutic actions of antidepressants are the consequence of changes in programs of gene expression that determine the intensities of incoming signals, the sensitivities of neuronal systems to those signals, and the nature, amplitude, and duration of CNS responses *(78)*. One can envision the possibility that abnormal behavior patterns—affective, cognitive, and somatosensory—might be the consequences of a disarray in the regulation of programs of gene expression in response to internal (neurohormonal, neuroendocrine) and external (environmental) stimuli, which, in turn, could make an individual vulnerable to psychiatric disorders. Recent studies from Michael Greenberg's laboratory have provided strong support for such a proposition. Brown et al. *(83)* have shown that female mice with an inactivating mutation in the *fos*B gene are "profoundly deficient" in their ability to nurture their young, but are "normal with respect to other cognitive and sensory functions." These truly exciting studies suggest that a transcription factor (*fos*B) can control a complex behavior through the activation of a program of gene expression, mediated via olfactory cues. Although the target genes that are activated, e.g., in the preoptic area of the hypothalamus by *fos*B (or its dimer with a member of the *jun* family), are unknown, the results of these experiments provide pertinent information on the molecular basis of behavior.

Insofar as antidepressant-induced changes in gene expression are concerned, results from our laboratory demonstrate that these changes can occur via two fundamentally different mechanisms: via agonist–receptor-mediated transduction cascades (Fig. 2); and by a mechanism that is independent of agonist–receptor-mediated signal transduction processes. Rossby et al. *(78)* have shown that the fluoxetine-induced increase in the steady-state level of preproenkephalin mRNA in the rat amygdala is dependent on the availability of synaptic 5-HT, since the increase in preproenkephalin mRNA by fluoxetine is abolished in brains depleted of central 5-HT. It has been suggested that the fluoxetine-induced serotonergic regulation of preproenkephalin gene expression in the rat amygdala could be mediated via phosphorylation of the nuclear transcription factor

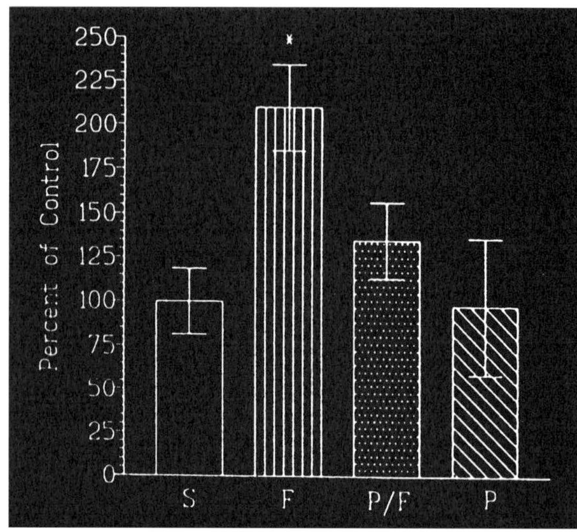


**Fig. 2.** Steady-state levels of PPE mRNA in rat amygdala determined by Northern blot hybridization analysis. Rats were treated for 13 d with saline (S, $n = 9$), fluoxetine (F, $n = 8$), PCPA plus fluoxetine (P/F, $n = 5$), or PCPA (P, $n = 6$). Values are expressed as percentage of control ± SEM. *$p = 0.0027$ vs saline. From ref. *78*.

CREB by the CaCAM-dependent protein kinase, which is assumed to be stimulated as a consequence of 5-HT/5-HT$_{2A}$ or 5-HT/5-HT$_{2C}$ receptor interactions (IP$_3$ formation and calcium mobilization).

Some tricyclic antidepressants have been reported to increase brain GR immunore-activity *(20)*, GR density *(21)*, and GR mRNA *(74,84)*. Since many antidepressant drugs increase the synaptic availability of NE and/or 5-HT, monoaminergic neuronal systems have been implicated in GR regulation. Since neither acute nor chronic fluox-etine or citalopram cause significant changes in steady-state levels of hippocampal GR mRNA or GR density *(21,77,84,85)*, we were curious to ascertain whether the increases in hippocampal GR mRNA induced by DMI in vivo are dependent on the synaptic availability of NE. In the virtual absence of noradrenergic input (DSP4 lesioned rats), chronic treatment with DMI increased the steady-state levels of hippocampal GR mRNA to the same degree in lesioned as in nonlesioned animals (Fig. 3). Moreover, both (+)-oxaprotiline (NE reuptake blocker) and (−)-oxaprotiline (no blockade of NE uptake) increased hippocampal GR mRNA following chronic administration of the drugs (unpublished results from this laboratory), thus demonstrating convincingly that an increased synaptic availability of NE is not a prerequisite for the effects of some antidepressants on gene expression. This agonist–receptor-independent action is rather surprising, because other biochemical effects of DMI-like drugs are dependent on the increased synaptic availability of NE, e.g., β-adrenoceptor downregulation and desen-sitization of the β-adrenoceptor-coupled adenylate cyclase system *(30,86,87)*. These in vivo results are compatible, however, with data demonstrating increased glucocorticoid receptor promoter activity after DMI treatment (Fig. 4) in transfected LTK$^-$ or neurob-lastoma cells in vitro *(88)*. It is conceivable that cytoplasmic concentrations of DMI are high enough after chronic administration to affect the activities and/or nuclear transport

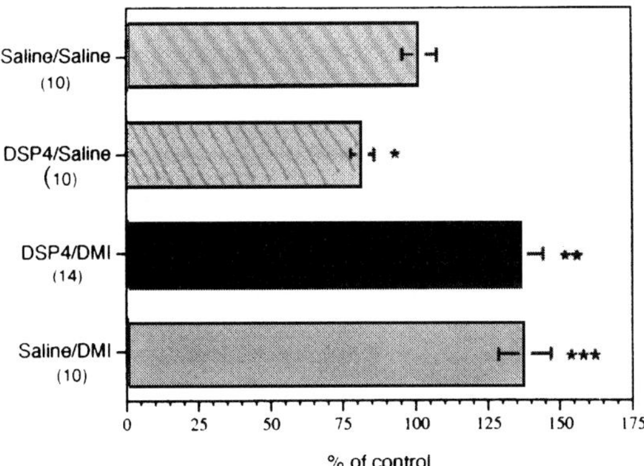


**Fig. 3.** Steady-state levels of glucocorticoid type II receptor (GRII) mRNA. Rats were treated with DSP4/saline ($n$ = 10), DSP4/DMI ($n$ = 14), and saline/DMI ($n$ = 10). The hippocampal GRII mRNA levels are expressed as a percent of controls treated with saline/saline. Significance was determined by ordinary ANOVA, followed by a Newman-Keuls test. *$p$ < 0.05 (DSP4/saline vs saline/saline); **$p$ < 0.001 (DSP4/DMI vs DSP4/saline); ***$p$ < 0.01 (saline/DMI vs saline/saline). From ref. *77*.

of gene-specific transcription factors, as discussed in the previous section. It has been reported that the transcription factor AP2 functions as a positive regulator of GR promoter activity *(89)*. This agonist–receptor cascade-independent regulation of gene expression by some antidepressants illustrates the great potential to affect programs of gene expression by influencing targets beyond the receptors. Table 3 shows data compiled from the recent literature on the effects of antidepressants on gene expression.

## 6. FUTURE PERSPECTIVES

According to Milner and Sutcliffe *(90)*, approx 20,000 genes are expressed exclusively in the CNS. It is reasonable to presume that psychic and emotional phenomena involve the expression of some or all of these genes (plus those that are also expressed in the periphery) in extremely complex temporal patterns. Perhaps the best analogy is a symphony orchestra composed of >20,000 instruments. Utilizing this analogy, the pathogenesis of any CNS disorder would involve some number of instruments (genes) producing dissonance by being played (transcribed and translated) at the wrong times (phase shifts) and/or for the wrong durations and/or amplitudes (numbers of protein and peptide molecules synthesized), compared with the same genes in a healthy CNS.

The notes played by all of the instruments in an orchestra are written linearly across the pages of the conductor's score. By examining the score vertically one can also see which notes are played simultaneously. This vertical aspect of the analogy implies that the state of the CNS, and thus the state of consciousness of an individual, at each moment in time is determined by synchronicity (*see* Acknowledgments) at the level of gene expression.

This concept is diametrically opposed to that of causality. Causality, based on the linear progressions of gene expression, is a matter of statistical probabilities ($p$ < 0.05) and is thus not absolute, whereas synchronicity takes the actual coincidence of events in space and time as meaning something more than mere chance. Accordingly, if we

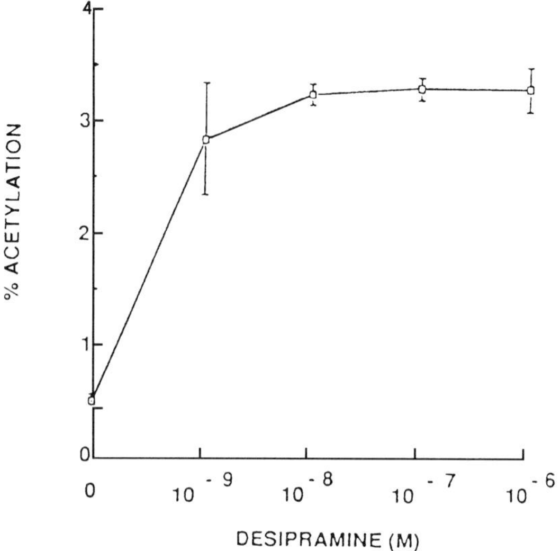


**Fig. 4 .** Desipramine dose–response curve for glucocorticoid receptor promoter activity measured with the reporter plasmid pHGR2.7 CAT. LTK⁻ cells were incubated with increasing concentrations of desipramine for 24 h before transient transfection with the pHGR2.7 CAT plasmid. The reporter plasmid pHGR2.7 CAT (5 µg) was precipitated with pRSV-*LacZ* (2.5 µg) for 5 h, on $2 \times 10^6$ cells, which were harvested 24 h later for assay of CAT activity (at constant β-galactosidase activity for each CAT assay). Results are shown as the mean ± SEM (four experiments). From ref. *88*.

**Table 3**
**Changes in Steady-State Levels of mRNAs by Chronic Administration of Some Antidepressants**

| mRNA | Tissue | Incr./decr. | Ref. |
|---|---|---|---|
| Tyrosine hydroxylase | Rat brain | ⇓⇓ | *72* |
| | Rat brain | ⇑⇑ | *73* |
| CRH | Rat hippocampus | ⇓⇓ | *73* |
| β₁-Adrenoceptor | Rat cortex | ⇑⇑⇓⇓ | *75* |
| Preproenkephalin | Rat amygdala | ⇑⇑ | *78* |
| Glucocorticoid receptors | Primary neuronal cultures | ⇑⇑ | *22* |
| | Rat hippocampus | ⇑⇑ | *74* |
| | Rat hippocampus | ⇑⇑ | *84* |
| | Rat hippocampus | ⇑⇑ | *77* |
| Mineralocorticoid receptor | Rat hippocampus | ⇑⇑ | *73* |
| c-*fos* induction by restraint stress | Rat cortex | ⇓⇓ | *82* |
| NGFI-A induction by restraint stress | Rat cortex | ⇓⇓ | *82* |
| CREB | Rat hippocampus | ⇑⇑ | *81* |
| BDNF and trk B | Rat hippocampus and frontal cortex | ⇑⇑ | *82* |

The data in the Table have been compiled from the recent literature (1989–1996). From ref. *41*.

could reconstruct and compare the scores of the various symphonies of gene expression as played in normal vs dysfunctional CNSs, we could determine which genes are involved in the pathogenesis of psychiatric disorders and how they are involved. From a clinical perspective, this should enable us to formulate pharmacologic therapies designed to restore plasticity in the CNS by regulating not only the levels of gene expression but also the rhythms (waveforms and phase positions).

## ACKNOWLEDGMENTS

The original studies from our laboratory have been supported by USPHS grant MH 29228. We thank Doris Head for the expert typing of this manuscript.

1. [Cf. "Synchronicity: An Acausal Connecting Principle," The Structure and Dynamics of the Psyche (Coll. Works of C.G. Jung, vol. 8).]

## REFERENCES


1. Goelet, P., Castellucci, V. F., Schacher, S., and Kandel, E. R. (1986) The long and the short of long-term memory—a molecular framework. *Nature* **322,** 419–422.
2. Comb, M., Hyman, S. E., and Goodman, H. M. (1987) Mechanisms of trans-synaptic regulation of gene expression. *Trends Neurosci.* **10,** 473–478.
3. Morgan, J. I. and Curran, T. (1988) Calcium as a modulator of the immediate-early gene cascade in neurons. *Cell Calcium* **9,** 303–311.
4. Vetulani, J. and Sulser, F. (1975) Action of various antidepressant treatment reduces reactivity of noradrenergic cyclic AMP generating system in limbic forebrain. *Nature* **257,** 495–496.
5. Sachar, E. J., Hellman, L., Roffwarg, H. P., Halpern, F. S., Fukushima, D. K., and Gallagher, T. F. (1973) Disrupted 24-hour patterns of cortisol secretion in psychotic depression. *Arch. Gen. Psychiatry* **28,** 19–24.
6. Duman, R. S. (1995) Regulation of intracellular signal transduction and gene expression by stress, in *Neurobiological and Clinical Consequences of Stress: From Normal Adaptation to PTSD* (Friedman, M. J., Charney, D. S., and Deutch, A. Y., eds.). Lippincott-Raven, Philadelphia.
7. Sheng, M. and Greenberg, M. E. (1990) The regulation and function of c-fos and other immediate early genes in the nervous system. *Neuron* **4,** 477–485.
8. Morgan, J. I. and Curran, T. (1991) Stimulus-transcription coupling in the nervous system: involvement of the inducible proto-oncogenes fos and jun. *Ann. Rev. Neurosci.* **14,** 421–451.
9. Yamamoto, K. R. (1985) Steroid receptor regulated transcription of specific genes and gene networks. *Ann. Rev. Genet.* **19,** 209–252.
10. Goodwin, F. K., Wirz-Justice, A., and Wehr, T. A. (1982) Evidence that the pathophysiology of depression and the mechanism of action of antidepressant drugs both involve alterations in circadian rhythms, in *Typical and Atypical Antidepressants: Clinical Practice* (Costa, E. and Racagni, G., eds.), Raven, New York.
11. Mendlewicz, J. (1991) Sleep-related chronobiological markers of affective illness. *Int. J. Psychophysiol.* **10,** 245–252.
12. Linkowski, P., Mendlewicz, J., Kerkofs, M., Leclercq, R., Golstein, J., Brasseur, M., Copinschi, G., and Van Cauter, E. (1987) 24-hour profiles of adrenocor-ticotropin, cortisol, and growth hormone in major depressive illness: effect of antidepressant treatment. *J. Clin. Endocrinol. Metab.* **65,** 141–152.
13. Goetz, U. and Tolle, R. (1987) Circadian rhythm of free urinary cortisol, temperature and heart rate in endogenous depressives and under antidepressant therapy. *Neuropsychobiology* **18,** 175–184.

14. Nicholson, S., Lin, J.-H., Mahmoud, S., Campbell, E., Gillham, B., and Jones, M. (1985) Diurnal variations in responsiveness of the hypothalamo-pituitary-adrenal axis in the rat. *Neuroendocrinology* **40,** 217–224.

15. Ozaki, N., Duncan, W. C., Jr., Johnson, K. A., and Wehr, T. A. (1993) Diurnal variations in serotonin and dopamine levels in discrete brain regions of Syrian hamsters and their modification by chronic clorgyline treatment. *Brain Res.* **627,** 41–48.

16. Green, C. B., Cahill, G. M., and Bexharse, J. C. (1995) Regulation of tryptophan hydroxylase expression by retinal circadian oscillator in vitro. *Brain Res.* **677,** 283–290.

17. Kwak, S. P., Morano, M. I., Young, E. A., Watson, S. J., and Akil, H. (1993) Diurnal CRH mRNA rhythm in the hypothalamus: decreased expression in the evening is not dependent on endogenous glucocorticoids. *Neuroendocrinology* **57,** 96–105.

18. Holmes, M. C., French, K. L., and Seckl, J. R. (1995) Modulation of serotonin and corticosteroid receptor gene expression in the rat hippocampus with circadian rhythm and stress. *Br. Res. Mol. Br. Res.* **28,** 186–192.

19. Rusak, B., McNaughton, L., Robertson, H. A., and Hunt, S. P. (1992) Circadian variation in photic regulation of immediate-early gene mRNAs in rat suprachiasmatic nucleus cells. *Br. Res. Mol. Br. Res.* **14,** 124–130.

20. Kitayama, I., Janson, A. M., Cintra, A., Fuxe, K., Agnati, L. F., Ögren, S. O., Harfstrand, A., Eneroth, P., and Gustafsson, J. A. (1988) Effects of chronic imipramine treatment on glucocorticoid receptor immunoreactivity in various regions of the rat brain. *J. Neural. Transm.* **73,** 191–203.

21. Budziszewska, B., Siwanowicz, J., and Przegaliñski, E. (1994) The effect of chronic treatment with antidepressant drugs on the corticosteroid receptor levels in the rat hippocampus. *Pol. J. Pharmacol.* **46,** 147–152.

22. Pepin, M. C., Beanlien, S., and Barden, N. (1989) Antidepressants regulate glucocorticoid receptor messenger RNA concentrations in primary neural cultures. *Mol. Brain Res.* **6,** 77–83.

23. Barden, N., Reul, J. M. H. M., and Holsboer, F. (1995) Do antidepressants stabilize mood through actions on the hypothalamic-pituitary-adrenocortical system? *TINS* **18,** 6–11.

24. Doig, R. J., Mummery, R. V., Wills, M. R., and Elkes, A. (1966) Plasma cortisol levels in depression. *Br. J. Psychiatry* **112,** 1263–1267.

25. Fullerton, D. T., Wenzel, F. J., Ohrenz, F. N., and Fahs, H. (1968) Circadian rhythm of adrenal cortical activity in depression. *Arch. Gen. Psychiatry* **19,** 674–688.

26. Conroy, R. T. W. L., Hughes, B. D., and Mills, J. N. (1968) Circadian rhythms of plasma 11-hydroxycorticosteroids in psychiatric disorders. *Br. J. Med.* **3,** 405–407.

27. Riederer, P., Birkmayer, W., Neumeyer, E., Ambrozi, L., and Linauer, W. (1974) The daily rhythm of HVA, VMA (VA) and 5-HIAA in depression syndrome. *J. Neural Trans.* **35,** 23–45.

28. Wehr, T. A., Muscettola, G., and Goodwin, F. K. (1980) Urinary methoxy-4-hydroxyphenylglycol circadian rhythm. *Arch. Gen. Psychiatry* **37,** 254–263.

29. Vetulani, J., Stawarz, R. J., Dingell, J. V., and Sulser, F. (1976) A possible common mechanism of action of antidepressant treatments. Reduction in the sensitivity of the noradrenergic cyclic AMP generating system in the rat limbic forebrain. *Naunyn-Schmiedeberg's Arch. Pharmacol.* **293,** 109–114.

30. Wolfe, B. B., Harden, T. K., Sporn, J. R., and Molinoff, P. B. (1978) Presynaptic modulation of β-adrenergic receptors in rat cerebral cortex after treatment with antidepressants. *J. Pharmacol. Exp. Ther.* **207,** 446–457.

31. Sellinger, M. D., Mendels, J., and Frazer, A. (1980) The effect of psychoactive drugs on β-adrenergic receptor binding sites in rat brain. *Neuropharmacol.* **19,** 447–454.

32. Banerjee, S. P., Kung, L. S., Riggi, S. J., and Chanda, S. K. (1977) Development of β adrenergic receptor subsensitivity by antidepressants. *Nature* **268,** 455–456.
33. Pryor, J. C. and Sulser, F. (1991) Evolution of monoamine hypotheses of depression, in *Biological Aspects of Affective Disorders* (Horton, R. W. and Katona, C., eds.), Academic, London, pp. 77–94.
34. Wong, D. T., Bymaster, F. P., and Engleman, E. A. (1996) Prozac (fluoxetine, Lilly 110140), the first selective serotonin uptake inhibitor and an antidepressant drug: twenty years since its first publication. *Life Sci.* **57,** 411–441.
35. Eison, A., Yocca, F. D., and Gianutsos, G. (1991) Effect of chronic administration of anti-depressant drugs on 5HT2 mediated behavior in the rat following noradrenergic or sero-tonergic denervation. *J. Neural Transm.* **8,** 19–32.
36. Todd, K. G., McManus, D. J., and Baker, G. B. (1995) Chronic administration of the anti-depressants phenelzine, desipramine, clomipramine or maprotiline decreases binding of 5HT2A receptors without affecting benzodiazepine binding sites in rat brain. *Cell Mol. Neu-robiol.* **15,** 361–370.
37. Goodnough, D. B. and Baker, G. B. (1994) 5HT2 and β-adrenergic receptor regulation in rat brain following chronic treatment with desipramine and fluoxetine alone and in combina-tion. *J. Neurochem.* **62,** 2262–2268.
38. Hyttel, J., Fredricson, O. K., and Arnt, J. (1984) Biochemical effects and drug levels in rats after long-term treatment with the specific 5-HT-uptake inhibitor, citalopram. *Psychopharmacology* **83,** 20–27.
39. Kellar, K. J., Cascio, C. S., Butler, J. A., and Kurtzke, R. N. (1981) Differential effects of electroconvulsive shock and antidepressant drugs on serotonin-2 receptors in rat brain. *Eur. J. Pharmacol.* **69,** 515–518.
40. Briley, M. and Moret, C. H. (1993) Neurobiological mechanisms involved in antidepressant therapies. *Clin. Neuropharmacol.* **16,** 387–400.
41. Nalepa, I., Manier, D. H., Gillespie, D. G., Rossby, S. P., Schmidt, D. E., and Sulser, F. (1997) Dual signaling by venlafaxine: I. The β adrenoceptor desensitization hypothesis revisited. *Eur. Neuropsychopharmacol.*, submitted.
42. Westphal, R. S., Backstrom, J. R., and Sanders-Bush, E. (1995) Increased basal phospho-rylation of the constitutively active serotonin 2C receptor accompanies agonist-mediated desensitization. *Mol. Pharmacol.* **48,** 200–205.
43. Sibley, D. R., Strasser, R. H., Benovic, J. L., Daniel, K., and Lefkowitz, R. J. (1986) Phosphorylation/dephosphorylation of the b-adrenergic receptor regulates its functional coupling to adenylate cyclase and subcellular distribution. *Proc. Natl. Acad. Sci. USA* **83,** 9408–9412.
44. Benovic, J. L., Strasser, R. H., Caron, M. G., and Lefkowitz, R. J. (1986) B-adrenergic receptor kinase: identification of a novel protein kinase that phosphorylates the agonist-occupied form of the receptor. *Proc. Natl. Acad. Sci. USA* **83,** 2797–2801.
45. Hausdorff, W. P., Caron, M. G., and Lefkowitz, R. J. (1990) Turing off the signal-desensi-tization of β-adrenergic receptor function. *FASEB J.* **4,** 2881–2888.
46. Premont, R. T., Inglese, J., and Lefkowitz, R. J. (1995) Protein kinases that phosphorylate activated G protein-coupled receptors. *FASEB J.* **9,** 175–182.
47. Ferguson, S. G., Menard, L., Barak, L. S., Koch, W. J., Colapietro, A. M., and Caron, M. G. (1995) Role of phosphorylation in agonist-promoted β-adrenergic receptor sequestration. *J. Biol. Chem.* **270,** 24,782–24,789.
48. Honegger, U. E., Roscher, A. A., and Wiesmann, U. N. (1983) Evidence for lysosomotropic action of desipramine in cultured human fibroblasts. *J. Pharmacol. Exp. Ther.* **225,** 436–441.

49. Duman, R. S., Terwilliger, R. Z., and Nestler, E. J. (1989) Chronic antidepressant regulation of GSα and cyclic AMP-dependent protein kinase. *Pharmacologist* **31,** 182.

50. Lesch, K. P. and Manji, H. K. (1992) Signal-transducing G proteins and antidepressant drugs: evidence of modulation of α subunit gene expression in rat brain. *Biol. Psychiatry* **32,** 549–579.

51. Li, P. P., Young, L. T., and Warsh, J. J. (1994) Effects of antibipolar and antidepressant drugs on the levels of signal transducing G proteins and their messenger ribonucleic acid transcripts. *Neuropsychopharmacology* **10,** 380S.

52. Rasenick, M. M. (1994) G proteins as the molecular target of antidepressant action: chronic treatment increases coupling between Gs and adenylate cyclase. *Neuropsychopharmacology* **10,** 580S.

53. Dwivedi, Y., Pandey, S. C., and Pandey, G. N. (1995) Effect of chronic administration of antidepressants on the levels of various subtypes of G-proteins in rat brain. *Soc. Neurosci.* **21,** 731.8 Abstract.

54. Emamghoreishi, M., Warsh, J. J., Sibony, D., and Li, P. P. (1996) Lack of effect of chronic antidepressant treatment on Gs and Gi α-subunit protein and mRNA levels in the rat cerebral cortex. *Neuropsychopharmacology* **15,** 281–287.

55. Chen, J. and Rasenick, M. M. (1995) Chronic antidepressant treatment facilitates G protein activation of adenylyl cyclase without altering G protein content. *J. Pharmacol. Exp. Ther.* **275,** 509–517.

56. Menkes, D. B., Rasenick, M. M., Wheeler, M. A., and Bitensky, M. W. (1983) Guanosine triphosphate activation of brain adenylate cyclase: enhancement by long-term antidepressant treatment. *Science* **219,** 65–67.

57. Avissar, S. and Schreiber, G. (1992) Interaction of antibipolar and antidepressant treatments with receptor-coupled G proteins. *Pharmacopsychiatry* **25,** 44–50.

58. Manji, H. K., Potter, W. Z., and Lenox, R. H. (1995) Signal transduction pathways. Molecular targets for lithium actions. *Arch. Gen. Psychiatry* **52,** 531–543.

59. Coleman, D. E. and Sprang, St. R. (1996) How G proteins work: continuing story. *TIBS* **21,** 41–44.

60. Shih, M. and Malbon, C. C. (1994) Oligodeoxygnucleotides antisense to mRNA encoding protein kinase A, protein kinase C and β-adrenergic receptor kinase reveal distinctive cell-type specific roles in agonist-induced desensitization. *Proc. Natl. Acad. Sci. USA* **91,** 12,193–12,197.

61. Döbbeling, U. and Berchtold, M. W. (1996) Down-regulation of the protein kinase A pathway by activators of protein kinase C and intracellular $Ca^{2+}$ fibroblast cells. *FEBS Lett.* **391,** 131–133.

62. Assie, M. B., Broadhurst, A., and Briley, M. (1988) Is down-regulation of β-adrenoceptors necessary for antidepressant activity? in *New Concepts in Depression* (Briley, M. and Fillion, G., eds.). Macmillan, London, pp. 161–166.

63. Hoeffler, J. P., Deutsch, P. J., Lin, J., and Habener, J. F. (1989) Cyclic adenosine monophosphate and phorbol esters—responsive signal transduction pathways converge at the level of transcriptional activation by the interactions of DNA-binding proteins. *Mol. Endocrinol.* **3,** 868–880.

64. Nishizuka, Y. (1992) Intracellular signaling by hydrolysis phospholipids and activation of protein kinase C. *Science* **258,** 607–614.

65. Faisst, S. and Meyers, S. (1992) Compilation of vertebrate-encoded transcription factors. *Nucleic Acids Res.* **20,** 3–26.

66. Hunter, T. and Karin, M.(1992) The regulation of transcription by phosphorylation. *Cell* **70,** 375–387.

67. Jans, D. A. (1995) The regulation of protein transport to the nucleus by phosphorylation. *Biochem. J.* **311,** 705–716.

68. Vandromme, M., Gauthier-Rouvière, C., Lamb, N., and Fernandez, A. (1996) Regulation of transcription factor localization: fine tuning of gene expression. *TIBS* **21,** 59–64.

69. Hagiwara, M., Brindle, P., Harootunian, A., Armstrong R., Rivier, J., Vale, W., Tsien, R., and Montminy, M. R. (1993) Coupling of hormonal stimulation and transcription via the cyclic AMP-responsive factor CREB is rate limited by nuclear entry of protein kinase A. *Mol. Cell. Biol.* **13,** 4852–4859.

70. Nestler, E. J., Terwilliger, R. Z., and Duman, R. S. (1989) Chronic antidepressant administration alters the subcellular distribution of cyclic AMP-dependent protein kinase in rat frontal cortex. *J. Neurochem.* **53,** 1644–1647.

71. Lee, K. A. W. and Masson, N. (1993) Transcriptional regulation by CREB and its relatives. *Biochem Biophys. Acta* **1174,** 221–233.

72. Nestler, E. J., McMahon, A., Sabban, E. L., Tallman, J. T., and Duman, R. S. (1990) Chronic antidepressant administration decreases the expression of tyrosine hydroxylase in the rat locus coeruleus. *Proc. Natl. Acad. Sci. USA* **87,** 7522–7526.

73. Brady, L. S., Whitfield, H. J., Jr., Fox, R. J., Gold, P. W., and Herkenham, M. (1991) Long-term antidepressant administration alters corticotropin releasing hormone, tyrosine hydroxylase and mineralocorticoid receptor gene expression in rat brain. *J. Clin. Invest.* **87,** 831–837.

74. Peiffer, A., Veilleaux, S., and Barden, N. (1991) Antidepressant and other centrally acting drugs regulate glucocorticoid receptor messenger RNA levels in rat brain. *Psychoneuroendocrinology* **16,** 505–515.

75. Hosoda, K. and Duman, R. S. (1993) Regulation of $\beta_1$-adrenergic receptor mRNA and ligand binding by antidepressant treatments and norepinephrine depletion in rat frontal cortex. *J. Neurochem.* **69,** 1335–1343.

76. Toth, M. and Shenk, T. (1994) Antagonist-mediated down-regulation of 5-hydroxytryptamine type 2 receptor gene expression: modulation of transcription. *Mol. Pharmacol.* **45,** 1095–1100.

77. Rossby, S. P., Nalepa, I., Huang, M., Burt, A., Perrin, C., Schmidt, D. E., and Sulser, F. (1995) Norepinephrine-independent regulation of GRII mRNA *in vivo* by a tricyclic antidepressant. *Brain Res.* **687,** 79–82.

78. Rossby, S. P., Perrin, C., Burt, A., Nalepa, I., Schmidt, D., and Sulser, F. (1996) Fluoxetine increases steady-state levels of preproenkephalin mRNA in rat amygdala by a serotonin dependent mechanism. *J. Serotonin Res.* **3,** 69–74.

79. Rossby, S. P. and Sulser, F. (1993) Die Wirkmechanismen von Antidepressiva: ein historischer Rückblick und neue neurobiologische Aspekte. *ZNS Journal, Forum für Psychiatrie und Neurologie* **1,** 10–19.

80. Schwaninger, M., Schöfl, C., Blume, R., Rössig, L., and Knepel, W. (1995) Inhibition by antidepressant drugs of cyclic AMP response element-directed gene transcription. *Mol. Pharmacol.* **47,** 1112–1118.

81. Nibuya, M., Nestler, E. J., and Duman, R. S. (1996) Chronic antidepressant administration increases the expression of CREB in rat hippocampus. *J. Neurosci.* **16,** 2365–2372.

82. Nibuya, M., Morinobu, S., and Duman, R. S. (1995) Regulation of BDNF and trkB mRNA in rat brain by chronic electroconvulsive seizure and antidepressant drug treatment. *J. Neurosci.* **15,** 7539–7547.

83. Brown, J. R., Ye, H., Bronson, R. T., Dikkes, P., and Greenberg, M. E. (1996) A defect in nurturing in mice lacking the immediate early gene fos B. *Cell* **86,** 297–309.

84. Seckl, J. R. and Fink, G. (1992) Antidepressants increase glucocorticoid and mineralocorticoid receptor mRNA expression in rat hippocampus in vivo. *Neuroendocrinology* **55,** 621–626.

85. Brady, L. S., Gold, P. W., Herkenham, M., Lynn, A. B., and Whitfield, H. J. (1992) The antidepressants fluoxetine, idazoxan and phenelzine alter corticotropin-releasing hormone and tyrosine hydroxylase mRNA levels in rat brain: therapeutic implications. *Brain Res.* **572,** 117–125.
86. Schweitzer, J. W., Schwartz, R., and Friedhoff, A. J. (1979) Intact presynaptic terminals required for β-adrenergic receptor regulation by desipramine. *J. Neurochem.* **33,** 377–379.
87. Janowsky, A. J., Steranka, L. R., Gillespie, D. D., and Sulser, F. (1982) Role of neuronal signal input in the down-regulation of central noradrenergic receptor function by antidepressant drugs. *J. Neurochem.* **39,** 290–292.
88. Pepin, M. C., Govindan, M. V., and Barden, N. (1992) Increased glucocorticoid receptor gene promoter activity after antidepressant treatment. *Mol. Pharmacol.* **41,** 1016–1022.
89. Nobukuni, Y., Smith, C. L., Hager, G. L., and Detera-Wadleigh, S. D. (1995) Characterization of the human glucocorticoid receptor promoter. *Biochemistry* **34,** 8207–8214.
90. Milner, R. J. and Sutcliffe, J. G. (1983) Gene expression in rat brain. *Nucleic Acid Res.* **11,** 5497–5520.
91. Hyman, St. E. and Nestler, E. J. (1993) *The Molecular Foundation of Psychiatry.* American Psychiatric, Washington, D.C.

**12**

# Animal Models to Detect Antidepressants

## *Are New Strategies Necessary to Detect New Agents?*

**Paul Willner and Mariusz Papp**

## 1. STRATEGIC TARGETS FOR ANTIDEPRESSANT DRUG DEVELOPMENT

Rational strategies for the development of novel antidepressant agents are based on an analysis of perceived therapeutic needs, which in turn derive from the inadequacies of the therapies that are currently available. The traditional (first-generation) antidepressants suffer from three major drawbacks: They have unacceptable side effects; their efficacy is low; and their therapeutic effects develop slowly, typically requiring 3–4 wk of treatment before clear improvements in mood are apparent. Since the 1970s, these three problems have formed the major therapeutic targets in antidepressant drug development, since any newly developed agent could only be marketed on the basis of an improvement over traditional agents in one or more of these areas.

Although a range of second-generation antidepressants have been developed and are widely used in the clinic, their success in meeting these therapeutic objectives is largely limited to the achievement of a more benign side effect profile. Tricyclic antidepressants (TCAs) have significant anticholinergic effects, which, in addition to being unpleasant, also encourage patients to discontinue treatment; this feature is weak or absent in all of the more recently developed drugs, with no loss of efficacy, and accounts in large measure for the successful market penetration of the specific serotonin reuptake inhibitors (SSRIs). Another unacceptable feature of tricyclics is their low safety margin, which leads to their frequent use by depressed patients as suicide agents. Many newer antidepressants are relatively safe in overdose (e.g., mianserin), and this must be counted as a significant improvement. The other class of first-generation antidepressants, the monoamine oxidase inhibitors (MAOIs), also produce unacceptable side effects, most notably their well-known interactions with tyramine-containing foods (the "cheese effect"). These unwanted effects derive from inhibition of MAO-B, and are absent in the recently developed specific reversible inhibitors of MAO-A (RIMAs), again, with no loss of efficacy. None of the newer antidepressants is entirely free of side effects, but these drugs have been extremely successful in overcoming the most serious of the unwanted effects of tricyclics and MAOIs.



*From: Antidepressants: New Pharmacological Strategies*
*Edited by: P. Skolnick, Humana Press Inc., Totowa, NJ*

Improving the efficacy of antidepressants has proved more problematic. The efficacy of TCAs is usually estimated to be around 65–70%. It is possible that the SSRIs might prove superior to tricyclics when applied for indications other than depression, for example, in panic or obsessive–compulsive disorder. However, there is little evidence that SSRIs or, indeed, any other of the numerous classes of novel agents under development, are more efficacious than tricyclics as antidepressants (though the RIMAs do improve on the older MAOIs, which are less effective than tricyclics). Thus around one-third of depressed patients are failing to benefit from the available treatments.

The reasons for this limited efficacy are far from clear. One obvious potential cause of treatment failure is noncompliance with treatment, and this factor has been assumed to account for a significant proportion of apparent treatment failures. However, if this were so, then efficacy should be higher for newer, better-tolerated antidepressants; and although it is sometimes claimed that SSRIs are indeed superior in efficacy to tricyclics, this probably reflects improved anxiolytic, rather than antidepressant, activity *(1)*. It has also been suggested that more vigorous pharmacotherapy leading to higher plasma drug levels should achieve higher success rates, but no such effect has been reliably documented. Alternatively, a hypothesis of different biochemical etiologies for depression has been advanced under which patients nonresponsive to one type of antidepressant (e.g., an SSRI) might prove more responsive to a neurochemically different agent (e.g., a specific noradrenaline uptake inhibitor [SNRI]); but, again, no evidence reliably supports this view. More successful have been a number of adjunctive therapies, such as the addition of thyroid hormone *(2)* or lithium *(3)* to the antidepressant treatment regime, which appear to be effective in a proportion of antidepressant-resistant patients. Even so, a high proportion of depressed patients remain in need of treatment, and the causes of this need remain uncertain.

How could this unmet need be addressed? Three general strategies are used in antidepressant development research. One is to focus on a known property of existing antidepressants, with a view to maximizing that property and minimizing others. This strategy underlay the development of the SSRIs, which maximize the serotonergic effects of tricyclics, while minimizing their noradrenergic effects, and, more recently, the development of specific 5-HT$_{1A}$ agonists, which target a single subset of the many 5-HT receptor populations stimulated by SSRIs. A second strategy is to target neurochemical effects that are identified as common to a variety of antidepressants. For example, the observation that most antidepressant drugs downregulate β-adrenergic receptors led to the development of rolipram, which achieves the same functional effect by an action beyond the receptor *(4)*, and the more recent observation that many antidepressants modulate *N*-methyl-D-aspartate (NMDA) receptors has led to the proposal that NMDA receptor antagonists might be developed as potential antidepressants *(5; see* Chapter 7). The third strategy is to identify potential antidepressant agents by screening novel compounds in behavioral tests predictive of antidepressant activity. In practice, behavioral tests are used primarily to bolster leads derived from the neurochemical strategies. However, this approach requires no preconceptions as to neurochemical mechanisms of action, and can generate novel neurochemical hypotheses for antidepressant action (e.g., cholecystokinin [CCK] antagonists *[6]* or enkephalinase inhibitors *[7]*).

Returning to the question of therapy-resistant depression, it is clear that the first two strategies rely on known properties of existing antidepressants and so are unlikely to improve significantly on their efficacy. Only the third strategy, the use of animal models of depression, generates truly novel neurochemical targets, with the potential for treating a group of patients neurochemically distinct from those who respond to conventional treatments. Even here, however, the likelihood of nonaccidentally discovering treatments for resistant depressions is low, since behavioral screening tests, by definition, identify compounds that resemble existing antidepressants. It follows that compounds so identified are likely to overlap substantially with existing antidepressants in their spectrum of action; there seems little rational basis to expect that agents discovered in this way should act differently in populations of depressed patients from those used initially to validate the test. A better strategy would be to use a model specifically developed to detect antidepressant effects in therapy-resistant subjects. Such a model would respond to conventional antidepressants, but would reliably produce a proportion of subjects who failed to do so, which could then be used in a second stage to test novel compounds. No such model exists at present; and, although this strategy is attractive in principle, the practical difficulties are so formidable that it is unlikely that a model of this kind will be developed, or, if developed, that it would be widely adopted. This means that the discovery of treatments for resistant depression is likely to remain serendipitous for the forseeable future.

The third major problem, the slow onset of antidepressant action, may prove more tractable. In general, the newer antidepressants do not act faster than the old and, indeed, the SSRIs may be a little slower *(8)*. There have, however, been frequent claims of more rapid onsets; most of these claims have not survived the rigors of blind controlled trials, but the claim does appear to be upheld in some cases, such as amineptine *(9)* and venlafaxine *(10)*. However, although a shortening of antidepressant onset latency from 3 to 2 wk is clinically meaningful, the "holy grail" of an acute antidepressant action has not yet been realized.

Why are antidepressants slow to act? To some extent, the answer to this question may involve pharmacokinetic variables that contribute to a slow achievement of steady-state plasma levels, and cognitive factors, such as a need to relearn what it means not to be depressed. However, the fact that antidepressants are known to cause a multiplicity of biochemical effects that develop slowly, on a time-scale comparable to the onset of therapeutic action, suggests that the slow onset of action may well be, in large measure, of pharmacodynamic origin. In fact, the majority of the neuroreceptor systems that have been examined show slow changes during the course of chronic antidepressant treatment. In some cases, such as the classic $\beta$-receptor downregulation, these changes can be viewed as homeostatic adaptations to changes in neurotransmitter concentration; in other cases, the changes are better viewed as a process of readjustment between the component parts of a more extensive system. A clear picture has recently emerged of how such adaptations may occur within the 5-HT system during chronic treatment with SSRIs. Early in the course of treatment, SSRIs fail to elevate 5-HT levels in prefrontal cortex, because an elevation of 5-HT levels in cell body regions acts at inhibitory 5-HT autoreceptors to inhibit cell firing, consequently decreasing 5-HT release in the forebrain. However, as treatment continues, cell body autoreceptors desensitize, firing rate increases, and 5-HT release in the forebrain normalizes; in these circumstances, blockade of reuptake by an SSRI now causes the expected increase in levels of extracellular 5-HT. Thus, the slow onset of action of SSRIs may reflect the time necessary to cause a desensitization of 5-HT autoreceptors located within the Raphe nuclei *(11)*. This suggests that it might be possible to accelerate the therapeutic action of SSRIs by blocking 5-HT autoreceptors, and preliminary clinical data support this hypothesis *(12; see* Chapter 1). In short, there are many indications that the achievement of a rapid onset of antidepressant action should prove a feasible objective.

## 2. STRATEGIC ISSUES FOR ANIMAL MODELS OF DEPRESSION

It has been argued that animal models of depression represent the best strategy for detecting truly novel pharmacotherapies for depression. However, animal models are also important for a second reason: Antidepressant drugs are devoid of mood-elevating effects in normal (i.e., nondepressed) human subjects. This fact raises serious doubts as to the clinical relevance of the numerous biochemical changes that are known to develop during the course of chronic antidepressant administration in normal animals. Additionally, the absence of clinically relevant behavioral correlates of antidepressant action in normal animals causes serious difficulty in interpreting the neurochemical effects, as is well illustrated by the longstanding and still ongoing controversy over whether antidepressants increase activity in NA systems (as suggested, inter alia, by the NA-uptake-inhibiting action of tricyclics), or decrease NA activity (as suggested, inter alia, by antidepressant-induced β-receptor downregulation). Clearly, any attempt to predict clinical actions from antidepressant effects in normal animals is fraught with uncertainty.

However, the use of animal models of depression to address this issue raises a further set of problems, since data derived from animal models are likely to be of value only to the extent that the models are valid. The procedures for validating animal models of psychiatric disorders have been discussed in detail elsewhere *(13,14)*; they include consideration of predictive validity (which concerns primarily the correspondence between drug actions in the model and in the clinic), face validity (phenomenological similarities between the model and the disorder), and construct validity (a sound theoretical rationale). Some desirable features in an animal model of depression are that the model should respond appropriately to antidepressant drugs, should employ realistic inducing conditions, and should model a core symptom of the disorder. Several of the available models have a reasonable pharmacological profile, with relatively few false positives and false negatives, but very few models perform well against all three sets of validating criteria *(13,14)*.

Furthermore, the majority of animal models of depression employ procedures that are of brief duration; this causes serious difficulties in using the models to investigate the time-course of antidepressant action. Indeed, it is difficult to see how a convincing study might be designed to reveal a rapid onset of antidepressant action in, for example, the Porsolt forced-swim test *(15)*. In this test, rats are forced to swim in a cylinder of water and, after an initial period of active escape attempts, they adopt an immobile posture; on a second exposure, the immobile posture is entered more rapidly. Antidepressants delay the onset of immobility in the second test and are active in this procedure after acute treatment at very high doses, after subacute treatment at moderate doses (typically, three treatments over a 24-h period), or after chronic treatment at low doses *(15)*. It has occasionally been claimed that an acute antidepressant-like action of a novel agent indicates a potential for rapid-onset clinical effects (e.g., ref. *16*), but, since the test responds acutely to conventional antidepressants (and is most widely used in the subacute-treatment modality), this conclusion is clearly unwarranted.

The fact that acute or subacute tests cannot, in principle, be used to generate reliable and interpretable estimates of the speed of onset of antidepressant action is by no means the only problem with these tests. If the study involves chronic antidepressant treatment, as is frequently the case with the forced-swim test, the period of chronic drug treatment must occur in advance of the test. However, in this case, it is difficult or impossible to dissociate therapeutic drug effects from prophylactic effects. It is quite possible that anxiolytic drugs could provide prophylactic protection against depression, and this may well explain why prophylactic treatment with benzodiazepines is occasionally reported to produce antidepressant-like effects in the forced-swim test *(17)*. Similar prophylactic effects of benzodiazepines have also been reported in another subacute test, the learned-helplessness model *(18)*. Finally, because acute tests necessarily involve an acute change of behavior, it can be very difficult to rule out drug-induced amnesia, or other sensorimotor or cognitive impairments, as the explanation of prophylactic drug effects *(19)*.

Together, these considerations argue strongly for chronicity as an important feature of animal models of depression: The models should not only be of demonstrable validity, but should also maintain an abnormal state for a prolonged period (weeks or months), during which therapy may be administered. It should be noted in passing that models meeting these criteria are of value not only for drug development, but also as experimental tools for investigating the physiological mechanisms underlying the depression-like behavior and the therapeutic action of antidepressant drugs, as well as nonphysiological (e.g., social) influences on the behavior and nonpharmacological (e.g., behavioral) therapies; and, because the model simulates depression in a relatively realistic and valid manner, the conclusions of such studies are likely to generalize to the clinic. It should also be noted that the arguments in favor of chronic models conflict with the traditional approach to the design of antidepressant screening procedures, which, for logistical reasons, requires that they respond to acute or subacute drug administration. However, such tests are incapable, by virtue of their design, of discovering new antidepressants that have a shorter onset of action. Because this represents the major current challenge in antidepressant drug development, the inability of traditional screening tests to repond to it represents a fundamental limitation on their continued usefulness.

## 3. THE CHRONIC MILD STRESS MODEL

Of the very few models that meet the criteria outlined above, the chronic mild stress (CMS) model has been most extensively validated and investigated. In the CMS model, rats or mice are exposed sequentially to a variety of extremely mild stressors (e.g., overnight illumination, cage tilt, change of cage mate), which change every few hours over a period of weeks or months. This procedure causes a decrease in sensitivity to rewards, which is usually monitored by a substantial decrease in the consumption of and/or preference for a palatable, weak (1%) sucrose solution *(20–22)*. No single element of the CMS schedule is either necessary or sufficient for these effects; variety is essential *(23)*. CMS-induced behavioral deficits may be maintained for several months; however, normal behavior is restored, during continued application of CMS, by chronic treatment with tricyclic or atypical antidepressants. The validation of the model will first be described, followed by its various applications.

### 3.1. Construct Validity

The theoretical rationale for the CMS model is that this procedure simulates anhedonia, a loss of responsiveness to pleasant events, which is a core symptom of depression and the defining feature of melancholia *(24)*. This rationale rests on two assumptions: that sucrose drinking is a valid measure of sensitivity to reward; and that CMS causes a generalized decrease in reward sensitivity, rather than a specific effect on responses to sweet tastes.

Decreases in sucrose drinking cannot be explained by nonspecific changes in fluid consumption (e.g., decreased thirst), since the intake of plain water is unaffected by CMS *(23)*, and the effects of CMS are seen in both single-bottle tests and in two-bottle (sucrose–water) preference tests *(20,25)*. CMS-induced decreases in sucrose drinking are seen in the ascending portion of the sucrose concentration-intake curve *(26)*, where intake and preference are monotonically related *(27)*, but not on the descending limb of the concentration-intake curve *(26)*, where intake is dissociated from preference *(27)*. The calorie content of the sucrose is also unimportant, since similar effects are seen in animals consuming calorie-free saccharin solutions *(20,25)*, and in both food-deprived and nondeprived animals *(23)*; and food intake is not decreased by CMS *(28,29)*. Finally, it has recently been suggested that CMS-induced decreases in sucrose consumption may be secondary to changes in body weight *(30)*, but this relationship could not be confirmed by five other laboratories; large decreases in body weight do not, in themselves, decrease sucrose intake *(31)*. The failure of these alternative accounts of the effects of CMS on sucrose drinking supports the conclusion that they do indeed reflect a decrease in the rewarding properties of the sucrose.

Further evidence for this conclusion comes from studies demonstrating that behavioral deficits are apparent in rewarded paradigms that do not depend on consummatory behavior. Thus CMS causes an increase in the threshold current required to support intracranial self-stimulation (brain stimulation reward) and attenuates or abolishes the ability to associate rewards with a distinctive environment (place conditioning). The latter effect has been demonstrated with a variety of different natural or drug reinforcers but does not extend to aversive place conditioning *(28,29,32–34)*. The most parsimonious explanation of all these findings is that CMS causes a generalized decrease in sensitivity to rewards (anhedonia).

### 3.2. Face Validity

In addition to inducing a state of anhedonia, CMS also causes the appearance of many other symptoms of major depressive disorder. Behavioral changes in animals exposed to CMS include decreases in sexual and aggressive behaviors *(35)*, and decreases in locomotor activity during the dark (waking) phase of the light–dark cycle *(36)*. In contrast, CMS did not cause the appearance of an "anxious" profile in two animal models of anxiety, the elevated plus-maze and the social interaction test *(35)*, suggesting that the behavioral changes are specific for depression. Animals exposed to CMS show an advanced phase shift of diurnal rhythms *(36)*, and a variety of sleep disorders characteristic of depression, including decreased rapid eye movement (REM) sleep latency and an increased number of REM sleep episodes *(36a,37)*. They also gain weight more slowly, leading to a relative loss of body weight *(23,31)*, and show signs of increased activity in the hypothalamus-pituitary-adrenal (HPA) axis, including adrenal hypertrophy *(23)* and corticosterone hypersecretion *(25)*. Abnormalities have also been detected in the immune system, including an increase in serum complement *(25)*, decreases in thymus weight, natural killer-cell activity, and reactivity to T-cell mitogens *(38,39)*, and an increase in acute phase proteins that was reversed by chronic antidepressant treatment *(40)*. Taken together with the generalized decrease in responsiveness to rewards, these parallels to the symptoms of depression are both extensive and comprehensive. Indeed, it is arguable that the only symptoms of depression that have not been demonstrated in animals exposed to CMS are those uniquely human symptoms that are only accessible to verbal enquiry *(41)*.

### 3.3. Predictive Validity

The reversal of CMS-induced anhedonia typically requires 3–4 wk of treatment, which closely resembles the clinical time-course of antidepressant action; a second parallel with the clinic is that antidepressants act specifically in animals exposed to CMS, but do not alter rewarded behavior in nonstressed control animals. These effects are illustrated in Fig. 1 for the prototypical antidepressant drug, imipramine.

Studies have been conducted in the CMS model with a wide range of antidepressant and nonantidepressant agents, in addition to a number of putative novel antidepressants. Ineffective agents in the CMS model include chlordiazepoxide *(29)*, D-amphetamine, and the neuroleptics chlorprothixene and haloperidol *(42)*; none of these drugs are effective as antidepressants. Drugs shown to be effective in reversing CMS-induced anhedonia include the tricyclics imipramine, desipramine, and amitriptyline *(20,42,43,43a)*; the SSRIs fluoxetine, fluvoxamine, and citalopram *(29,44,45)*; the SNRI maprotiline *(29)*, the monoamine oxidase inhibitors moclobemide *(34)* and brofaromine *(42)*; and the atypical antidepressant mianserin *(46,47)*. Other less conventional, but clinically effective, antidepressants that are also effective in the CMS model include the antimanic agents lithium *(48)* and carbamazepine *(49)*, and the 5-HT$_{1A}$ partial agonist buspirone *(42,44)*. Additionally, activity in the CMS model has been reported for the corticosterone synthesis inhibitor ketoconazole *(49)*, which has been reported to have clinical antidepressant activity in a recent open study *(50)*. Finally, electroconvulsive therapy (ECT) has also been shown to restore normal responsiveness to reward in animals exposed to CMS, and, unlike all of the drug effects listed above, this response was present after a single week of treatment *(37)*.

In addition to these clear and appropriate positive and negative responses, there are also a number of questionable findings. For example, morphine was effective early in treatment at a low dose (1 mg/kg), but the effects were not sustained *(50a)*, and no activity was seen at a higher dose (an escalating regime, rising from 10 to 90 mg/kg) *(42)*; morphine has not been shown to be an effective antidepressant in properly con-

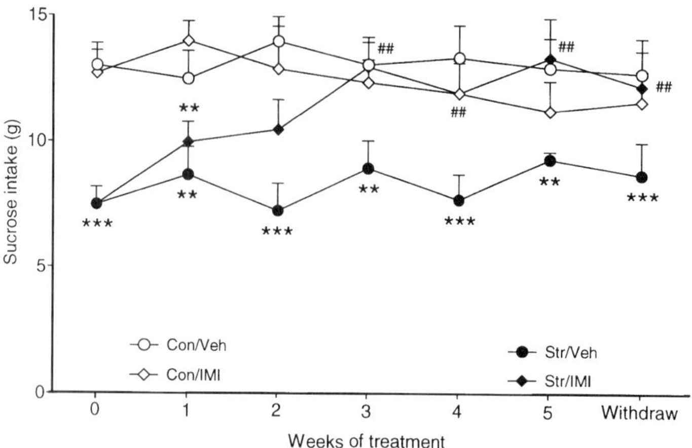



**Fig. 1.** The antianhedonic effect of imipramine. Rats were subjected for 3 wk to chronic mild stress (Str), which resulted in significant decreases in sucrose intake (measured once weekly) relative to control (Con) animals; these data are shown at wk 0. Subsequently, CMS continued during 5 wk of treatment with imipramine (IMI: 10 mg/kg/d) or vehicle (Veh), and 1 wk of drug withdrawal. Stars represent significant differences between CMS and control animals (** $p < 0.01$, *** $p < 0.001$); hatches represent significant increases in sucrose drinking in the CMS-IMI group relative to its own predrug (week 0) intake (##$p < 0.01$). Note that full recovery of sucrose drinking occurred after 3 wk of treatment and that imipramine was without effect in the nonstressed control group.

ducted clinical trials, but was widely used for this purpose in the early part of this century *(51)*. Both mepyramine, an antihistamine, and atropine, an anticholinergic, showed antidepressant-like activity, and would appear to be false positives; however, it is not entirely clear that these drugs would not show clinical antidepressant activity if formally tested and, in the case of atropine, the argument that it might possess clinical antidepressant activity is quite compelling *(42)*. Finally, unlike buspirone, the more specific 5-HT$_{1A}$ partial agonist ipsapirone was inactive in the CMS model, and this may represent a false-negative response *(44)*. However, although ipsapirone has clear anxiolytic activity *(52)*, there are as yet no published studies claiming that ipsapirone is effective in major depressive disorder. Indeed, another 5-HT$_{1A}$ partial agonist, gepirone, has been reported to be an effective antidepressant in nonmelancholic patients, but to be ineffective in melancholia, of which anhedonia is the core symptom *(53)*. From these data, it would not be predicted that ipsapirone should reverse anhedonia.

To summarize, a wide variety of antidepressant drugs, as well as ECT, are active in increasing responsiveness to rewards in animals exposed to CMS (but not in control animals), and the time-course of the therapeutic improvements closely mirrors the clinical action of these agents. Conversely, a number of nonantidepressants are inactive in the CMS model, as predicted. There are a few drugs that appear to behave in an inappropriate manner, but some of these apparent failures may reflect inadequacies in the clinical literature. At present, there are no unequivocal discrepancies between the model and the clinic. This suggests that the CMS model provides a basis for drug development that could be used with a fair degree of confidence.

## 4. DOPAMINERGIC MECHANISMS IN THE CMS MODEL

Because the CMS paradigm provides a relatively valid animal model of depression, investigation of the neurochemical mechanisms underlying CMS-induced behavioral and physiological changes may provide leads to the pathophysiology of depression and the mechanisms of antidepressant drug action. A variety of neurochemical systems have been examined as a potential basis for the behavioral sequelae of CMS; of these, the most interesting changes, which form the starting point for an analysis of antidepressant action and potential novel approaches, are found within the mesolimbic dopamine (DA) system. This is not a traditional starting point for thinking about antidepressants, which has tended to focus primarily on 5-HT or NA systems. However, the mesolimbic DA system does represent a logical starting point for thinking about anhedonia, because this system has been the major focus for studies of mechanisms underlying responsiveness to rewards *(54)*; and it is now well established that chronic treatment with antidepressant drugs (in normal animals) leads to an increase in the locomotor stimulant response to stimulation of D2/D3 receptors within the nucleus accumbens *(55)*.

CMS causes a decrease in D2/D3-receptor binding in the nucleus accumbens, which is reversed by chronic treatment with imipramine. D1 receptors, and D2/D3 receptors in the dorsal striatum, are unaffected by CMS *(56)*. The decrease in D2/D3 receptor binding is accompanied by a pronounced functional subsensitivity to the rewarding and locomotor stimulant effects of the D2/D3 agonist quinpirole, administered systemically or within the nucleus accumbens *(57)*. CMS also causes presynaptic changes in DA function, including an increase in concentrations of DA and its metabolite (DOPAC) *(54)*, and an increase in electrically-stimulated DA release *(58)*, which, again, are confined to the nucleus accumbens. However, recent studies using brain microdialysis have demonstrated a decrease in DA release in the nucleus accumbens, following CMS, in response to a palatable food (G. Di Chiara, personal communication). Thus, both presynaptic and postsynaptic markers indicate a decrease in DA transmission, specific to the nucleus accumbens, in animals exposed to CMS.

There is also evidence that these changes are functionally important in mediating the behavioral effects of CMS and antidepressant treatment. In animals successfully treated with antidepressants (including tricyclics, specific 5-HT or NA uptake inhibitors, or mianserin), behavioral recovery is reversed by acute administration of D2/D3 receptor antagonists, at low doses that are without effect in nonstressed animals or in untreated stressed animals *(43,43a,46)*. Chronic stress also causes an antidepressant-reversible decrease in aggressive behavior, and this effect of chronic antidepressant treatment is also reversed by acute administration of DA antagonists *(59)*. These data argue strongly that a sensitization of D2/D3 receptors may be responsible for the therapeutic action of antidepressants in this model.

One test of this hypothesis is to ask whether sensitization of this receptor population is sufficient for antidepressant action. In a series of studies, anhedonic animals were treated with directly acting D2/D3 agonists (quinpirole, bromocriptine, pramipexole) administered twice weekly. This treatment schedule causes a sensitization of the locomotor-stimulant response comparable to that observed following chronic antidepressant treatment, and also caused a complete recovery of responsiveness to rewards, as measured by intake of a dilute sucrose solution and by place-preference conditioning

*(60–62)*. Thus, not only are sensitized responses of D2/D3 receptors necessary for antidepressant action in the CMS model (effects are reversed by low doses of D2/D3 antagonists), but this effect also appears to be sufficient for an antidepressant-like action.

These findings point to the mesolimbic DA system as a potential target for antidepressant drug development. DA-uptake inhibitors represent one strategy to increase activity in this system. DA-uptake inhibition is a feature of several recent antidepressants (e.g., nomifensine, buproprion, amineptine, minaprine), but these agents have not been tested in the CMS model and, as noted above, amphetamine is ineffective both clinically and in the model *(63)*. However, tolcapone, which raises intrasynaptic DA levels by a different mechanism—inhibition of the catabolic enzyme COMT—did show antidepressant-like activity in the CMS model *(64)*. Clinical antidepressant activity has also been reported with a variety of directly acting D2/D3 receptor agonists *(63)*. The novel D2/D3 agonist pramipexole *(65)* reversed the effects of CMS in two tests of rewarded behavior (sucrose intake, place conditioning), exactly as observed with conventional antidepressants. The data suggested that pramipexole might have a faster onset of antianhedonic action than conventional antidepressants, but the dose tested was somewhat too high, and some nonspecific drug effects made it impossible to draw a definitive conclusion concerning the onset latency *(62)*. Phase 3 trials of pramipexole in depression are currently in progress.

## 5. OTHER NEUROCHEMICAL SYSTEMS IN THE CMS MODEL

The fact that low doses of D2/D3 receptor antagonists reverse the actions of antidepressants with widely divergent primary mechanisms of action suggests that D2/D3 receptors in the nucleus accumbens may represent a final common pathway, through which chronically administered antidepressant drugs influence behavioral output systems. However, this does not imply that antidepressants directly alter DA receptor function within the nucleus accumbens. On the contrary, it seems more likely that antidepressants act in a more traditional manner (e.g., at 5-HT or NA synapses), and at more traditional sites. For example, the primary action of SSRIs is likely to be at 5-HT (rather than DA) synapses, located within limbic forebrain regions, such as amygdala, hippocampus, or prefrontal cortex (rather than nucleus accumbens). All of these traditional sites of antidepressant action *(51)* project into the nucleus accumbens, via glutamatergic afferents, providing a route through which actions at distant 5-HT (or NA) synapses could influence DA function in the accumbens *(66)*. The slow onset of antidepressant action presumably reflects slow processes of adaptation at one or more sites within this circuitry.

### 5.1. Serotonin

Several interactions of CMS with the 5-HT system have been described. In initial studies, CMS was found to cause increased tissue levels of 5-HT and its metabolite 5-HIAA in the nucleus accumbens (but not the dorsal striatum) *(26)*. Subsequent receptor-binding studies found that CMS increased the density of both 5-HT$_{1A}$ and 5-HT$_2$ receptors in the cerebral cortex. Chronic treatment with imipramine decreased the density of 5-HT$_2$ receptors in control animals and normalized 5-HT$_2$ receptor density in animals exposed to CMS. However, in the case of the 5-HT$_{1A}$ receptor, the effect of imipramine in control animals was similar to that of CMS (increased receptor density), and imipramine failed to normalize the effect of CMS *(67)*. In an electrophysiological study, CMS had no effect on the 5-HT$_{1A}$-mediated inhibitory effect of 5-HT in hippocampal slices, and chronic treatment with imipramine inhibited this effect to a similar extent in control and stressed animals *(68)*. These studies suggest that postsynaptic 5-HT$_{1A}$ receptors are not involved in the therapeutic effects of imipramine in the CMS model.

As noted in Section 4., one 5-HT$_{1A}$ partial agonist, buspirone, reversed CMS-induced anhedonia, but another 5-HT$_{1A}$ partial agonist, ipsapirone, did not. The 5-HT$_{1A}$ full agonist 8-OH-DPAT was also ineffective. However, antidepressant-like activity was seen with the 5-HT$_{1A}$ antagonist WAY 100135 *(44)*. Whether this effect is presynaptically or postsynaptically mediated remains to be determined. A presynaptic site of action for WAY 100135 is suggested by the effects of another novel agent, BIMT-17, which has postsynaptically mediated antidepressant-like effects in the forced-swim test. BIMT-17 has 5-HT$_{1A}$ agonist and 5-HT$_2$ antagonist properties, and has postsynaptic effects in prefrontal cortex similar to those of 5-HT. Unusually, these effects of BIMT-17 are present on acute treatment, but other 5-HT agonists, or SSRIs, require chronic treatment to produce 5-HT-like effects *(69,70)*. In mice exposed to CMS, BIMT-17 showed a unique acute antidepressant-like effect, which was present after a single injection, and was then sustained over chronic treatment *(71)*.

There is clinical evidence that the speed of onset of antidepressant action can be accelerated by adjunctive treatment with either lithium, which potentiates 5-HT function through mechanisms that remain uncertain *(72)*, or pindolol, which, in addition to its β-adrenergic antagonist properties, is also a 5-HT$_{1A}$ receptor antagonist *(12; see* Chapter 1). Both lithium and pindolol have recently been reported to accelerate the onset of antidepressant action in the CMS model: Lithium was shown to potentiate imipramine and fluoxetine; pindolol was shown to potentiate fluvoxamine and buspirone. In all cases, significant improvements were seen after a single week of treatment *(45,48)*. These studies serve to support clinical observations, rather than to introduce new treatments, but they do dramatically illustrate the potential of the CMS model to detect a rapid onset of antidepressant action.

### 5.2. Noradrenaline

In common with most other recent antidepressant research, studies of the NA system in the CMS model have been less extensive than those of the 5-HT system. CMS has been found to increase the density of β-adrenergic receptors in cerebral cortex, and also to increase the cyclic AMP response to NA in cortical slices. These effects are opposite to those seen after chronic antidepressant treatment in normal animals and, as predicted, chronic treatment with imipramine reversed the effects of CMS *(67,73)*. By contrast, CMS had no effect on the α$_1$-receptor-mediated inhibitory effect of phenylephrine on cell firing in hippocampal slices; this effect was inhibited by chronic imipramine, but to a similar extent in control and stressed animals *(68)*. There are even fewer studies of antidepressant effects mediated via the NA system. As noted above, the SNRI maprotiline was active in the CMS model *(29)*; the very specific α$_2$-receptor antagonist RX-811059 (ethoxy-idazoxan) was not *(74)*.

### 5.3. Glutamate

In addition to their classical effects on monaminergic neurotransmission, antidepressant drugs have also been found to interact significantly with glutamatergic systems. Thus, in normal animals, antidepressants interact acutely with NMDA receptors, and after chronic treatment cause a variety of adaptive changes at the NMDA receptor that decrease its functional activity *(5)*. It should be noted that these effects were observed in cerebral cortex, but not in hippocampus, striatum, or basal forebrain [5; *see* Chapter 7]. Thus, if the effects of antidepressants on NMDA receptors are involved in generating the adaptive changes in DA functioning that appear to represent the final common pathway for antidepressant action in the CMS model, such an interaction is more likely to derive from a change in the activity of glutamatergic afferents to the nucleus accumbens, rather than within the accumbens itself. It has recently been demonstrated that CMS increased the potency of glycine to bind to strychnine-insensitive sites on the NMDA receptor, an effect opposite to that of chronic antidepressant treatment in normal animals; imipramine reversed this effect *(75)*.

These neurochemical changes suggest that NMDA receptor antagonists might have antidepressant properties, and, indeed, antagonists acting at a variety of sites on the NMDA receptor complex have been found to display antidepressant-like effects in acute preclinical tests (e.g., forced-swim and tail suspension) predictive of antidepressant action, and to downregulate β-adrenoceptors in rodent cortex *(5)*. Recent studies in the CMS model have demonstrated antidepressant-like effects of four ligands acting at different loci on the NMDA receptor complex; the uncompetitive (MK-801) and competitive (CGP 37849, CGP 40116) NMDA antagonists, and the high-affinity, partial agonist at strychnine-insensitive glycine sites, ACPC. The magnitude and time-course of the effects of MK-801, CGP 37849, and CGP 40116 were comparable to those observed following a similar treatment regime with imipramine. However, the ACPC-treated animals recovered significantly faster (1–2 wk) than those treated with imipramine and the three NMDA antagonists *(76,77)*. This finding suggests that ACPC may act clinically as a rapid-onset antidepressant, and again demonstrates the ability of the CMS model to detect rapid onset (Fig. 2).

### 5.4. Enkephalin

Finally, two recent studies have reported enkephalin changes in the nucleus accumbens of rats exposed to CMS. One study found that CMS decreased tissue levels of met-enkephalin; chronic imipramine treatment had the opposite effect in control animals and blocked the effect of CMS *(78)*. The second study found no difference in basal levels of met-enkephalin release, measured by microdialysis, but control animals showed an increase in met-enkephalin release during social interaction, which was absent in animals exposed to CMS *(79)*. Both studies suggest that CMS decreases the responsiveness of enkephalinergic systems in the nucleus accumbens, which could play an important role in CMS-induced anhedonia, and could provide another target for antidepressant drug development. However, the enkephalinase inhibitor RB 101 was ineffective in reversing the suppression of sucrose drinking by CMS *(50a)*.

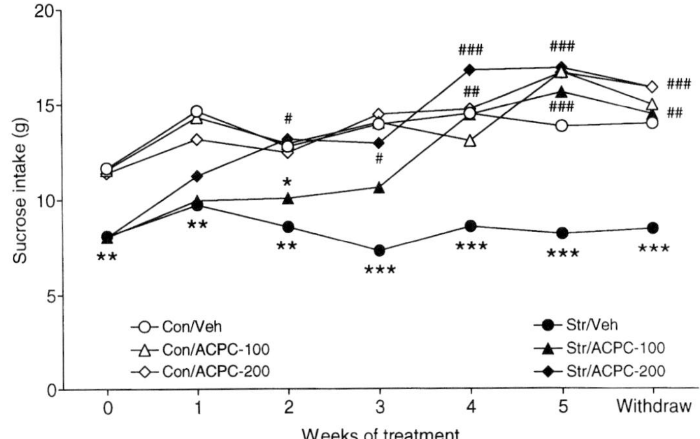

**Fig. 2.** The antianhedonic effect of the glycine site antagonist ACPC. Rats were subjected for 3 wk to chronic mild stress (Str), which resulted in significant decreases in sucrose intake (measured once weekly) relative to control (Con) animals; these data are shown at wk 0. Subsequently, CMS continued during 5 wk of treatment with ACPC (100 or 200 mg/kg/d) or vehicle (Veh), and 1 wk of drug withdrawal. Stars represent significant differences between CMS and control animals (*$p < 0.05$, **$p < 0.01$, ***$p < 0.001$); hatches represent significant increases in sucrose drinking in the CMS-IMI group relative to its own pre-drug (week 0) intake (#$p < 0.05$, ##$p < 0.01$, ###$p < 0.001$). Note that at the higher dose of ACPC the difference between control and stressed animals was no longer significant after 1 wk of treatment, and full recovery of sucrose drinking occurred after 2 wk of treatment. These data should be compared with the effects of imipramine as illustrated in Fig. 1.

## 6. ALTERNATIVE STRATEGIES

The CMS model has proved extremely successful in meeting the objectives set out at the start of this chapter: It is a chronic model with a high degree of validity, which has proved effective both in identifying potential novel antidepressants and in discriminating different rates of onset. However, although the model incorporates many features of depression, virtually all of the pharmacological studies have focused on the primary target symptom, around which the model was designed: anhedonia. This strategy is justified by the productivity of the model, but it is important not to overlook the fact that depression is a symptomatically heterogeneous disorder, in which two patients bearing the diagnostic label "major depressive disorder" could have totally distinct and nonoverlapping symptoms *(80)*. This argues for a diversity of animal models that focus on different target symptoms.

This view is supported empirically by studies demonstrating different neurochemical mechanisms of antidepressant action in different animal models of depression. As noted above, antidepressant action in the CMS model appears to be channeled through a final common pathway at the level of DA synapses, and can be reversed by blockade of DA receptors. This is also true of, for example, the Porsolt forced-swim test *(15)*. However, in other models, 5-HT mechanisms appear to predominate. For example, prolonged social isolation of rats causes an impairment of cooperative social behavior (a putative model of poor social skills), which responds to chronic antidepressant treatment. However, in this case, the action of imipramine was reversed by the 5-HT receptor antagonist metergoline *(81)*, which was ineffective in reversing the action of imipramine in rats receiving an almost identical treatment regime in the CMS model *(43)*. Similarly, 5-HT mechanisms also predominate in the olfactory bulbectomy model. In this procedure, bilateral lesions of the olfactory bulb cause a chronic but antidepressant-reversible syndrome, which differs from that observed in almost all other animal models of depression in being characterized by locomotor hyperactivity and heightened aggression. Again, antidepressant effects in this model are blocked by metergoline *(82)*. Antidepressant effects in the olfactory bulbectomy model usually require chronic treatment, but several studies have reported that 5-HT agonists and SSRIs act acutely *(see* ref. *14)* (though some other studies have found SSRIs to be inactive, even after chronic treatment *[83]*). Hence the ability of this model to detect rapid onset of antidepressant action is questionable. Nevertheless, the uniqueness of the atypical symptom pattern suggests that this model might in principle detect novel agents overlooked by more symptomatically conventional models.

In order to broaden the range of available models, several recent studies have attempted to model depression using rodent social behavior. These models have not been extensively validated, but they do involve chronicity. They are based on the parallels that have frequently been drawn between the behavior of depressed people and patterns of submissive behavior in animals *(84)*. In one procedure, groups of male rats are observed repeatedly at the start of the dark cycle, when they are at their most active, in order to determine the dominance hierarchy; chronic antidepressant treatment of the subdominant animal increases its hierarchical status *(85)*. In a related procedure, defeat of the dominant animal by a rat from a more aggressive strain causes a loss of dominant status, which can be maintained over long periods by repeated exposure to defeat, and can be reversed by chronic antidepressant treatment *(86)*. Defeat, in rats, also increases immobility in the Porsolt forced-swim test, and causes a variety of other behavioral changes potentially relevant to depression *(87)*. Similar effects of defeat have been described in mice *(88)*. Although these ethological procedures are attractive in principle, they do have some significant disadvantages. Foremost among these is that the social requirement multiplies the number of animals used in each experiment and the complexity of the experimental design. The workload involved in processing video recordings to establish dominance status should also not be underestimated, and this may limit the practical usefulness of these models and the extent of the validation they receive. Finally, although the experience of occupying a nondominant position within a social hierarchy is chronically stressful *(89)*, it is not clear whether the consequences of repeated social defeat are significantly different from those of other types of stressors, as observed, for example, in the CMS model. A similarity of these two procedures is suggested by the observations that, on the one hand, exposure to CMS increases submissive behavior *(35)*, but on the other, defeat causes a decrease in responsiveness to rewards, as measured by sucrose consumption and place-conditioning methods *(86,90)*.

A different approach to modelling depression in animals focuses on vulnerability, and, to this end, a number of neurodevelopmental strategies have been proposed. An extensive literature describes abnormalities in later life resulting from neonatal administration of antidepressants to rats *(91)*. Some of these effects, such as decreases in responsiveness to rewards *(91)* or immobility in the forced-swim test *(92)*, are consistent with what would be expected in an animal model of depression, but others, such as a decreased HPA response to stress *(93)*, are not. The pharmacological validation of this model is minimal *(91,94)*, and its theoretical rationale is obscure. A less extensively documented, but more promising, procedure is based on maternal separation of preweaning rats, which leads to persistent increases in HPA activity *(95)* and a decreased behavioral response to incentive stimuli in the adult animal *(96)*. No pharmacological studies using this procedure have been published.

Genetic strategies offer an alternative, and possibly superior, approach to modelling individual differences in vulnerability to depression. Flinders Sensitive Line (FSL) rats were bred for sensitivity to cholinergic agonists, and show a number of features consistent with an animal model of depression *(97)*. FSL rats are more submissive than their control strain, and, although the two strains do not differ in tests of rewarded behavior, the FSL rats are sensitive to the anhedonic effects of CMS, but their control strain are not *(98)*. However, FSL rats show reduced, rather than increased, HPA activity *(97)*. FSL rats also show increased immobility in the forced-swim test, and these effects are reversed by antidepressant drugs, but not by scopolamine or amphetamine *(99)*. This brief survey illustrates the status of the FSL model: It may prove useful for investigating neurochemical mechanisms relevant to depression, but it does not have any obvious advantages as a drug development tool. The same is likely to be true of a number of other selectively inbred strains, such as the apomorphine-susceptible *(100)* and Fawn hooded *(101)* rats, which have been less extensively characterized as animal models of depression.

Finally, the potential of quantitative and molecular genetic techniques should not be overlooked. One strategy that could profitably be developed is the use of recombinant inbred strains to identify candidate genes for depressive behaviors, using quantitative trait loci (QTL) analysis. This technique has been widely used in alcohol research *(102)*, but has not yet been applied to animal models of depression. Alternatively, selective gene targeting can now be used to remove (knock-outs) or insert (transgenics) specific genes *(103)*. Increases in aggressive behavior have been described in mice following a variety of gene knock-outs *(103)*, but alterations in responsiveness to rewards, or other behavioral changes directly relevant to depression, have not yet been reported. However, two transgenes have been described that cause changes in immobility in the forced-swim test. These effects are seen in TGF-α mice (which overexpress TGF-α) *(104)*, and in mice expressing glucocorticoid receptor antisense, which causes HPA hyperactivity *(105)*. Unfortunately, the changes in immobility are opposite in the two models (increased and decreased, respectively), but the limited pharmacological evidence suggests that both abnormalities are corrected by antidepressant drugs *(106,107)*. The pace of molecular biology is such that rapid advances in the application of genetic techniques to animal models of depression are inevitable; but it seems less likely that these developments will find immediate application in the search for novel antidepressant agents.

## 7. CONCLUSIONS

Traditional methods for developing antidepressant drugs have proved relatively successful in producing a variety of novel antidepressants that act at a diversity of neurochemical targets. The major achievement of these newer agents has been a reduction in the adverse side effects associated with tricylics and MAOIs, with no loss of efficacy. However, actual improvements in efficacy have not been forthcoming, and little progress has been made in accelerating the onset of antidepressant action; this constitutes a strong case for the development of novel models. If such models are to be of value for investigating onset latency, then chronicity is an essential design feature. Validity is also an important consideration: Increasingly, novel putative antidepressants are progressed to clinical testing only if they show evidence of activity in behavioral models that display at least a modicum of validity.

Of the relatively few chronic models that have been proposed, the CMS model has been most extensively validated. This model has identified a variety of potential novel antidepressants, including a number of agents that appear to show rapid onset of action, and has also been used to elucidate neurochemical mechanisms that may be relevant to depression and its treatment. However, the diversity of depressive disorders argues for a diversity of models, rather than an exclusive reliance on a single model, however effective. The success of the CMS model demonstrates that novel strategies are possible, and should encourage their further development.

## REFERENCES


1. Sitsen, J. M. A. and Montgomery, S. A. (1994) The pharmacological treatment of depression and its problems, in: *Handbook of Depression and Anxiety* (Den Boer, J. A. and Sitsen, J. M. A., eds.). Marcel Dekker, New York, pp. 349–377.
2. Prange, A. (1987) L-triiodothyronine (T3): its place in the treatment of TCA-resistant depressed patients, in: *Treating Resistant Depression* (Zohar, J. and Belmaker, R. H., eds.). PMA, New York, pp. 147–162.
3. De Montigny, C. and Cournoyer, G. (1987) Lithium addition in treatment-resistant major depression, in: *Treating Resistant Depression* (Zohar, J. and Belmaker, R. H., eds.). PMA, New York, pp. 147–162.
4. Wachtel, H. (1983) Potential antidepressant activity of rolipram and other selective cyclic adenosine 3',5'-monophosphate phosphodiesterase inhibitors. *Neuropharmacology* **22,** 267–272.
5. Layer, R. T., Popik, P., Nowak, G., Paul, I. A., Trullas, R., and Skolnick, P. (1996) A unified theory of antidepressant action: evidence for adaptation of the N-methyl-D-aspartate (NMDA) receptor following chronic antidepressant treatments, in: *Ion Channel Pharmacology* (Cena, V. and Soria, B., eds.). Oxford University Press, London, in press.
6. Smadja, C., Maldonado, R., Turcaud, S., Fournie-Zaluski, M. C., and Roques, B. P. (1995) Opposite effects of CCK-A and CCK-B 95 receptors in the modulation of endogenous enkephalin antidepressant-like effects. *Psychopharmacology* **120,** 400–408.
7. Tejedor-Real, P., Mico, J. A., Maldonado, R., Roques, B. P., and Gibert-Rahola, J. (1993) Effect of mixed (RB 38A) and selective (RB 38B) inhibitors of enkephalin degrading enzymes on a model of depression in the rat. *Biol. Psychiatry* **34,** 100–107.
8. Andrews, J. M. and Nemeroff, C. B. (1994) Contemporary management of depression. *Am. J. Med.* **97,** 24S–32S.

9. Rampello, L., Nicoletti, G., and Raffaele, R. (1991) Dopaminergic hypothesis of retarded depression: a symptom profile for predicting therapeutical responses. *Acta Psychiatrica Scand.* **84,** 552–554.

10. Derivan, A., Entsuah, A. R., and Kikta, D. (1995) Venlafaxine: measuring the onset of antidepressant action. *Psychopharmacol. Bull.* **31,** 439–447.

11. De Montigny, C., Chaput, Y., and Blier, P. (1990) Modification of serotonergic neuron properties by long-term treatment with serotonin reuptake blockers. *J. Clin. Psychiatry* **51(Suppl),** 4–8.

12. Artigas, F., Perez, V., and Alvarez, E. (1994) Pindolol induces a rapid improvement of depressed patients treated with serotonin reuptake inhibitors. *Arch. Gen. Psychiatry* **51,** 248–251.

13. Willner, P. (1984) The validity of animal models of depression. *Psychopharmacology* **83,** 1–16.

14. Willner, P. (1990) Animal models of depression: an overview. *Pharmacol. Ther.* **45,** 425–455.

15. Borsini. F. and Meli, A. (1990) The forced swimming test: its contribution to the understanding of the mechanisms of action of antidepressants, in: *Dopamine and Mental Depression* (Gessa, G. L. and Serra, G., eds.). Pergamon, Oxford, pp. 63–76.

16. Sarges, R., Howard, H. R., Browne, R. G., Lebel, L. A., Seymour, P.A., and Koe, B. K. (1990) 4-Amino[1,2,4]triazolo[4,3-a]quinoxalines. A novel class of potent adenosine receptor antagonists and potential rapid-onset antidepressants. *J. Med. Chem.* **33,** 2240–2254.

17. Flugy, A., Gagliano, M., Cannizzaro, C., Novara, V., and Cannizzaro, G. (1992) Antidepressant and anxiolytic effects of alprazolam versus the conventional antidepressant desipramine and the anxiolytic diazepam in the forced swim test in rats. *Eur. J. Pharmacol.* **214,** 233–238.

18. Drugan, R., Ryan, S. M., Minor, T. R., and Maier, S. F. (1984) Librium prevents the analgesia and shuttle-box escape deficits typically observed following inescapable shock. *Pharmacol. Biochem. Behav.* **21,** 749–754.

19. De Pablo, J. M., Ortiz-Caro, J., Sanchez-Santed, F., and Guillamon, A. (1991) Effects of diazepam, pentobarbital, scopolamine and the timing of saline injection on learned immobility in rats. *Physiol. Behav.* **50,** 895–899.

20. Willner, P., Towell, A., Sampson, D., Sophokleous, S., and Muscat, R. (1987) Reduction of sucrose preference by chronic mild stress and its restoration by a tricyclic antidepressant. *Psychopharmacology* **93,** 358–364.

21. Willner, P., Muscat, R., and Papp, M. (1992) Chronic mild stress-induced anhedonia: a realistic animal model of depression. *Neurosci. Biobehav. Rev.* **16,** 525–534.

22. Monleon, S., D'Aquila, P., Parra, A., Simon, V. M., Brain, P. F., and Willner, P. (1994) Attenuation of sucrose consumption in mice by chronic mild stress and its restoration by imipramine. *Psychopharmacology* **117,** 453–457.

23. Muscat, R. and Willner, P. (1992) Suppression of sucrose drinking by chronic mild unpredictable stress: a methodological analysis. *Neurosci. Biobehav. Rev.* **16,** 507–517.

24. American Psychiatric Association (1994) *DSM IV-Diagnostic and Statistical Manual of Psychiatric Disorders*, 4th ed. American Psychiatric Association, Washington DC.

25. Ayensu, W.K., Pucilowski, O., Mason, G. A., Overstreet, D. H., Revzani, A. H., and Janowsky, D. S. (1995) Effects of chronic mild stress on serum complement activity, saccharin preference and corticosterone levels in Flinders lines of rats. *Physiol. Behav.* **57,** 165–169.

26. Willner, P., Klimek, V., Golembiowska, K., and Muscat, R. (1991) Changes in mesolimbic dopamine may explain stress-induced anhedonia. *Psychobiology* **19,** 79–84.

27. Muscat, R., Kyprianou, T., Osman, M., Phillips, G., and Willner, P. (1991) Sweetness-dependent facilitation of sucrose drinking by raclopride is unrelated to calorie content. *Pharmacol. Biochem. Behav.* **40**, 209–213.
28. Papp, M., Willner, P., and Muscat, R. (1991) An animal model of anhedonia: attenuation of sucrose consumption and place preference conditioning by chronic unpredictable mild stress. *Psychopharmacology* **104**, 255–259.
29. Muscat, R., Papp, M., and Willner, P. (1992) Reversal of stress-induced anhedonia by the atypical antidepressants, fluoxetine and maprotiline. *Psychopharmacology* **109**, 433–438.
30. Matthews, K., Forbes, N., and Reid, I. C. (1995) Sucrose consumption as a hedonic measure following chronic unpredictable mild streess. *Physiol. Behav.* **57**, 241–248.
31. Willner, P., Moreau, J.-M., Nielsen, C., Papp, M., and Sluzewska, A. (1996) Decreased hedonic responsiveness following chronic mild stress is not secondary to loss of body weight. *Physiol. Behav.* **60**, 129–134.
32. Papp, M., Lappas, S., Muscat, R., and Willner, P. (1992) Attenuation of place preference conditioning but not place aversion conditioning by chronic mild stress. *J. Psychopharmacol.* **6**, 352–356.
33. Moreau, J.-L., Jenck, F., Martin, J. R., Mortas, P., and Haefely, W. E. (1992) Antidepressant treatment prevents chronic mild stress-induced anhedonia as assessed by ventral tegmentum self-stimulation behaviour in rats. *Eur. Neuropsychopharmacol.* **2**, 43–49.
34. Moreau, J.-L., Jenck, F., Martin, J. R., Mortas, P., and Haefely, W. E. (1993) Effects of moclobemide, a new generation reversible MAO-A inhibitor, in a novel animal model of depression. *Pharmacopsychiatry,* **26**, 30–33.
35. D'Aquila, P., Brain, P. F., and Willner, P. (1994) Effects of chronic mild stress in behavioural tests relevant to anxiety and depression. *Physiol. Behav.* **56**, 861–867.
36. Gorka, Z., Moryl, E., and Papp, M. (1996) The effect of chronic mild stress on circadian rhythms in the locomotor activity of rats. *Pharmacol. Biochem. Behav.* **54**, 229–234.
36a. Cheeta, S., Ruigt, G., van Proosdij, J., and Willner, P. (1997) Changes in sleep architecture following chronic mild stress. *Biol. Psychiatry,* in press.
37. Moreau, J.-L., Scherschlict, R., Jenck, F., and Martin, J. R. (1997) Chronic mild stress-induced anhedonia model of depression: sleep abnormalities and curative effects of electroshock treatment. *Behav. Pharmacol.* **6**, 682–687.
38. Kubera, M., Basta-Kaim, A., Papp, M., and Skowron-Cendrzak, A. (1994) Immunological changes after chronic mild stress and psychotropic drugs administration. 12th European Immunology Meeting (abstracts), p. 37.
39. Kubera, M., Basta-Kaim, A., and Papp, M. (1995) The effect of chronic treatment with imipramine on the immunoreactivity of animals subjected to chronic mild stress model of depression. *Immunopharmacology* **30**, 225–230.
40. Sluzewska, A., Gryska, K., and Mackiewicz, A. (1996) Acute phase proteins in chronic mild stress model of depression. *Behav. Pharmacol.* **7(Suppl 1)**, 105–106.
41. Willner, P. (1991) Animal models as simulations of depression. *TIPS* **12**, 131–136.
42. Papp, M., Moryl, E., and Willner, P. (1996) Pharmacological validation of the chronic mild stress model of depression. *Eur. J. Pharmacol.* **296**, 129–136.
43. Muscat, R., Sampson, D., and Willner, P. (1990) Dopaminergic mechanisms of imipramine action in an animal model of depression. *Biol. Psychiatry* **28**, 223–230.
43a. Sampson, D., Muscat, R., and Willner, P. (1991) Reversal of antidepressant action by dopamine antagonists in an animal model of depression. *Psychopharmacology* **104**, 491–495.
44. Przegalinski, E., Moryl, E., and Papp, M. (1995) The effect of 5-HT1A receptor ligands in a chronic mild stress model of depression. *Neuropharmacology* **34**, 1305–1310.

45. Sluzewska, A. and Szczawinska, K. (1996) The effects of pindolol addition to fluvoxamine and buspirone in chronic mild stress model of depression. *Behav. Pharmacol.* **7(Suppl 1)**, 105.

46. Cheeta, S., Broekkamp, C., and Willner, P. (1994) Stereospecific reversal of stress-induced anhedonia by mianserin and its (+)-enantiomer. *Psychopharmacology* **116**, 523–528.

47. Moreau, J.-L., Jenck, F., Martin, J. R., and Mortas, P. (1994) Curative effects of the atypical antidepressant mianserin in the chronic mild stress-induced anhedonia model of depression. *J. Psychiatr. Neurosci.* **19**, 51–56.

48. Sluzewska, A. and Szczawinska, K. (1996) Lithium potentiation of antidepressants in chronic mild stress model of depression in rats. *Behav. Pharmacol.* **7(Suppl 1)**, 105.

49. Sluzewska, A. and Nowakowska, E. (1994) The effects of carbamazepine, lithium and ketoconazole in chronic mild stress model of depression in rats. *Behav. Pharmacol.* **5(Suppl 1)**, 86.

50. Thakore, J. H. and Dinan, T. G. (1995) Cortisol synthesis inhibition: a new treatment strategy for the clinical and endocrine manifestations of depression. *Biol. Psychiatry* **37**, 364–368.

50a. Smadja, C., Ruiz, F., Turcaud, S., Roques, B. P., and Maldonado, R. (1995) Effects induced by endogenous and exogenous opioids in different animal models of depression: modulation by the CCK system. *Analgesia* **1**, 742–745.

51. Willner, P. (1985) *Depression: A Psychobiological Synthesis.* Wiley, New York.

52. Cutler, N. R., Keppel Hesselink, J. M., and Sramek, J. J. (1994) A phase II multicenter dose-finding, efficacy and safety trial of ipsapirone in outpatients with generalized anxiety disorder. *Prog. Neuro-Psychopharmacol. Biol. Psychiatry* **18**, 447–463.

53. Amsterdam, J. D. (1992) Gepirone, a selective serotonin (5HT(1A)) partial agonist in the treatment of major depression. *Prog. Neuropsychopharmacol. Biol. Psychiatr.* **16**, 271–280.

54. Willner, P. and Scheel-Kruger, J., eds. (1991) *The Mesolimbic Dopamine System: From Motivation to Action.* Wiley, New York.

55. Maj, J. (1990) Behavioral effects of antidepressant drugs given repeatedly on the dopaminergic system, in: *Dopamine and Mental Depression* (Gessa, G. L. and Serra, G., eds.). Pergamon, Oxford, 139–146.

56. Papp, M., Klimek, V., and Willner, P. (1994) Parallel changes in dopamine D2 receptor binding in limbic forebrain associated with chronic mild stress-induced anhedonia and its reversal by imipramine. *Psychopharmacology* **115**, 441–446.

57. Papp, M., Muscat, R., and Willner, P. (1993) Subsensitivity to rewarding and locomotor stimulant effects of a dopamine agonist following chronic mild stress. *Psychopharmacology* **110**, 152–158.

58. Stamford, J. A., Muscat, R., O'Connor, J. J., Patel, J., Trout, S. J., Wieczorek, W. J., Zruk, Z. L., and Willner, P. (1991) Subsensitivity to reward following chronic mild stress is associated with increased release of mesolimbic dopamine. *Psychopharmacology* **105**, 275–282.

59. Zebrowska-Lupina, I., Ossowska, G., and Klenk-Majewska, B. (1992) The influence of antidepressants on aggressive behavior in stressed rats: the role of dopamine. *Pol. J. Pharmacol. Pharm.* **44**, 325–335.

60. Muscat, R., Papp, M., and Willner, P. (1992) Antidepressant-like actions of dopamine agonists in an animal model of depression. *Biol. Psychiatry* **31**, 937–946.

61. Papp, M., Muscat, R., and Willner, P. (1993) Behavioural sensitization to a dopamine agonist is associated with recovery from stress-induced anhedonia. *Psychopharmacology* **110**, 159–164.

62. Willner, P., Lappas, S., Cheeta, S., and Muscat, R. (1994) Reversal of stress-induced anhedonia by the dopamine receptor agonist, pramipexole. *Psychopharmacology* **115**, 454–462.

63. Willner, P. (1994) Dopaminergic mechanisms in depression and mania, in: *Psychopharmacology: The Fourth Generation of Progress* (Bloom, F. E. and Kupfer, D. J., eds.). Raven, New York, pp. 921–931.

64. Moreau, J.-L., Borgulya, J., Jenck, F., and Martin, J. R. (1994) Tolcapone: a potential new antidepressant detected in a novel animal model of depression. *Behav. Pharmacol.* **5,** 344–351.

65. Mierau, J., and Schingnitz, G. (1992) Biochemical and pharmacological studies on pramipexole, a potent and selective dopamine D2 receptor agonist. *Eur. J. Pharmacol.* **215,** 161–170.

66. Cador, M., Robbins, T. W., Everitt, B. E., Simon, H., Le Moal, M., and Stinus, L. (1991) Limbic-striatal interactions in reward processes: modulation by the dopaminergic system, in: *The Mesolimbic Dopamine System: From Motivation to Action.* (Willner, P. and Scheel-Kruger, J., eds.), Wiley, New York, pp. 225–250.

67. Papp, M., Klimek, V., and Willner, P. (1994) Effects of imipramine on serotonergic and beta-adrenergic receptor binding in a realistic animal model of depression. *Psychopharmacology* **114,** 309–314.

68. Bijak, M. and Papp, M. (1995) The effect of chronic treatment with imipramine on the responsiveness of hippocampal CA1 neurons to phenylephrine and serotonin in a chronic mild stress model of depression. *Eur. Neuropsychopharmacol.* **5,** 43–48.

69. Cesana, R., Ciprandi, C., and Borsini, F. (1995) The effect of BIMT 17, a new potential antidepressant, in the forced swimming test in mice. *Behav. Pharmacol.* **7,** 688–694.

70. Borsini, F. (1996) Antidepressants: behavioral consequences of 5-HT postsynaptic agonism. *Behav. Pharmacol.* **7(Suppl. 1),** 10.

71. Willner, P. (1995) Animal models of depression: Validity and applications. In: *Depression and Mania: From Neurobiology to Treatment* (Gessa, G. L., Fratta, W., Pani, L., and Serra, G., eds.). Raven, New York, pp. 19–41.

72. McCance-Katz, E., Price, L. H., Charney, D. S., and Heninger, G. R. (1992) Serotonergic function during lithium augmentation of refractory depression. *Psychopharmacology* **108,** 93–97.

73. Papp, M., Nalepa, I., and Vetulani, J. (1994) Reversal by impramine of beta-adrenoceptor up-regulation induced in a chronic mild stress model of depression. *Eur. J. Pharmacol.* **261,** 141–147.

74. Cheeta, S. (1995) A model of depression: assessment of the validity of the chronic mild stress procedure. PhD thesis, University of Wales.

75. Nowak, G., Papp, M., and Paul, I. A. (1995) Adaptation of NMDA receptors in animal models of depression: chronic mild stress (CMS). *Soc. Neurosci. Abstracts* **21,** 2107.

76. Papp, M. and Moryl, E. (1994) Antidepressant activity of non-competitive and competitive NMDA receptor antagonists in a chronic mild stress model of depression. *Eur. J. Pharmacol.* **263,** 1–7.

77. Papp, M. and Moryl, E. (1997) Antidepressant-like effects of l-aminocyclopropanecarboxylic acid and d-cycloserine in an animal model of depression. *Eur. J. Pharmacol.,* in press.

78. Dziedzicka-Wasylewska, M. and Papp, M. (1996) Effect of chronic mild stress and prolonged treatment with imipramine on the levels of endogenous met-enkephalin in the rat dopaminergic mesolimbic system. *Pol. J. Pharmacol.* **48,** 53–56.

79. Dauge, V., Bertrand, E., Smadja, C., Fournie-Zaluska, M. C., and Roques, B. P. (1996) Analysis of extracellular levels of met-enkephaline in the nucleus accumbens of rats submitted to a chronic unpredictable mild stress regimen. *Behav. Pharmacol.* **7(Suppl. 1),** 24.

80. Fibiger, H. C. (1991) The dopamine hypotheses of schizophrenia and depression: contradictions and speculations, in: *The Mesolimbic Dopamine System: From Motivation to Action* (Willner, P. and Scheel-Kruger, J., eds.). John Wiley, Chichester, pp. 615–637.

81. Willner, P., Sampson, D., Phillips, G., Fichera, R., Foxlow, P., and Muscat, R. (1989) Effects of isolated housing and chronic antidepressant treatment on cooperative social behaviour in rats. *Behav. Pharmacol.* **1,** 85–90.

82. Broekkamp, C. L., Garrigou, D., and Lloyd, K. G. (1980) Serotonin-mimetic and antidepressant drugs on passive avoidance learning by olfactory bulbectomized rats. *Pharmacol. Biochem. Behav.* **13,** 643–646.

83. Redmond, A. M., Kelly, J. P., and Leonard, B. E. (1994) Effect of paroxetine and fluvoxamine on behavioural changes in a number of paradigms in the olfactory bulbectomized rat model of depression. *J. Serotonin Res.* **3,** 199–205.

84. Gardner, R. (1982) Mechanisms in manic-depressive disorder: an evolutionary model. *Arch. Gen. Psychiatry* **39,** 1436–1441.

85. Mitchell, P. J. and Redfern, P. H. (1992) Chronic treatment with chlomipramine and mianserin increases the hierarchical position of subdominants rats housed in triads. *Behav. Pharmacol.* **3,** 239–247.

86. Willner, P., D'Aquila, P., Coventry, T., and Brain, P. (1995) Loss of social status: preliminary evaluation of a novel animal model of depression. *J. Psychopharmacol.* **9,** 207–213.

87. Koolhaas, J. M., Hermann, P. M., Kemperman, C., Bohus, B., van den Hoofdakker, R. H., and Beersma, D. G. M. (1990) Single social defeat in male rats induces a gradual but long lasting behavioral change: a model of depression? *Neurosci. Res. Comm.* **7,** 35–41.

88. Kudryatseva, N. N., Bakshtanowskaya, I. V., and Koryakina, L. A. (1991) Social model of depression in mice of C57BL/6J strain. *Pharmacol. Biochem. Behav.* **38,** 315–320.

89. Blanchard, D. C., Spencer, R. L., Weiss, S. M., Blanchard, R. J., McEwen, B., and Sakai, R. R. (1995) Visible burrow system as a model of chronic social stress: behavioral and neuroendocrine correlates. *Psychoneuroendocrinology* **20,** 117–134.

90. Coventry, T., D'Aquila, P., Brain, P., and Willner, P. (1997) Social influences on morphine conditioned place preference. *Behav. Pharmacol.,* in press.

91. Vogel, G. W., Neill, D., Hagler, M., and Kors, D. (1990) A new animal model of depression: a summary of present findings. *Neurosci. Biobehav. Rev.* **14,** 49–63.

92. Hilakivi, L. A., Taira, R., Hilakivi, I., and Loikas, P. (1988) Neonatal treatment with monoamine uptake inhibitors alters late response in behavioral 'despair' test to beta and GABA-B receptor agonists. *Pharmacol. Toxicol.* **63,** 57–61.

93. Ogawa, T., Mikuni, M., Kuroda, Y., Muneoka, K., Mori, K. J., and Takahashi, K. (1994) Effects of the altered serotonergic signalling by neonatal treatment with 5,7-dihydroxytryptamine, ritanserin or clomipramine on the adrenocortical stress response and the glucocorticoid receptor binding in the hippocampus in adult rats. *J. Neural Transm.* **96,** 113–123.

94. Dwyer, K. D. and Roy, E. J. (1993) Juvenile desipramine reduces adult sensitivity to imipramine in two behavioral tests. *Pharmacol. Biochem. Behav.* **45,** 201–207.

95. Ladd, C. O., Owens, M. J., and Nemeroff, C. B. (1996) Persistent changes in corticotropin-releasing factor neuronal systems induced by maternal deprivation. *Endocrinology* **137,** 1212–1218.

96. Matthews, K., Wilkinson, L. S., and Robbins, T. W. (1996) Repeated maternal separation of preweanling rats attenuates behavioral responses to primary and conditioned incentives in adulthood. *Physiol. Behav.* **59,** 99–107.

97. Overstreet, D. H. (1993) The Flinders sensitive line rats: a genetic animal model of depression. *Neurosci. Biobehav. Rev.* **17,** 51–68.

98. Pucilowski, O., Overstreet, D, H., Rezvani, A. H., and Janowsky, D. S. (1993) Chronic mild stress-induced anhedonia: greater effect in a genetic rat model of depression. *Physiol. Behav.* **54,** 1215–1220.

99. Overstreet, D. H., Pucilowski, O., and Rezvani, A. (1995) Administration of antidepressants, diazepam and psychomotor stimulants further confirms the utility of Flinders Sensitive Line rats as an animal model of depression. *Psychopharmacology* **121,** 27–37.
100. Cools, A. R., Brachten, R., Heeren, D., Willemen, A., and Ellenbroek, B. (1990) Search after neurobiological profile of individual-specific features of Wistar rats. *Brain Res. Bull.* **24,** 49–69.
101. Overstreet, D. H., Rezvani, A., and Janowsky, D. S. (1992) Genetic animal models of depression and ethanol preference provide support for cholinergic and serotonergic involvement in depression and alcoholism. *Biol. Psychiatry* **31,** 919–936.
102. Plomin, R. and McClearn, G. E. (1993) Quantitative trait loci (QTL) analyses and alcohol-related behaviors. *Behav. Genet.* **23,** 197–211.
103. Gold, L. H. (1996) Integration of molecular biological techniques and behavioural pharmacology. *Behav. Pharmacol.* **7,** 589–615.
104. Hilakivi-Clarke, L. A., Arora, P. K., Sabol, M. B., Clarke, R., Dickson, R. B. and Lippman, M. E. (1992) Alterations in behavior, steroid hormones and natural killer cell activity in male transgenic TGF alpha mice. *Brain Res.* **588,** 97–103.
105. Beaulieu, S., Rousse, I., Gratton, A., Barden, N., and Rochford, J. (1994) Behavioral and endocrine impact of impaired type II glucocorticoid receptor function in a transgenic mouse model. *Ann. NY Acad. Sci.* **746,** 388–391.
106. Hilakivi-Clarke, L. A. and Goldberg, R. (1993) Effects of tryptophan and serotonin uptake inhibitors on behaviour in male transgenic transforming growth factor alpha mice. *Eur. J. Pharmacol.* **237,** 101–108.
107. Montkowski, A., Barden, N., Wotjak, C., Stec, I., Ganster, J., Meaney, M., Engelmann, M., Reul, J. M. H. M., Landgraf, R., and Holsboer, F. (1995) Long-term antidepressant treatment reduces behavioural deficits in transgenic mice with impaired glucocorticoid receptor function. *J. Neuroendocrinol.* **7,** 841–845.

# 13

# Molecular Strategies to Novel Antidepressant Discovery

## Steven M. Paul, Xin Wu, Yanbin Liang, and Edward I. Ginns



## 1. INTRODUCTION

Given the relatively recent introduction, established effectiveness, and widespread clinical use of selective serotonin reuptake inhibitors (SSRIs) *(1)*, one might legitimately ask, "Why do we need new antidepressant agents?" Indeed, the SSRIs and their less-selective predecessors, the tricyclic antidepressants and monoamine oxidase inhibitors, are effective in approx 65–70% of patients when systematically investigated in "double-blind" placebo-controlled trials *(2,3)*. Moreover, patients who fail to respond to one antidepressant often respond to another compound, either from the same or different "class" *(3)*. It has also been estimated that at least 80% of depressed patients will eventually respond to one or more antidepressants, and this may be greater than 90% for patients treated with electroconvulsive therapy (ECT) *(4)*.

There are several obvious reasons why the search for new and better antidepressants should continue. First, all currently available antidepressants have two major limitations: an unfortunate time-delay to full clinical response and an overall efficacy (% response) that is far less than desirable (only 65% in moderately to severely depressed patients) *(3,4)*. While it is generally recognized that antidepressants require several weeks (approx 2–4) of continuous administration before their mood-elevating properties are fully manifest, the exact time-course of response is still controversial *(5)*. Indeed, it has been argued that current antidepressants may actually improve mood relatively soon after steady-state plasma levels have been achieved, although a more complete response is not observed until several weeks later *(2,6)*. Is it even possible to achieve a rapid (several days) and robust (>90%) antidepressant response? In this regard, most investigators agree that ECT improves mood somewhat sooner (and perhaps somewhat better) than other antidepressants and, in fact, rather rapid responses are often observed after only a few treatments *(4,7)*. Nevertheless, the best evidence that a rapid antidepressant response is possible comes from studies of sleep deprivation, in which it has been consistently found that greater than 50% of depressed patients will display a rapid and (in many cases) dramatic "antidepressant" response to a single night



*From: Antidepressants: New Pharmacological Strategies*
*Edited by: P. Skolnick, Humana Press Inc., Totowa, NJ*

of sleep deprivation *(8)*. Unfortunately, these often dramatic responses are short lived, as patients relapse within a day or so *(8)*. It is conceivable however that similar, but sustained, responses could be achieved with pharmacotherapy.

Second, it has been increasingly recognized that for many depressed patients maintenance treatment is required to prevent frequent relapses and (or) to maintain a euthymic state. Moreover, in relative long-term prospective studies of treatment outcomes gathered from large cohorts of depressed patients, a surprisingly low (in our view) percentage of patients remain symptom free even while taking maintenance antidepressant therapy *(9)*. These data again underscore the need for better (faster-acting and more efficacious) antidepressant drugs.

How then will new antidepressants be discovered and, specifically, what approaches will be most effective in finding these newer agents? Indeed, this volume highlights the current excitement in the field and identifies several scientific strategies that are likely to prove successful. In this chapter we focus our discussion on several related but as yet relatively unexplored approaches to antidepressant drug discovery—namely the use of genomics (broadly defined) to fully delineate the downstream mechanisms underlying antidepressant drug action and to find the genes (and their protein products) that determine the now-well-accepted genetic vulnerability to develop major affective disorder. The latter, if modifiable by pharmacotherapy, could provide the most effective antidepressants of all, as their use may alter the underlying disease process itself.

## 2. GENOMICS AS AN APPROACH TO UNDERSTANDING THE TIME-LAG OF ANTIDEPRESSANT RESPONSE

### 2.1. Antidepressants Work On First, Second, and Third Messengers to Alter Gene Expression

#### 2.1.1. First Messengers

The vast majority of research on the neurochemical substrates of antidepressant action has focused primarily on delineating the acute effects of these drugs on the function(s) of various neurotransmitters (i.e., first messengers). These studies, dating back over 30 yr, have shown that most (if not all) antidepressants enhance neurotransmitter function by either inhibiting their catabolic enzymes or their inactivation via reuptake blockade. Tricyclic antidepressants, for example, inhibit biogenic amine reuptake and in some cases have direct actions (agonist or antagonist) at various neurotransmitter receptors. With respect to SSRIs, it seems highly probable that each member of this structurally diverse class of compounds initiates its beneficial pharmacological actions by increasing synaptic serotonin (5-hydroxytryptamine, or 5-HT) concentration, since their one common neuropharmacological feature is potent inhibition of 5-HT reuptake *(1,10)*. The latter, however, occurs almost immediately following drug treatment as documented by in vivo microdialysis in a variety of laboratory animal studies *(10)*. Since acute 5-HT reuptake blockade does not result in an acute mood-elevating or antidepressant action, how does reuptake blockade eventually lead to the antidepressant actions of SSRIs? To explain the time-lag of therapeutic response, both the acute and chronic effects of SSRIs on serotonergic neurotransmission have been explored in some detail *(11,12)*. Acute administration of SSRIs and the resultant increase in terminal and somatodendritic 5-HT concentration result in a "feedback inhibition" of the firing rate or activity of 5-HT neurons of the midbrain and an attenuated release of 5-HT from terminals that project from these cell bodies to forebrain regions, including the hippocampus and cerebral cortex *(11,12)*. This feedback inhibition is the result of an acute activation of somatodendritic and terminal (prejunctional) 5-HT autoreceptors of the 5-HT$_{1A}$ and 5-HT$_{1B}$ subtypes, respectively *(13,14)*. Following repeated administration of SSRIs, these autoreceptors (particularly the somatodendritic 5-HT$_{1A}$ autoreceptor) become "subsensitive" and, after several weeks, the levels of "synaptically available" 5-HT are elevated over those seen prior to or following acute SSRI administration *(15)*. The time-lag for developing this functional subsensitivity of 5-HT autoreceptor responses in laboratory animals parallels the time-lag to clinical response, and has therefore been postulated to mediate the therapeutic effects of SSRIs and perhaps other antidepressants as well *(16)*. In fact, strategies to prevent (i.e., block) the acute activation of somatodendritic 5-HT$_{1A}$ autoreceptors following SSRI administration have been devised; and preliminary clinical data *(17)* suggest that such approaches may reduce the time-lag of antidepressant response (*see also* Romero, et al., this volume). However, even if one accepts that the enhanced synaptically available 5-HT (which is observed after chronic but not acute antidepressant treatment) is critical to the mood-elevating properties of these drugs, the exact receptor and postreceptor targets of SSRI antidepressant drug action still need to be identified.

### 2.1.2. Alterations in Second and Third Messengers Lead to Changes in Gene Expression

Antidepressants, by "upregulating" neurotransmitter levels either acutely or chronically, subsequently alter the activity of intracellular second messengers such as cAMP and products of the polyphosphoinositide cascade. Of the 14 identified 5-HT receptors, for example, all but one are coupled by G proteins to adenylate cyclase or phospholipase C (the 5-HT$_3$ receptor represents a ligand-gated ion channel) *(18)*. Most of these 5-HT receptors are linked to inhibition of adenylate cyclase through Gi but three (5-HT$_{4,6,7}$), like β-adrenoceptors, are positively coupled to adenylate cyclase through Gs *(18)*. The 5-HT$_2$ receptors (5-HT$_{2A,B,C}$) are linked to activation of phospholipase C, resulting in the production of several intracellular second messengers that serve to mobilize intracellular Ca$^{2+}$ and activate various protein kinases (e.g., protein kinase C) *(18)*. These second messengers can in turn activate third messengers via protein phosphorylation ultimately resulting in changes in transcription factor-dependent gene expression. Thus, at 5-HT synapses, enhanced 5-HT levels are likely to result in multiple changes in second-messenger function. The situation is all the more complex for "less-specific" antidepressants such as the tricyclic antidepressants and ECT, since multiple first and second messengers are undoubtedly affected by treatment with these agents. Nonetheless, tricyclic antidepressants, MAOIs, SSRIs, and ECT have been shown to affect certain common second-messenger systems, especially after chronic administration *(19)*. Perhaps the best characterized of these antidepressant-responsive second messenger systems is the cAMP cascade, where changes in both adenylate cyclase and cyclic AMP-dependent protein kinase (PKA) activity have been demonstrated *(19)*.

Activation of the cAMP cascade results in the PKA-mediated phosphorylation (via this third messenger) of various proteins, including the nuclear transcription factor, cAMP response element-binding protein (CREB). Interestingly, in a recent study by Nibuya and colleagues *(20)*, the levels of CREB mRNA, CREB protein, and CRE-binding activity were all upregulated in the hippocampus of rats chronically treated with various antidepressants *(20)*. In that study, the prototypic tricyclic antidepressants imipramine and desipramine, the MAOI tranylcypromine, the SSRIs fluoxetine and sertraline, as well as ECS, all upregulated hippocampal CREB after chronic but not acute administration *(20)*. The upregulation of CREB after chronic antidepressant treatment is consistent with the time-course of therapeutic response. Moreover, chronic treatment with morphine, cocaine, or haloperidol (nonantidepressant psychotropic drugs) failed to affect CREB levels in the hippocampus *(20)*.

Significantly, the levels of hippocampal CRE-binding activity as measured by gel-shift analysis are also upregulated by chronic (but not acute) administration of antidepressants *(20)*. Thus it seems clear that CREB-responsive gene expression is altered by chronic administration of chemically and pharmacologically dissimilar antidepressants. However, given the number of receptors, second, and third messengers involved (vide supra), it seems likely to us that the tissue levels of other transcription factors will also be altered following chronic antidepressant administration. These drug-induced and time-dependent changes in gene expression could underlie the delayed time-course of therapeutic response and may ultimately provide clues to help identify target genes mediating the therapeutic response to antidepressants.

## 2.2. Antidepressant-Induced Changes in Gene Expression

Antidepressant treatment has recently been shown to alter the level of brain mRNAs encoding neurotransmitter-related proteins, including biosynthetic enzymes *(21)* and receptors *(22)*. Transcription of these genes was predictably affected since their protein products had been previously shown to be altered by chronic antidepressant administration. For example, "downregulation" of $\beta_1$-adrenoceptors by chronic administration of antidepressants is accompanied by a similar reduction in $\beta_1$-adrenoceptor mRNA *(22)*. Given the effects of antidepressants in altering the tissue levels of transcription factors like CREB, it seems likely that other less-obvious changes in gene expression will ultimately be delineated. One might speculate that virtually all genes whose expression is regulated via a CRE will be affected, i.e., in those brain regions where CREB is upregulated. In fact, Nibuya and colleagues *(23)* have also recently shown that two CRE-containing (and potential target) genes for CREB (namely BDNF [brain-derived neurotrophic factor] and its receptor trkB) are also upregulated following chronic (but not acute) antidepressant administration (*see* chapter by Duman, et al.). The upregulation of genes encoding a neurotrophic factor and its receptor were unanticipated and suggest that chronic antidepressant treatment may increase the survival of hippocampal neurons or promote the sprouting of neurons that innervate the hippocampus, such as those 5-HT and NE neurons that project from the midbrain raphe and locus coeruleus, respectively. Interestingly, repeated environmental "stressors" have been shown to cause atrophy and, in severe stress paradigms, death of hippocampal neurons *(24)*. Smith and colleagues *(25)* have also recently shown that exposure of rats to stress reduces the levels of BDNF in the hippocampus, which may account for the morphological changes in these neurons observed following stress. Thus, chronic antidepressant treatment may prevent or reverse stress-induced changes in hippocampal neuron structure/function by upregulating CRE-responsive gene products like BDNF and trkB. Would drugs that directly activate BDNF expression or that directly stimulate trkB receptors be rapidly acting antidepressants?

### 2.2.1. Using Differential Display to Find Antidepressant-Responsive Genes

Given that the expression of multiple genes is likely to be affected by chronic antidepressant treatment, how can such genes be systematically identified and how can their expression be linked to the therapeutic action of these drugs? Over the past few years, several powerful techniques have emerged to allow the identification of mRNAs that are "differentially" expressed in one tissue compared to another. Earlier studies employing subtractive hybridization proved successful in isolating genes in which expression was altered by drug treatment; but this technique is rather time consuming, requires relatively large amounts of RNA, and is often not very repro- ducible from experiment to experiment. In 1992, Liang and Pardee *(26)* described a novel method for identifying differentially expressed mRNAs by "differential dis- play." In this technique, RNA from a tissue is reverse transcribed (RT) into cDNA and amplified by the polymerase chain reaction (PCR) using two different primers. The first anchored oligo-dT primer anneals to the 3' poly(A) tail to generate cDNAs from a subpopulation of mRNAs. The second primer, an arbitrary primer comprised of approx 10 nucleotides of defined sequence, is then used to amplify the cDNAs from the mRNA pool. The reaction products are subsequently separated by size on a dena- turing polyacrylamide gel. DNA fragments that are present in one sample and absent in another (or vice versa) can be readily detected by visual inspection. Once a specific fragment is identified as being differentially expressed, the DNA can be reamplified, cloned, sequenced, and used to isolate full-length cDNA. If the appropriate number of arbitrary 5' primers are employed, it is theoretically possible to examine all genes expressed in a given cell or tissue. Moreover, the differential display method allows for simultaneous detection of genes that are either upregulated or downregulated in the same sample.

The RT-PCR based differential display method is very sensitive and requires mini- mal amounts of mRNA, so it is quite useful for studying gene expression in relatively small brain regions. Since "false-positive" DNA fragments are frequently generated both within a given experiment (same RNA sample) as well as between experiments (different RNA samples but from the same tissue/treatment conditions, and so on), dis- play patterns should be reproduced and differential expression of a given transcript must be confirmed using an alternate method (e.g., "reverse" Southern or Northern analysis). Also, because the RT-PCR differential display technique is only semiquanti- tative, it may not be sufficiently sensitive to detect small but physiologically (or phar- macologically) meaningful changes in gene expression (e.g., $<100\%$ or so). For example, it is not clear that a 40–50% change in hippocampal CREB mRNA observed following chronic antidepressant administration *(20)* would be detected by differential display. However, despite these limitations, the differential display technique has proven

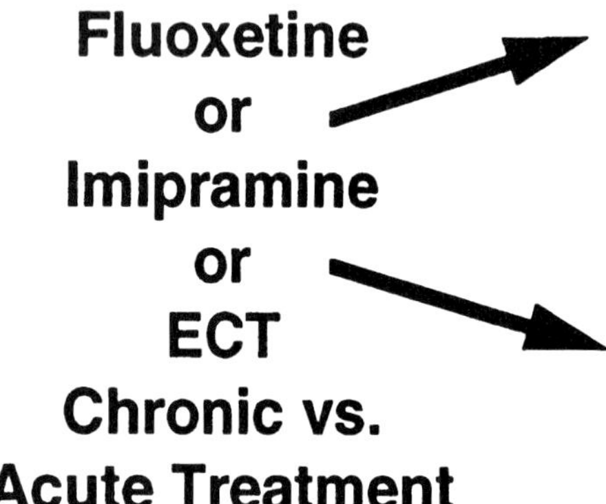



**Fig. 1.** *(see also opposite page)* The use of differential display to characterize antidepressant-induced changes in gene expression. Animals are treated acutely and chronically with antidepressants such as fluoxetine, imipramine, and ECT *(see text* for details). RT-PCR based differential display using various primer pairs is used to effectively sample most, if not all, mRNAs expressed in different brain regions. The radioactively labeled reaction products can be separated using a denaturing polyacrylamide gel and the pattern visualized autoradiographically. Changes in gene expression (transcription) induced by drug treatment are readily identified (*see* Fig. 2) and the DNA fragments can be reamplified and used to obtain full-length cDNAs for sequence determination and other characterization.

quite useful in identifying differentially expressed genes in a number of experimental paradigms *(27,28)*.

We have used the differential display method as described above to identify genes that are differentially expressed following antidepressant administration. Rats were treated acutely (3 d) or chronically (21 d) with fluoxetine (5 mg/kg/d) or imipramine (5 mg/kg/d) or vehicle (Fig. 1). Another group of animals was administered electron-convulsive treatment (ECT) either acutely (one treatment) or chronically (once daily for 10 d). In all treatment groups, animals were sacrificed 24 h following their last treatment and RNA was prepared from various brain regions using standard methods (Fig. 1). Differential display of RNA isolated from the hippocampus (a brain region implicated in the pathophysiology of depression and where the nuclear transcription factor

Gene Discovery
Differentially Expressed Genes Following
Chronic Antidepressant Treatment
DIFFERENTIAL DISPLAY
cxaaaaa_n
txaaaaa_n
gxaaaaa_n
axaaaaa_n
5'_n tttttxc3'
5'_n tttttxg3'
5'_n tttttxa3'
5'_n tttttxg3'
Reverse transcribe
cxaaaaa_n
gxttttttttt_n
forward primers:
random 10 -mers
i.e.
5' tgc tga cct g 3'
reverse primers:
5'_n tttttxc3'
5'_n tttttxg3'
5'_n tttttxa3'
5'_n tttttxg3'
PCR Amplification
α 33P dATP
tgctgacctg
gxttttttttt_n
Electrophoresis

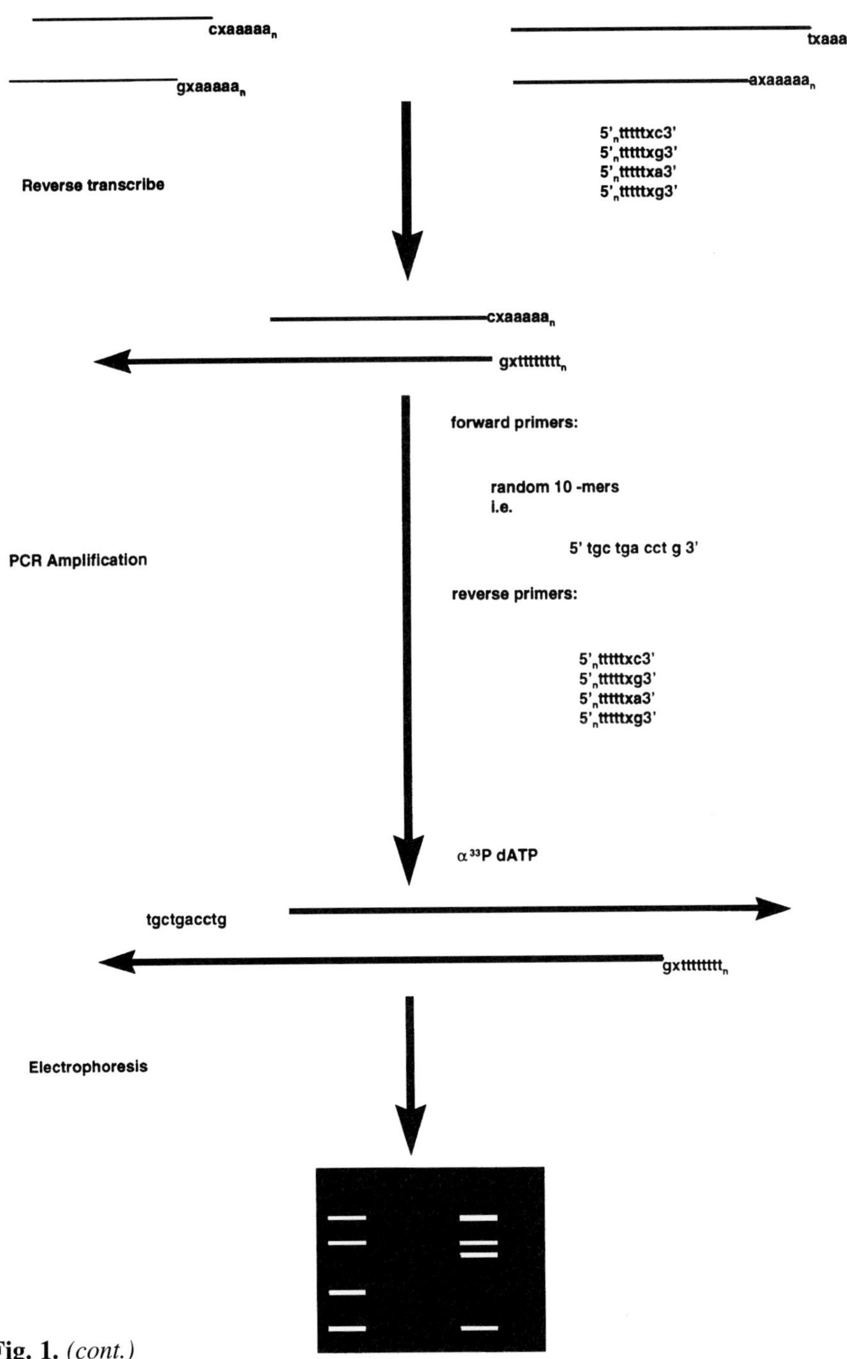

**Fig. 1.** *(cont.)*

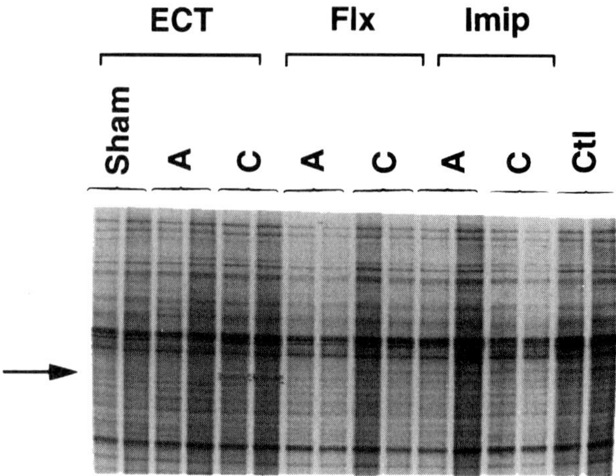



**Fig. 2.** Differential display reveals a novel transcript upregulated by chronic (but not acute) ECT. RT-PCR differential display was used to study the effects of acute (A) and chronic (C) antidepressants on gene expression in the hippocampus of adult rats. Note transcript (arrow) upregulated by chronic ECT but not by the other antidepressant treatments. Similar experiments have revealed transcripts upregulated by the other antidepressant treatments as well (data not shown). Flx, fluoxetine; ECT, electroconvulsive therapy; Imip, imipramine; Ctl, control (untreated).

CREB is upregulated after chronic, but not acute, antidepressant treatment) *(20)* revealed several differentially expressed genes. However, as anticipated, antidepressant treatment also resulted in a number of differentially expressed genes in the cerebral cortex, where alterations in CREB are not observed *(20)*. The most obvious and reproducible changes in gene expression were of transcripts upregulated by chronic ECT (*see* Fig. 2 for example). These transcripts are now being sequenced and full-length cDNAs are being isolated so they can be more completely characterized. Since ECT is generally considered to be a more rapid and more effective antidepressant treatment than drug therapies, genes altered by ECT may prove useful in discovering more effective small molecule antidepressants. It is possible that target genes upregulated by all three antidepressant treatments (chronic but not acute) will be found. However, we must also entertain the possibility that each of these three antidepressant treatments work by altering the expression of different genes and that no common pattern of gene expression will emerge. Moreover, different patterns of gene expression may occur, depending on the brain region examined. In this regard Wong and colleagues *(29)*, using differential display, have recently reported the presence of two transcripts that are upregulated in the hypothalamus of rats chronically treated (8 wk) with either imipramine or fluoxetine. These two transcripts (out of 23 initially identified by differential display!) were confirmed by reverse Northern analysis *(29)*. Unfortunately, Wong et al. *(29)*. did not provide any information as to the sequences or possible identity of these transcripts. Once more fully characterized, it will be important to determine whether such genes and their protein products are important (i.e., necessary) for the antidepressant actions of these drugs since changes in transcription of these genes may be unrelated to their beneficial pharmacological properties.

## 3. GENOMICS AS AN APPROACH TO FINDING SUSCEPTIBILITY GENES FOR DEPRESSION

### 3.1. The Role of Genetic Factors in the Etiology of Depression

Depression is a common clinical disorder affecting approx 4–5% of the general population *(30)* and, as such, is likely to be etiologically quite heterogeneous. Like many other common medical conditions (e.g., hypertension or diabetes), depression appears to be the result of a complex interplay between environmental and constitutional (e.g., genetic) factors. Decades of family, twin, and adoption studies have confirmed an important contribution of genetic factors to the etiology of affective disorders *(31)*. The relative risk of developing an affective disorder, for example, increases in direct proportion to genetic "relatedness" so that the more related a given family member is to an affected proband, the greater is his/her chance of becoming ill. For instance, a first-degree relative of a patient with bipolar affective disorder is 17–20 times more likely to develop the illness than is a first-degree relative of a healthy control *(31)*.

Studies comparing the concordance rates between monozygotic (MZ) and dizygotic (DZ) twin pairs have found much higher concordance rates in MZ compared to DZ pairs *(31,32)*. However, despite sharing an identical complement of genes, in all studies the concordance rates for MZ twin pairs is always well below 100%. This latter observation can be interpreted in several ways—including the possibilities that not all forms of affective disorder are inherited (i.e., genetic) or that the genes responsible are not fully penetrant, or that the phenotype can be modified by either genetic and (or) nongenetic factors. Moreover, in most affected families, depression is inherited as a complex trait in which the mode of inheritance does not follow a simple Mendelian pattern *(31,33)*. Thus multiple predisposing genes may be required for manifestation of the disease phenotype and, even then, perhaps only as sequelae to the necessary environmental precipitants *(34)*. Based on these data, the genes that contribute to the pathogenesis of affective disorders are best viewed as "predisposing or susceptibility" genes whose expression can be modified by genetic and environmental factors *(34)*. It is hoped that the identification of such susceptibility genes may lead to new and better targets for antidepressant drug discovery.

### 3.1.1. Finding Susceptibility Genes for Bipolar Affective Disorder

Clinically, bipolar affective disorder or manic-depressive illness is characterized by alternating episodes of mania, hypomania, or depression *(35)*. For most bipolar patients the age of onset (15–35 yr old) is relatively early compared to other forms of affective disorder, and the natural history of the disorder, if untreated, is often severe (the suicide rate for untreated patients is approx 20%!). Bipolar affective disorder has an estimated lifetime prevalence rate of approx 1% and is one of the most "inherited" of all psychiatric disorders, with MZ twin concordance rates approaching 70% in most studies *(32)*. Consequently, the vast majority of recent and current research on the genetics of affective disorders has focused on delineating the chromosomal loci harboring susceptibility genes for bipolar affective disorder *(see below)*. To carry out such studies, patients with bipolar affective disorder (probands) are identified and their family members are examined to ascertain affection status (presence or absence of illness) of first-degree relatives (e.g., parents, siblings, and so on). A critical element to these genetic linkage studies is the availability of sufficient numbers of well-characterized patients and family members. In association studies, a group of affected individuals is usually compared to a "matched" group of unaffected subjects and thus family members *per se* are not required. This approach generally requires large numbers of patients and a control group of identical ethnic background, or the rate of false-positive associations (type 1 errors) becomes problematic.

For positional cloning of disease susceptibility genes *(36)*, investigators have generally utilized one of the following approaches. The first approach involves the study of a relatively large number of small nuclear families in which one or more affected sib pairs and at least one affected parent can be identified and samples for DNA analysis can be collected. The second approach involves the study of so-called "genetic isolates"—relatively large and related families or pedigrees with a relatively high percentage of affected individuals, and in which the illness can be traced to only a few (ideally one) progenitor(s). Each of these approaches has advantages and disadvantages but, for common and complex disorders caused by multiple genes and where genetic heterogeneity is likely, the use of genetic isolates can reduce the number of disease susceptibility genes, facilitating linkage analysis (*see below*). Nevertheless, if sufficient numbers of affected sib pairs from small unrelated families can be collected, it is still possible to identify susceptibility loci even if the disorder is characterized by considerable genetic heterogeneity.

Once suitable numbers of well-characterized pedigrees containing as many affected and unaffected members as possible are collected and DNA marker genotyping is completed, linkage analysis can be used to delineate the chromosomal regions in which susceptibility genes for bipolar affective disorder reside *(36)* (Fig. 3). Initially, a genome screen using DNA markers spaced at approx 10–20 cM (10–20 million bp) intervals is performed in order to identify regions of the genome that appear to be segregating with illness, i.e., from affected grandparent or parent to affected offspring (Fig. 3). Thus every available member of the pedigree is genotyped for markers spaced as evenly as possible over all regions of the genome. In this way, "linkage" between a DNA marker at a known locus and the illness phenotype can provide an approximate location of a susceptibility gene. If evidence for linkage is obtained in the initial genome screen, additional markers can be genotyped in the interesting regions to confirm (or refute) the initial linkage findings and to further narrow the location of the disease locus. The large number of highly polymorphic DNA markers now available has greatly facilitated genetic linkage studies and greatly improved the informativeness of almost all meiotic events *(36)*. In other words, for any given member of the pedigree, it is usually possible to ascertain whether a specific marker allele is derived from the maternal or paternal chromosome. Using a variety of statistical computational programs now available, linkage between DNA markers and the presence or absence of the phenotype can be tested. One of the more traditional tests for linkage, the LOD score (logarithm to the base 10 of the ratio of the probability of the pedigree assuming linkage at a given recombination fraction divided by the probability assuming no linkage) is useful for the study of rare disorders that have Mendelian (single gene) inheritance. However, for bipolar affective disorder (as well as for other forms of depression), the evidence strongly suggests polygenic, oligogenic, or even multigenic transmission *(37)*. Unfortunately, virtually all early genetic linkage studies of bipolar affective disorder assumed

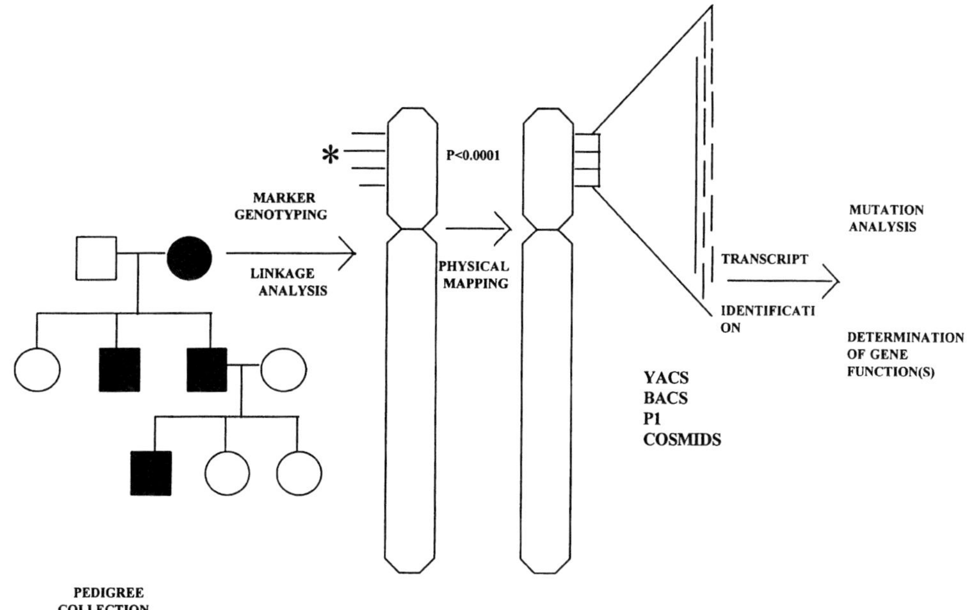


**Fig. 3.** Positional cloning strategy for finding a disease gene. DNA is prepared from blood or tissue samples. Transformed lymphoblast lines can be established for each subject to provide a replenishable source of DNA for genotyping and other genomic studies. Linkage analyses using various statistical algorithms can be used to identify chromosome regions containing susceptibility genes. Once a candidate locus is found, a physical map of the chromosome interval is constructed and expressed sequences (cDNAs) are isolated and analyzed for mutations. Putative mutations in cDNA are studied to rule out "silent" polymorphisms by comparing DNA and, if necessary, protein from normal and affected individuals. Transgenic and knockout mice can be utilized to help characterize gene function (*see text* for discussion).

single-gene or simple Mendelian inheritance; the results of these studies have been difficult to replicate, especially in unrelated families (*see* ref. *37* for a review of the various problems and pitfalls associated with these studies). Consequently, it has become more useful to perform nonparametric linkage analyses using "model-independent" allele-sharing methods in which it is not necessary to make assumptions about the mode of inheritance *(37)*.

3.1.1.1. Chromosomal Loci Linked to Bipolar Affective Disorder

Recent genome screens carried out using nonparametric linkage methods have revealed evidence for linkage of bipolar affective disorder to a number of chromosomal regions *(38,39)*. Notably, different loci on chromosome 18 have been identified and there have been claims of replication by some *(40)* but not all *(41,42)* investigators. In our collaborative study of bipolar affective disorder in the Old Order Amish (a well-characterized genetic isolate), we have found evidence for susceptibility loci on chromosomes 6, 13, and 15 *(43)*. Still, the statistical evidence for linkage to these loci as evaluated using the criteria of Lander and Kruglyak *(44)*, while suggestive, are not definitive. Therefore these regions need to be studied in greater detail by examining additional highly informative DNA markers and by including additional (particularly other affected) members of these and related pedigrees (thus increasing the statistical power to establish linkage). In our view, while none of the linkage findings of bipolar affective disorder published to date have been convincing, a number of highly suggestive regions have been identified and could be confirmed in the near future. It is also possible given the likely oligogenic nature of the disorder that two-gene (or multigene) models will be necessary to detect linkage. Additional allele sharing methods, such as the extended haplotype method, should also prove useful *(45)*. Finally, we suggest that susceptibility genes that actually reduce the relative risk for developing bipolar affective disorder similar to that recently shown for the apolipoprotein E2 allele and late-onset Alzheimer's disease *(46)* should be considered, and the presence of these genes may further complicate genetic linkage studies of bipolar affective disorder.

### 3.1.2. From Chromosomal Locus to Gene to New Drug Target

Once chromosomal loci containing susceptibility genes for bipolar affective disorder are definitively identified, the disease susceptibility genes themselves can be identified in several ways. Often, a physical map of the region can be constructed utilizing yeast artificial chromosomes (YAC) or other vectors containing relatively large fragments of DNA—covering the smallest possible region of interest (Fig. 3). Next, novel expressed sequences (transcripts) are isolated, or sequences are previously mapped to the region in question are obtained from genomic databases. The great progress that has been made in sequencing and mapping human transcripts has greatly facilitated identification of candidate susceptibility genes. Once these candidate expressed sequences (cDNAs) are identified, the transcription level and sequence of these genes can be analyzed. Mutations causing changes in protein function and that lead to disease can be characterized (Fig. 3). Allelic heterogeneity commonly seen in many inherited neuropsychiatric disorders can also be readily characterized and the impact of variations in protein function on illness presentation can be studied in a variety of ways. Overexpression of a mutant transgene in mice, or targeted disruption (knock-out) of the wild-type (normal) gene, may lead to a phenotype mimicking some aspects of the human disorder and provide important information on the protein's function. The generation of an animal model of the disease could subsequently be used to test new drugs for therapeutic activity. This approach has already proven useful for single-gene (and thus relatively simple) genetic disorders such as early onset Alzheimer's disease *(47)*, in which reliable neuropathological end points can be readily measured to validate the transgenic model and to follow disease progression *(47)*. However, for behavioral disorders such as bipolar affective disorder (in which few, if any, structural abnormalities are anticipated), validation will be much more challenging.

Finally, another potentially useful approach will be to explore candidate genes of known function that map to the susceptibility loci identified from the genome screen and linkage analysis. Genes encoding neurotransmitter-related proteins (biosynthetic or catabolic enzymes, receptors, transporters, and so on) that map to these loci can then be quickly evaluated as potential disease or susceptibility genes (*see* ref. *48* for example). Although many candidate genes have been postulated to be involved in the etiology of psychiatric disorders, their association with a particular form of psychopathology has unfortunately been difficult to replicate. Future studies combining the positional cloning and candidate gene approaches may prove more productive, especially as more "candidate genes" are mapped and included in the genomic databases.

## 4. SUMMARY AND CONCLUSION

Whereas traditional approaches to antidepressant drug discovery have resulted in an efficacious and useful generation of therapeutic agents, it is clear that new and better antidepressants are still needed. Revolutionary advances in molecular biology should permit a more complete characterization of the changes in gene expression, which may account for the well-recognized time-delay of antidepressant responses and which underlie the ultimate mechanism of action of these agents. This, in turn, may reveal a host of new targets for drug discovery. Given the important contribution of genetic factors to the etiology of many forms of depression, the identification of susceptibility genes will likely provide novel therapeutic approaches to disease prevention and more effective management of even the most severe and malignant forms of the disorder.

## ACKNOWLEDGMENTS

The authors wish to thank Dr. Binhui Ni and Ms. Kelly Bales for their careful review and comments on our manuscript.

## REFERENCES


1. Montgomery, S. A. (1995) Selective serotonin reuptake inhibitors in the acute treatment of depression, in *Psychopharmacology: The Fourth Generation of Progress* (Bloom, F. E. and Kupfer, D. J., eds.), Raven, New York, ch. 89, pp. 1043–1051.
2. Burke, M. J. and Preskorn, S. H. (1995) Short-term treatment of mood disorders with standard antidepressants, in *Psychopharmacology: The Fourth Generation of Progress* (Bloom, F. E. and Kupfer, D. J., eds.). Raven, New York, ch. 90, pp. 1053–1065.
3. Katz, M. M., Koslow, S. H., Maas, J., Frazer, A., Bowden, C. L., Casper, R., Croughan, J., Kocsis, J., and Redmond, E. (1987) The timing, specificity and clinical prediction of tricyclic drug effects in depression. *Psychol. Med.* **17,** 297–309.
4. Fink, M. (1979) *Convulsive Therapy: Theory and Practice.* Raven, New York.
5. Potter, W. Z. and Manji, H. K. (1994) Clinical onset of antidepressant action: implications for new drug development, in *Critical Issues in the Treatment of Affective Disorders* (Langer, S. Z., Brunello, N., Racagni, G., and Mendlewicz, J., eds.), Karger, S., Basel, pp. 118–126.
6. Haskell, D. S., DiMascio, A., and Prusoff, B. (1975) Rapidity of symptom reduction in depressions treated with amitriphyline. *J. Nerv. Ment. Dis.* **160,** 24–33.
7. Sackeim, H. A., Devanard, D. P., and Nobler, M. S. (1995) Electroconvulsive therapy, in *Psychopharmacology: The Fourth Generation of Progress,* (Bloom, F. E. and Kupfer, D. J., eds.) Raven, New York, ch. 95, pp. 1123–1141.
8. Wu, J. C. and Bunney, W. E. (1990) The biological basis of an antidepressant response to sleep deprivation and relapse: review and hypothesis. *Am. J. Psychiatry* **147,** 14–21.
9. Thase, M. E. and Kupfer, D. J. (1987) Characteristics of treatment-resistant depression, in *Treating Resistant Depression* (Zohar, J. and Belmaker, R. H., eds.) PMA, New York, pp. 23–45.
10. Fuller, R. W. and Wong, D. T. (1990) Serotonin uptake and serotonin uptake inhibition. *Ann. NY Acad. Sci.* **600,** 68–78.
11. Blier, P., de Montigny, C., and Chaput, Y. (1990) A role for the serotonin system in the mechanism of action of antidepressant treatments: preclinical evidence. *J. Clin. Psychiatry* **51,** 14–20.
12. Artigas, F. (1993) 5-HT and antidepressants: new views from microdialysis studies. *Trends Pharmacol. Sci.* **14,** 262.

13. Piñeyro, G., de Montigny, C., Weiss, M., and Blier, P. (1996) Autoregulatory properties of dorsal raphe 5-HT neurons: possible role of electrotonic coupling and 5-HT$_{1D}$ receptors in the rat brain. *Synapse* **22,** 54–62.

14. Blier, P., Chaput, Y., and de Montigny, C. (1988) Long-term 5-HT reuptake blockade, but not monoamine oxidase inhibition, decreases the function of terminal 5-HT autoreceptors: an electrophysiological study in the rat brain. *Naunyn-Schmiedeberg's Arch. Pharmacol.* **337,** 246–254.

15. Artigas, F., Romero, L., de Montigny, C., and Blier, P. (1996) Acceleration of the effect of selected antidepressant drugs in major depression by 5-HT$_{1A}$ antagonists. *TINS* **19,** 378–383.

16. Blier, P. and de Montigny, C. (1994) Current advances and trends in the treatment of depression. *TIPS* **15,** 220–226.

17. Perez, V., Gilaberte, I., Faries, D., Alvarez, E., and Artigas, F. (1997) Pindolol augments the antidepressant efficacy of fluoxetine: results of a double-blind, randomized trial. *Lancet,* in press.

18. Sanders-Bush, E. and Canton, H. (1995) Serotonin receptors: signal transduction pathways, in: *Psychopharmacology: The Fourth Generation of Progress* (Bloom, F. E. and Kupfer, D. J., eds.) Raven, New York, ch. 38, pp. 431–441.

19. Ozawa, H. and Rasenick, M. M. (1991) Chronic electroconvulsive treatment augments coupling of the GTP-binding protein Gs to the catalytic moiety of adenylyl cyclase in a manner similar to that seen with chronic antidepressant drugs. *J. Neurochem.* **56,** 330–338.

20. Nibuya, M., Nestler, E. J., and Duman, R. S. (1996) Chronic antidepressant administration increases the expression of cAMP response element binding protein (CREB) in rat hippocampus. *J. Neurosci.* **16(7),** 2365–2372.

21. Nestler, E. J., McMahon, A., Sabban, E. L., Tallman, J. F., and Duman, R. S. (1990) Chronic antidepressant administration decreases the expression of tyrosine hydroxylase in the rat locus coeculeus. *Proc. Natl. Acad. Sci. USA* **87,** 7522–7526.

22. Hosoda, K. and Duman, R. S. (1993) Regulation of β$_1$-adrenergic receptor mRNA and ligand binding by antidepressant treatments and norepinephrine depletion in rat frontal cortex. *J. Neurochem.* **60,** 1335–1343.

23. Nibuya, M., Morinobu, S., and Duman, R. S. (1995) Regulation of BDNF and trkB mRNA in rat brain by chronic electroconvulsive seizure and antidepressant drug treatments. *J. Neurosci.* **15,** 7539–7547.

24. Stein-Behrens, B., Mattson, M. P., Chang, I., Yeh, M., and Sapolsky, R. (1994) Stress, exacerbates neuron loss and cytoskeletal pathology in the hippocampus. *J. Neurosci.* **14,** 5373–5380.

25. Smith, M. A., Makino, S., Kvetnansky, R., and Post, R. M. (1995) Stress alters expression of brain-derived neurotrophic factor and neurotrophin-3 mRNAs in the hippocampus. *J. Neurosci.* **15,** 1768–1777.

26. Liang, P. and Pardee, A. B. (1992) Differential display of eukaryotic messenger RNA by means of the polymerase chain reaction. *Science* **257,** 967–971.

27. Blanchard, R. K. and Cousins R. J. (1996) Differential display of intestinal mRNAs regulated by dietary zinc. *Proc. Natl. Acad. Sci. USA* **93,** 6863–6868.

28. Zimmerman, J. W. and Schultz, R. M. (1994) Analysis of gene expression in the preimplantation mouse embryo: use of mRNA differential display. *Proc. Natl. Acad. Sci. USA* **91,** 5456–5460.

29. Wong, M.-L., Khatri, P., Licinio, J., Esposito, A., and Gold, P. W. (1996) Identification of hypothalamic transcripts upregulated by antidepressants. *Biochem. Biophys. Res. Commun.* **229,** 275–279.

30. Regier, D. A., Boyd, J. H., Burke, J. D., Rae, D. S., Myers, J. K., Kramer, M., Robins, L. N., George, L. K., Karno, M., and Locke, B. Z. (1988) One-month prevalence of mental disorders in the United States: Based on five epidemiological catchment area sites. *Arch. Gen. Psychiat.* **45,** 977–986.

31. Tsuang, M. T. and Farone, S. V. (1990) *The Genetics of Mood Disorders.* The Johns Hopkins University Press, Baltimore, MD.

32. Bertelsen, A., Harvald, B., and Hauge, M. (1977) A danish twin study of manic-depressive illness. *Br. J. Psychiat.* **130,** 330–351.

33. Plomin, R., Owen, M. J., and McGuffin, P. (1994) The genetic basis of complex human behaviors. *Science* **264,** 1733–1739.

34. Philibert, R., Egeland, J., Paul, S. M., and Ginns, E. I. (1997) The inheritance of bipolar affective disorder: abundant genes coming together. *J. Affective Dis.,* in press.

35. Winokur, G. (1991) *Mania and Depression: A Classification of Syndrome and Disease.* The Johns Hopkins University Press, Baltimore, MD.

36. Collins, F. S. (1995) Positional cloning moves from perditional to traditional. *Nature Genet.* **9,** 347–350.

37. Straub, R. E. and Gilliam, T. C. (1993) Genetic linkage studies of bipolar affective disorder, in *Genome Maps and Neurological Disorders* (Davies, K. E. and Tilghman, S. M., eds.) Cold Spring Harbor Laboratory, Cold Spring Harbor, NY, pp. 77–99.

38. Berrettini, W. H., Ferraro, T. N., Goldin, L. R., Weeks, D. E., Detera-Wadleigh, S., Nurnberger, J. I., and Gershon, E. S. (1994) Chromosome 18 DNA markers and manic-depressive illness: evidence for a susceptibility gene. *Proc. Natl. Acad. Sci. USA* **91,** 5918–5921.

39. Straub, R. E., Lehner, T., Luo, Y., Loth, J. E., Shao, W., Sharpe, L., Alexander, J. R., Das, K., Simon, R., Fieve, R. R., Lerer, B., Endicott, J., Ott, J., Gilliam, T. C., and Baron, M. (1994) A possible vulnerability locus for bipolar affective disorder on chromosome 21q22.3. *Nature Genet.* **8,** 291–296.

40. Stine, O. C., Xu, J. F., Koskcla, R., Mcmahon, F. J., Gschwend, M., Friddle, C., Clark, C. D., Mcinnis, M. G., Simpson, S. G., Breschel, T. S., Vishio, E., Riskin, K., Feilotter, H., Chen, E., Shen, S., Folstein, S., Meyers, D. A., Rotstein, D., Marr, T. G., and Depaulo, J. R. (1995) Evidence for linkage of bipolar disorder to chromosome-18 with a parent-of-origin effect. *Am. J. Hum. Genet.* **57,** 1384–1394.

41. Pauls, D. L., Ott, J., Paul, S. M., Allen, C. R., Fann, C. S. J., Carulli, J. P., Falls, K. M., Bouthillier, C. A., Gravius, T. C., Keith, T. P., Egeland, J. A., and Ginns, E. I. (1995) Linkage analysis of chromosome 18 markers do not identify a major susceptibility locus for bipolar affective disorder in the Old Order Amish. *Am. J. Hum. Genet.* **57,** 636–643.

42. Smyth, C., Kalsi, G., Brynjolfsson, J., O'Neill, J., Curtis, D., Rifkin, L., Moloney, E., Murphy, P., Sherrington, R. S., Petursson, H., and Gurling, H. M. D. (1996) Further tests for linkage at the tyrosine hydroxylase gene locus on chromosome 11p15 in a new sample of British multiplex manic depression (bipolar and unipolar affective disorder) families. *Am. J. Psychiatry* **153,** 271–274.

43. Ginns, E. I., Ott, J., Egeland, J. A., Allen, C. R., Fann, C. S. J., Pauls, D. L., Weissenbach, J., Carulli, J. P., Falls, K. M., Keith, T. P., and Paul, S. M. (1996) A genome-wide search for chromosomal loci linked to bipolar affective disorder in the Old Order Amish. *Nature Genet.* **12,** 431–435.

44. Lander, E. and Kruglyak, L. (1995) Genetic dissection of complex traits: guidelines for interpreting and reporting linkage results. *Nature Genet.* **11,** 241–247.

45. Houwen, R. H., Baharloo, S., Blankenship, K., Raeymaekers, P., Juyn, J., Sandkuijl, L. A., and Freimer, N. B. (1994) Genome screening by searching for shared segments: mapping a gene for benign recurrent intrahepatic cholestasis. *Nature Genet.* **8,** 380–386.

46. Corder, E. H., Saunders, A. M., Risch, N. J., Strittmatter, W. J., Schmechel, D. E., Gaskell, P. C., Rimmler, J. B., Locke, P. A., Conneally, P. M., Schmader, K. E., Small, C. W., Roses, A. D., Haines, J. L., and Pericak-Vance, M. A. (1995) Protective effect of apolipoprotein E type 2 allele for late onset Alzheimer's disease. *Nature Genet.* **7,** 180–184.

47. Games, D., Adams, D., Alessandrini, R., Barbour, R., Berthelette, P., Blackwell, C., Carr, T., Clemens, J., Donaldson, T., Gillespie, F., Guido, T., Hagoplan, S., Johnson-Wood, K., Khan, K., Lee, M., Leibowitz, P., Lieberburg, I., Little, S., Ellezer, M., McConlogue, L., Montoya-Zavala, M., Mucke, L., Paganini, L., Penniman, E., Power, M., Schenk, D., Seubert, P., Snyder, B., Soriano, F., Tan, H., Vitale, J., Wadsworth, S., Wolozin, B., and Zhao, J. (1995) Alzheimer-type neuropathology in transgenic mice overexpressing V717F β-amyloid precursor protein. *Nature* **373,** 523–527.

48. Lesch, K.-P., Bengel, D., Heils, A., Sabol, S. Z., Greenberg, B. D., Petri, S., Benjamin, J., Müller, C. R., Hamer, D. H., and Murphy, D. L. (1996) Association of anxiety-related traits with a polymorphism in the serotonin transporter gene regulator region. *Science* **274,** 1527–1531.